Clinical Magnetic Resonance Spectroscopy

Clinical Magnetic Resonance Spectroscopy

Ernest B. Cady

University College Hospital
London, United Kingdom

Plenum Press • New York and London

Library of Congress Cataloging-in-Publication Data

Cady, Ernest B.
    Clinical magnetic resonance spectroscopy / Ernest B. Cady.
        p.   cm.
    Includes bibliographical references.
    Includes index.
    ISBN-13: 978-1-4684-1335-9   e-ISBN-13: 978-1-4684-1333-5
    DOI: 10.1007/978-1-4684-1333-5

    1. Nuclear magnetic resonance spectroscopy--Diagnostic use.
I. Title.
    [DNLM: 1. Nuclear Magnetic Resonance.   QD 96.N8 C126c]
RC78.7.N83C33   1990
616.07'548--dc20
DNLM/DLC
for Library of Congress   90-7400
                                                                    CIP

Foreword

Nobody can know everything. For the successful application of techniques based on nuclear magnetic resonance to clinical problems, it is a vital necessity that individuals with widely different skills should learn a little of each others' trades by co-operation and communication. Ernest Cady has long proved himself a master of these arts to his colleagues at University College London, and by writing this excellent book he extends his experience to a wide circle of readers.

Although the nuclear magnetic resonance (NMR) phenomenon had been predicted theoretically (and to some degree demonstrated experimentally) appreciably earlier, it required the advances in electronics that took place during World War II to turn NMR into a practical technique, as demonstrated independently in 1946 by Bloch and Purcell. Since then, NMR has been used extensively and increasingly by chemists and physicists.

In the 1970s the first applications of NMR to animal organs yielded important advances in our knowledge of the biochemical and physiological processes as they occur in genuinely intact tissues. They showed incidentally that some conventional techniques introduce significant artifacts.

More importantly, when combined with massive technical advances in the design of superconducting magnets and in radiofrequency technology, this work laid the foundations for the explosive expansion of human clinical applications that has occurred within the past decade. One casualty of this expansion has been in terminology. The very word "nuclear" sent shudders down the spines of many patients even though the energy per quantum of the radiofrequencies employed is some 10^{10} times less than that of ionizing radiations. Nevertheless for clinical purposes the "N" has been dropped from the name; "MR" is used instead throughout this book. This may seem trivial but it is not; the mental equanimity of patients is just as important as their physical condition.

The change is distinctly confusing to other scientists, who wonder whether the clinicians are employing a totally different technique, especially since in some new developments they are! Another loss is that the name NMR is succinctly descriptive—*N*uclear because the fundamental phenomenon arises from the spin of those atomic nuclei that have it (all those containing uneven numbers of protons and/or neutrons); *M*agnetic because the spin and positive charge make the nucleus behave like a tiny magnet which will interact with externally applied magnetic fields; and *R*esonance because the use of the technique depends on first exciting and then detecting resonant frequencies.

This book deals almost exclusively with MR spectroscopy (MRS), one of whose main uses is the noninvasive chemical analysis of tissues. MRS has had some success in solving clinical problems, notably in assessing the condition of the brain and the prognosis in new-born human babies. Enzyme defects in muscle, which are easily detected by MRS, turn out on survey to be rare. The real and well-justified clinical success of the decade has been in MR imaging (MRI). In this technique, the resonances from the ^1H nuclei, so plentiful in the body, and also intrinsically the most sensitive, are manipulated in such a way that they produce pictures of high quality of sections in any plane within the body. Because radio-frequency technique is so versatile, these sections can be "stained" in different ways so as to show up contrasts that are impossible when using X-ray tomography.

Nevertheless, the appearance of a book on clinical MRS now is singularly well timed. MRS is increasingly being combined with MRI in the same wide-bore, high-field spectrometer. In this way a certain volume of tissue can be located using MRI and its chemical composition determined by MRS. The possibilities seem endless and the excellent reference listings not only describe the past but also put up many pointers to the future.

Since clinical MR is such a multifaceted subject, it is hard to know how to begin to present it optimally to readers whose own knowledge will also be very variable. My advice to them is that they need not necessarily start at the beginning of the book but should start from topics that they already know something about and are especially interested in. They can then work outward, colloquially speaking, to less familiar but very important territory. Anyone who wants to learn about MRS but feels daunted by *every* part of this book can take heart from the fact that Ernest Cady's first degree was in astronomy!

Prof. D. R. Wilkie MD FRCP FRS
Department of Physiology
University College London

Preface

I have found the production of this book an exhilarating and self-educative experience. The initial impetus for its production occurred during the production of a chapter on the same subject, which was coauthored with S. R. Williams. It was apparent that enough material was available and that the subject had reached a sufficient degree of maturity to proceed with a more substantial work. It is hoped not only that the reader will find the text interesting as a description of the medical applications of magnetic resonance spectroscopy (MRS) but also that those who are actively involved in clinical investigations will find the work useful for reference purposes. Furthermore, it is anticipated that clinicians and other researchers entering the field without prior experience will find that the text gives useful information about both the practical implementation of MRS systems and the physiological information which can be derived from the spectra.

Over the last decade, the in vivo MRS field has always been exciting —from the first collections of the ^{31}P spectra of in situ human muscle using surface coils to the recent acquisition of highly resolved proton spectra with full three-dimensional ^{1}H image-guided localization from the living human brain. The field is still developing rapidly, with continuous publication of technical advances in data collection, localization, spectral editing, and spectral-analysis methods. I have attempted to ensure that as many recent "state of the art" publications as possible have been cited in the text.

My initial interest in the clinical aspects of MRS was aroused by D. R. Wilkie, E. O. R. Reynolds, and R. H. T. Edwards; for this, and for the leadership shown by these individuals, I am eternally thankful. I am also grateful for discussions with various colleagues concerning the application of MRS in the clinic. These include P. L. Hope, P. A. Hamilton, J. S. Wyatt, A. D. Edwards, A. M. de L. Costello, R. M. Gardiner, P. S. Tofts, S. Wray, M. J. Dawson, and D. T. Delpy. I must express my thanks to D. J. Newham, R. Aldridge, D. A. Jones, and J. K. Nicholson for assistance with collecting some of the spectra used in the present work and to H. Greenwood for secretarial assistance.

Finally, but by no means least, I must thank my wife, Lorna, and my sons, Henry and Douglas, for their tolerance and patience during my many periods of absence while preparing the manuscript.

Ernest B. Cady

Hornsey, London

Contents

Clinical Magnetic Resonance Spectroscopy

An Introduction to Medical Magnetic Resonance Spectroscopy

1.1. THE MAGNETIC RESONANCE PHENOMENON

All atomic nuclei contain protons, which are positively charged. Many nuclei also behave as if they were spinning. In the terminology of quantum mechanics, they are said to possess net spin angular momenta. The combination of charge and spin produces a nuclear magnetic dipole, and the nuclei can be likened to microscopic magnets. If an external, constant magnetic field is applied, the spin axis of the nucleus precesses about the applied field in a manner similar to that of a gyroscope in the earth's gravitational field (see Figure 1.1a). Theoretical analysis and experiment[1, 2] have shown that the precessional frequency is proportional to the strength of the applied magnetic field which is usually measured in units of tesla (1 tesla = 10,000 gauss). The constant of proportionality is known as the gyromagnetic ratio (γ), and for the hydrogen nucleus (a proton) this is 26.75×10^7 rad/sec per tesla, or 42.57 MHz/tesla. For an applied field of the order of 1 tesla, the precessional frequency is in the radio band of the electromagnetic spectrum. Radio frequency (RF) is conventionally measured in Hertz (Hz; 1 Hz is equivalent to 1 cycle/sec), but because the precessional, or resonant, frequency is proportional to the field strength, nuclei in magnetic fields with higher or lower strengths resonate at greater or lesser frequencies. In order to make data collected on different instruments com-

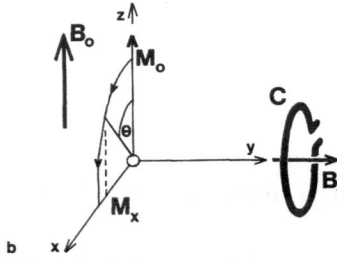

Figure 1.1. (a) A representation of the behavior of hydrogen nuclei (protons) in a static applied magnetic field B_0. The nuclei are shown as $\oplus$ with their precessing magnetic moments indicated by m; m↑ or m↓ represents the mean resultants of these precessing magnetic moments. M_0 is the net resultant magnetization, the vector sum of all the m↑ and m↓. (b) An external, applied magnetic field oscillating at the resonant frequency v_0 and of amplitude B_1 is generated by a brief radio pulse from the tuned coil C. In a frame of reference (x, y, z) which is rotating about z at v_0, the oscillating magnetic field B_1 appears to be constant, and B_0 seems to be nonexistent because in the spinning frame it has no effect. The only magnetic field which has an effect on M_0 is B_1, and this results in M_0 precessing about y for the duration of the pulse. It is often arranged that the pulse is of such a length that it flips M_0 by approximately 90°. In the non-rotating laboratory frame this produces a significant component of M_0 rotating about z at frequency v_0 in the xy plane. The tuned coil senses this varying magnetic field M_x, and the RF voltage generated (the FID) can be detected by a radio receiver. In this example the same coil is used both to perturb the system and to detect the magnetic resonance signal. It is possible for these functions to be performed by separate coils, and sometimes this can be advantageous.

parable, the resonant frequencies are always quoted in parts per million (ppm) of the applied magnetic field strength.

For the proton there are two possible orientations of the spin angular momentum with respect to the applied field, aligned with the field or against it; these orientations correspond to different energy levels. At room temperature a nucleus has a slightly greater chance of being in the lower energy state than in the higher one. Therefore, if one considers a large number of hydrogen nuclei in an applied field, a small net magnetization will result from the mutual cancellation of the effects of all but a few of the precessing nuclei, as shown in Figure 1.1a. The induced magnetization M_0 is aligned with the applied static field B_0 and can be perturbed by applying a varying magnetic field of the same frequency as the rate of precession either as a continuous wave or as a brief pulse. The perturbation is provided by RF energy transmitted from a suitably designed antenna and results in an angular flip of M_0. The subtlety of magnetic resonance spectroscopy (MRS) lies in the fact that nuclei in different molecules or located in various chemical groups in the same molecule experience slightly different magnetic field strengths. This is because the molecular electronic structure in the neighborhood of the nucleus produces a local magnetic field, which adds to the applied field. Hence the precessional frequencies for particular nuclei in various compounds depend on the nature of the molecule and the position of the nucleus in the molecule. Most samples therefore contain nuclei that resonate at a variety of frequencies, depending on the local chemical environment. A broadband RF pulse or a continuous RF wave that varies in frequency over a sufficiently wide range perturbs them all. The relative resonant frequency of the nucleus, is called chemical shift, is usually measured in ppm relative to a reference signal derived either from a suitable metabolite in the tissue or from an external sample.

For many reasons, most modern spectrometers use RF pulses to perturb the magnetization—a technique called "pulse Fourier transform spectroscopy." It is usually arranged that the net magnetizations are perturbed by the pulse or pulses in such a way that significant components are produced perpendicular to the main applied field, as shown in Figure 1.1b. These vectors continue precessing at frequencies proportional to the sum of the applied and local fields, and their precession can induce a current in an adjacent antenna. This signal, recorded by the RF receiver of the spectrometer as a function of voltage against time, is called the free induction decay (FID). By a process called Fourier transformation,[3] the FID can be analyzed mathematically for the frequencies that it contains, so as to produce a spectrum. Analysis of the spectrum gives information not only about molecular structure but also about metabolite concentrations. A more detailed account of the basic theory of MRS will be given in Chapter 2. Figure 1.2 shows an FID obtained from phosphorus (^{31}P) nuclei in resting human forearm muscle and the spectrum derived therefrom. The chemical shifts of the peaks are displayed in ppm relative to phosphocreatine (PCr). Due to various effects caused by the inhomogeneity of the magnetic field and by interactions with neighboring nuclei, the FID decays in a quasi-exponential manner. The different constituent signals from various metabolites may decay at faster or slower rates in the FID, producing broader or narrower peaks in the spectrum.

Two processes that affect the strengths and widths of spectral peaks are called spin–lattice and spin–spin relaxation, and these have characteristic exponential decay times T_1 and T_2, respectively (dealt with in greater detail in Chapter 2). The resonances in the spectrum (labeled 1 to 7 in Figure 1.2) are due to highly mobile molecules and are narrow (a half width of about 1 ppm) when compared with the broad, low-level feature due to the relatively immobile nuclei in phospholipids and bone (half width about 25 ppm). Because of the rapid molecular motion of the mobile species, the nuclei concerned experience an average field more homogeneous than that pertaining to the immobile species and hence

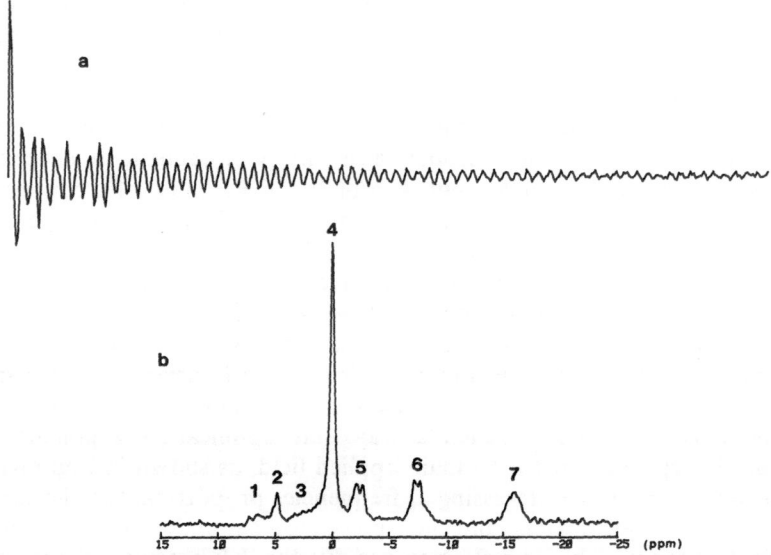

Figure 1.2. (a) A ^{31}P FID obtained from resting human forearm muscle at 32.5 MHz. The data were obtained using a 6-cm, single-turn surface coil with a 90° pulse at the coil's center. 512 scans were accumulated at 2.256-sec intervals. (b) The spectrum derived from the FID by Fourier transformation. The peaks are due largely to sugar phosphates (1), Pi (2), phosphodiesters (3), PCr (4), and mainly ATP (5–7), although peaks (5) and (6) contain contributions from ADP and peak (6) includes signal due to nicotinamide–adenine dinucleotides (NAD and NADH). (Spectrum acquired in collaboration with R. Aldridge, D. Jones, and G. Obletter.)

resonate over a much narrower frequency band. For this reason, in vivo biological MRS has been used mainly to investigate highly mobile metabolites, although there is potential for application to studies of bound molecules such as are found in cellular membranes.

1.2. MAGNETIC RESONANCE IN A CLINICAL CONTEXT

The first, pioneering magnetic resonance experiments on atomic nuclei in liquids and solids were carried out in the 1940s by Purcell, Torrey, and Pound[4] and by Bloch, Hansen, and Packard.[5] Theoretical analysis of the behavior of the spinning atomic nucleus in a magnetic field led to the development of the Bloch equations, which form a classical basis for a description of the phenomenon.[1, 6, 7] In the 40 years since the initial studies were made, MRS has blossomed into one of the most important research tools for the quantitative analysis of chemical composition and structure and has found application in many distinct fields for the investigation of gases, liquids, and solids. In the life sciences, biological application has naturally preceded clinical use, and research workers have, as always, had to work within the constraints of the technology available. For many years studies were made on test-tube samples small enough to insert in the narrowbore magnets of the day, either permanent or resistive. Fundamental to the application of MRS in clinical science has been the development of economical highfield superconducting magnets large enough to allow the insertion of all or part of the infant or adult body.

Magnetic resonance imaging (MRI) has found major application in medicine and vies for supremacy with its main competitor, X-ray computerized tomography (CAT). The main advantages of both MRI and MRS are their noninvasive nature and the flexibility with which they can be applied to obtain different sorts of information from various tissues and pathologies. The static magnetic fields and RF strengths currently in clinical use cause no known harmful effects, and studies can be repeated safely even on newborn infants. (Safety aspects of MRS will be dealt with in more detail in Chapter 4.) Other diagnostic modalities such as ultrasound, CAT, positron emission tomography, and nuclear medicine imaging techniques give mainly structural information and, except for ultrasound, involve the use of ionizing radiation. Magnetic resonance can give information on nuclear density, nuclear relaxation properties, metabolite and metallic ion concentrations, metabolite exchange rates, intracellular pH (pH_i), and 3-dimensional macroscopic tissue structure.

Both MRI and MRS depend on the same underlying physical principles. In imaging, the resonant frequency of the nucleus (which is proportional to the local magnetic field strength) is translated into positional information by the application of a magnetic field gradient. The structure of internal anatomical features is revealed by the density or relaxation properties of some constituent of the tissue, usually water protons. By the use of extremely homogeneous magnetic fields, biological spectroscopy strives to resolve the frequencies of signals from the vast multitude of metabolites, many found only in minute concentrations, present in living systems. For this reason MRS has developed as a discipline which requires relatively high magnetic fields (up to 10 tesla or more) in order to attain the resolution and signal strengths necessary to produce quantifiable results. Until recently, good quality clinical images have been obtained at relatively low field strengths (0.05 to 0.2 tesla), but with the development of large-bore, high-field superconducting magnets it has been found possible to get good imaging results at fields of 1.5 to 4 tesla. This has raised the possibility of manufacturing clinical systems with both spectroscopic and imaging capability and of applying imaging techniques to practical problems such as localizing the tissue from which the spectra are obtained. This development has given MRS the impetus to become a major clinical tool for the diagnosis, prognosis, and monitoring of a large variety of pathologies with metabolic as well as structural consequences.

1.3. NUCLEI WITH POTENTIAL CLINICAL APPLICATIONS

It should be remembered that only nuclei with spin can exhibit magnetic resonance. This limits the sensitivity of the technique to some extent because, in many biological situations, the isotopes that can be used have low natural abundance and may also have low intrinsic sensitivity. Consider some of the commonest nuclei in organic molecules: ^{12}C, ^{1}H, ^{16}O, ^{14}N, and ^{31}P. Only ^{1}H, ^{14}N, and ^{31}P have spin and are almost 100% naturally abundant. (^{14}N, however, has spin equal to 1 and possesses an electric quadrupole moment, which gives rise to very broad spectral peaks and hence limits the usefulness of this nucleus.) In order to study the biochemistry of the remaining nuclei, one would have to observe low-abundance isotopes and hence be content with a reduced signal. Table 1.1 lists some of the nuclei with potential for biological studies and other relevant data. It should be noted that there are over 100 nuclei which have spin and, although many of these have low natural abundance, sensitivity, and concentration under normal conditions, certain applications may be found for them as tracers using labeled compounds.

1.3.1. Phosphorus (^{31}P)

Although ^{31}P has an absolute sensitivity of only 7×10^{-2} when compared with ^{1}H, this nucleus has so far proved to be one of the most important in MRS investigations of living systems. This importance can be attributed largely to the presence of the nucleus in such metabolites as PCr, ATP, and Pi, which are important in energy metabolism. These are found in many biological tissues at millimolar concentrations, which can easily be detected by MRS. In particular, these metabolites are found in skeletal muscle at relatively high concentrations, and this tissue has proved to be very amenable to in vivo studies. The small amounts of phosphorus metabolites in surface tissues (skin and fat) have permitted objective investigations of skeletal muscle using only the most primitive signal localization methods (e.g., surface coils,[8] which usually consist of one or more coplanar, circular turns of highly conductive material and are positioned noninvasively so as to receive the RF signal from adjacent tissue). A great deal of the pioneering work using medium-bore magnets (up to 25 cm in diameter) has been done on muscle tissue in distal limbs, and this has led the way to more general investigations of the whole body using larger instruments. The ^{31}P resonances occur over a spectral band of about

Table 1.1. Some Nuclei with Potential for Clinical Studies

Nucleus	Spin	Relative sensitivity[a]	Natural abundance (%)	Absolute sensitivity[b]	Frequency at 1 tesla (MHz)
^{1}H	1/2	1.00	99.98	1.00	42.57
^{13}C	1/2	1.59×10^{-2}	1.11	1.76×10^{-4}	10.71
^{14}N	1	1.01×10^{-3}	99.63	1.01×10^{-3}	3.08
^{15}N	1/2	1.04×10^{-3}	0.37	3.85×10^{-6}	4.31
^{17}O	5/2	2.91×10^{-2}	0.04	1.08×10^{-5}	5.77
^{19}F	1/2	8.3×10^{-1}	100.00	8.3×10^{-1}	40.05
^{23}Na	3/2	9.25×10^{-2}	100.00	9.25×10^{-2}	11.26
^{31}P	1/2	6.63×10^{-2}	100.00	6.63×10^{-2}	17.23

[a] Field constant and equal numbers of nuclei.
[b] The product of the relative sensitivity and the natural abundance (normalized so that the ^{1}H absolute sensitivity = 1.00).

40 ppm, and spectra tend to be relatively simple because many metabolites contain only one phosphate group in each molecule and hence are represented in the spectrum by a single resonance. (^{1}H spectra, for instance, are more complicated, mainly because protons are found in many different situations in the same molecule, and hence a given metabolite may give several resonances at different chemical shifts.) Many tissue phosphates are highly mobile and give sharp resonances. Some phosphate is relatively immobile (mainly in bone and membrane phospholipids) and gives a broad underlying spectral "hump," but it is possible to remove this component from the spectra so as to facilitate quantitation. Other metabolites detected by ^{31}P MRS are sugar phosphates and phosphomono- and di-esters thought to be related to membrane development. Figure 1.3 shows the resonance bands in which different types of phosphorus metabolites are found in biological spectra.

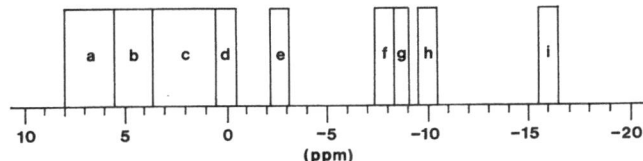

Figure 1.3. A schematic ^{31}P spectrum indicating the resonance bands in which are found signals from various phosphorylated metabolites commonly encountered in biological tissue: *a*, phosphomonoesters including sugar phosphates; *b*, inorganic orthophosphate; *c*, phosphodiesters; *d*, phosphocreatine and phosphoarginine; *e*, nucleoside triphosphates (e.g. ATP)—mainly ionized-end phosphate (γ) but including nucleoside polyphosphates (e.g. ADP); *f*, nucleoside triphosphates (mainly esterified-end phosphate (α) but including nucleoside polyphosphates); *g*, dinucleotides (e.g. nicotinamide–adenine dinucleotides); *h*, nucleoside diphospho-sugars; and *i*, nucleoside triphosphates—middle phosphate (β).

In addition to having the potential for determining metabolite concentrations, ^{31}P MRS can also give estimates of pH_i. This is because the chemical shifts of many ^{31}P resonances are pH dependent.[9] The Pi resonance has been found to be particularly sensitive over the pH range encountered in biological material and is commonly used for this purpose. The chemical shifts of the ATP resonances can be used similarly to estimate the pH,[10, 11] Mg^{2+} concentration, and degree of binding of Mg^{2+} to the ATP,[12] as well as temperature.[13]

1.3.2. Hydrogen (^1H)

The ^1H nucleus has the highest absolute abundance of those listed in Table 1.1 and is the commonest in biological systems. In vivo studies using ^1H are handicapped, however, by the very narrow range of chemical shifts encountered (covering about 8 ppm, so that magnetic field homogeneity is critical) and by the large number of metabolites which give many overlapping peaks and produce very complex spectra. Coupling between the spins of particular protons causes resonances to occur as doublets, triplets, or more complicated multiplets. This is called spin coupling and further increases interpretational problems. (Spin coupling will be dealt with in more detail in Chapter 2.) Large molecules, such as proteins, are relatively immobile and hence produce spectrally broad features which underlie the peaks due to smaller, more mobile species such as amino acids. Various techniques have been developed to simplify ^1H spectra by reducing the signals from immobile molecules and selecting remaining metabolites according to their relaxation constants (T_1 and T_2). In addition to the large number of resonances, many tissues have a high water content (typically 75%, giving a water-proton concentration of about 80 mole/kg wet wt.), and this produces further problems. Due to the finite digitization range of the spectrometer, difficulty will be encountered when trying to record adequately the signals from water and from metabolites with much smaller concentrations. The width of the water peak also implies that most signals from metabolites of interest are superimposed upon significant wings. For these reasons, it is usually necessary to adopt some method for suppressing the tissue-water signal in order to obtain quantifiable spectra. Techniques for spectral simplification and water suppression will be dealt with in Chapter 5. Figure 1.4 shows the metabolite resonance bands in biological ^1H spectra.

Certain ^1H resonances, such as those due to the C2 and C4 protons of histidine, possess pH-dependent chemical shifts.[14] The amide resonances of oligopeptides, including N-acetylaspartate (N-acasp), have chemical shifts that depend on tissue temperature.[15, 16]

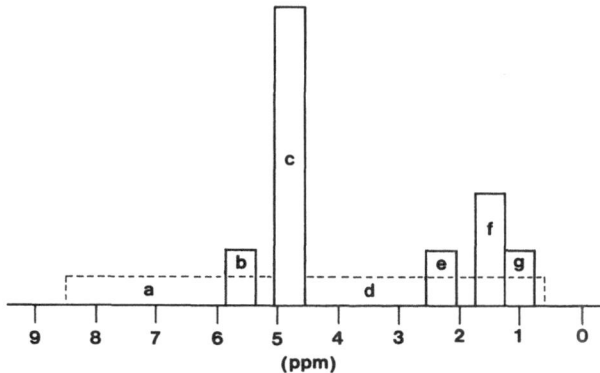

Figure 1.4. The resonance bands in which particular types of metabolite are detected in ^1H spectra obtained from biological samples: *a*, aromatic ring protons (ATP, ADP, phenylalanine, etc.); *b*, fatty acyl protons; *c*, water (from intracellular and other body-fluid sources); *d*, aliphatic chain protons (amino acids, lactate, creatine, etc.); *e*, *f*, and *g*, fatty acyl protons.

1.3.3. Carbon (^{13}C)

Due to the low natural abundance and low relative sensitivity of ^{13}C, the absolute sensitivity compared with ^1H is only about 2×10^{-4}. This limits the potential for applications using ^{13}C; however, important in vivo results have been obtained. A major impetus to the use of this nucleus arises from the large number of ^{13}C-enriched compounds readily available from commercial sources. This means that it is easy to carry out "tracer" studies of the chemical pathways utilized by a particular metabolite. Figure 1.5 displays the spectral bands in which ^{13}C resonances are found, and it

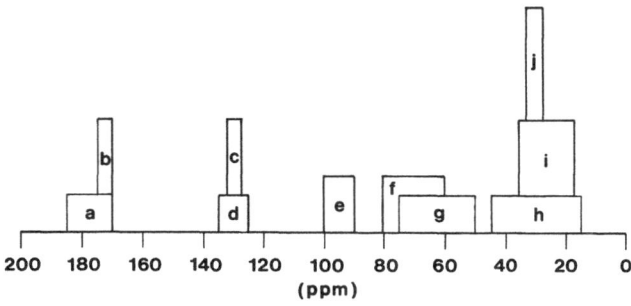

Figure 1.5. The regions of the ^{13}C spectrum in which signals from various types of tissue metabolite are found: *a*, amino acids, creatine, lactate, etc.; *b* and *c*, fatty acyl chains; *d*, membrane lipids; *e* and *f*, glycogen and glucose; *g* and *h*, amino acids, creatine, lactate, etc.; *i*, fatty acyl chains; and *j*, fatty acyl chains $-(CH_2)_n-$.

can be seen that the chemical shift range in which resonances are observed
is large (about 200 ppm). However, the usefulness of this nucleus is offset
to some extent by line splitting caused by adjacent protons. (The low ^{13}C
abundance means that there is only a very small chance of finding two
similar nuclei in close proximity, and therefore spin-coupling effects due
to adjacent carbon nuclei are not significant.) The line splitting has the
effect of producing doublet, triplet, and more complicated line structures
rather than single peaks. By simultaneously irradiating the sample at the
^{1}H frequency, the effect of proton coupling can be removed and a
significant increase in the ratio of signal to noise can be obtained due to the
nuclear Overhauser effect (NOE).[17] Great care must be taken when using
decoupling on living organs to ensure that no heating occurs, especially
when collecting data from organs close to tissues with low circulation such
as the lens of the eye.

1.3.4. Nitrogen (^{15}N)

The ^{15}N nucleus has low sensitivity and a low natural abundance,
giving an absolute sensitivity of about 4×10^{-6} when compared with ^{1}H.
Hence, one might expect this nucleus to be a poor research tool. However,
for moderately sized biomolecules (e.g., globular proteins with mol. wt.
about 14,000 in aqueous solution), the spin–lattice relaxation time T_1 of
the nucleus is very short, and this allows for rapid pulsing during data
accumulation.[18] An additional advantage is the 900-ppm chemical shift
range over which resonances are observed.

1.3.5. Oxygen (^{17}O)

The ^{17}O nucleus has an absolute sensitivity of about 10^{-5}, but if
labeled compounds are used a sensitivity about twice that of ^{13}C can be
achieved, although the lines are very broad.

1.3.6. Fluorine (^{19}F)

The ^{19}F nucleus has a high relative sensitivity and 100% natural
abundance but in biological systems occurs naturally only in low concentrations. Due to its small atomic size and low reactivity, ^{19}F is expected to
have little effect on molecular systems if it is introduced as a replacement
for ^{1}H and hence can be used for tracer experiments with fluorine-labeled
material including polypeptides, proteins, cytotoxics, and anesthetics.

1.3.7. Sodium (^{23}Na)

Although the ^{23}Na nucleus has a relative sensitivity better than that of ^{31}P, a natural abundance of 100%, an intracellular concentration of about 15 mM, and a concentration in blood plasma of over 100 mM, its use is limited because, unless a chemical shift reagent[19] is used (to separate resonances from intra- and extracellular sodium), only a single resonance is observed in biological systems.[20, 21]

1.4. THE DEVELOPMENT OF BIOLOGICAL MRS

The application of MRS in a medical environment has been dependent on a vast amount of research on mammalian tissues in vitro and in vivo. Almost within a decade of the pioneering work of the founders of the magnetic resonance technique,[1, 2, 4, 5] high-resolution proton (^{1}H) spectroscopy had been used to study ribonuclease,[16] and the potential for the investigation of biological systems was established. Various mammalian tissues have been excised, freeze-clamped, perfused, or studied in situ, and cellular extracts and suspensions have been prepared for high-resolution investigation using conventional MRS techniques. Undoubtedly, one of the major advances so far has been the in vivo application of surface coils[8] which can localize the tissue from which signals are collected. Subsequent clinical work, including the development of techniques and the interpretation of results, has depended to a large extent on these important preliminary studies.

^{31}P MRS is a main area of exploratory research, primarily because of the important biochemical role played by phosphorylated metabolites, such as PCr and ATP. In the early 1960s, the frequency dependences on pH and metal-ion concentration of ATP and ADP resonances were investigated by Cohn and Hughes.[10, 12] This was followed in the early 1970s by an important study of the pH dependence of several intracellular erythrocyte phosphates by Moon and Richards.[9] It was found that many ^{31}P resonances shifted to higher frequencies at alkaline pHs and to lower frequencies under acid conditions, and this demonstrated the use of the Pi resonance for the noninvasive monitoring of pH$_i$. [2,3-diphosphoglycerate (2,3-DPG) was found to be more sensitive to changes in pH although far less abundant in most tissues and hence of less use in general.]

The next section will describe some of the more important preliminary studies, which have provided the foundations for the clinical application of MRS. The rate at which publications appear is now very high, and it is beyond the scope of this book to provide a comprehensive review of the

literature on the biological applications of MRS. Some general works have
appeared on the subject,[23–26] but it is largely up to the individual researcher
to scour the relevant journals regularly for articles of specialized interest.

1.4.1. Skeletal Muscle

Of particular importance was the early application of ^{31}P MRS to the
study of phosphorylated muscle metabolites by Hoult et al.,[27] in which
signals from ATP, PCr, Pi, and sugar phosphates were detected. Freshly
dissected frog-muscle tissue was simply immersed, unperfused, in calcium-
free Ringer's solution at 20° C, and a gradual decrease in PCr and ATP,
accompanied by a rise in Pi, was observed over a 160-min period. During
the study, pH_i was observed to drop from 7.1 to 6.2. Following this work,
Dawson et al.[28] investigated the biochemical changes occurring during
contraction and recovery of frog and toad muscles. The tissues were
oxygenated by perfusion and maintained in physiological condition at 4° C.
Contractions were stimulated electrically, and force transducers were used
to measure muscle tension. Quantitations of muscle metabolites were
attempted using a substitution technique, and good agreements with conventional chemical analyses were obtained. Metabolite levels for resting
muscle were slightly different from those obtained previously[27] (sugar
phosphate and Pi were very low), and this difference was said to be related
to the maintenance of physiological condition during the studies. The spectra obtained were somewhat similar in appearance to the human in vivo
spectrum shown in Figure 1.2b. Large changes in metabolite levels and
intracellular acidosis were induced by prolonged (25-sec) electrical stimulation resulting in increased Pi and sugar phosphate, accompanied by a
reduction in PCr. On termination of the stimulation, the metabolites
recovered almost completely. Stimulations of 1-sec duration repeated every
125 sec did not reveal any significant changes. The same group has
extended its studies by investigating muscular fatigue using similar
methods.[29] They concluded that the development of muscular force is
closely correlated with metabolite levels and is proportional to the
hydrolysis rate of ATP.

The earliest pathological change in muscle detected by ^{31}P MRS was
the report, in 1977, of an unusual phosphodiester peak in spectra from
dystrophic chicken muscle.[30] The resonance was identified as L-serine
ethanolamine phosphodiester (SEP) by the use of perchloric acid extraction, barium and alcohol fractionation, and chromatographic isolation
techniques. The role of SEP in muscle tissue is unclear, although some relation with phospholipid metabolism is possible and recent theories on the
causes of the muscular dystrophies have focussed on cellular-membrane
defects.[31] Further work by the same group has indicated that another

phosphodiester, glycerol-3-phosphorylcholine (GPC), may be a marker for Duchenne muscular dystrophy.[32]

In 1980, the use of a surface coil to obtain ^{31}P spectra from in situ skeletal muscle was reported,[8] and this has paved the way for many subsequent in vivo studies of normal and pathological human muscle.[33]

^{1}H MRS was used in 1981 to obtain high-resolution spectra from intact frog and rat muscles.[34] The signal from tissue water was reduced by application of narrow-band saturation irradiation at the water-proton frequency. Comparison was made between spectra collected from intact muscle, the cytosolic fraction, and a model solution designed to simulate the results from tissue. In the spectra from intact frog muscle, seven resonances apart from water were detected. These were assigned to carnosine; PCr and creatine (Cr), both of which have peaks that resonate at almost the same ^{1}H frequency); possibly phosphorylcholine (PCh); and the CH_2 and CH_3 groups of alkyl chains in lipids. Four additional resonances were observed in spectra obtained from the cytosolic fraction. One of these was assigned to carnosine. It was also noticed that the cytosolic fraction resonance, in the position of the CH_2 lipid peak in the intact-muscle spectra, is a doublet, and the chemical shift and spin coupling (see Chapter 2) coincided with those of methyl lactate protons. A strong dependence of chemical shift on pH was found for one of the carnosine resonances. The investigation of rat muscle revealed much larger signals from lipid alkyl chains but, some of the resonances detected in frog muscle (notably carnosine) were not seen.

In 1984, ^{1}H spectra from dissected frog and chicken muscle were obtained at 20° C using an 11-tesla magnet.[35, 36] The signal from muscle-tissue water was suppressed by a technique called "jump–return." This first flips the water magnetization and then returns it to its starting orientation, while at the same time producing substantial transverse magnetization from metabolites of interest that resonate at other frequencies. (See Chapter 5 for details of this and other methods for suppressing the signal from tissue water.) In addition to the detection of resonances from the metabolites reported in the previous study, peaks due to anserine and carnitine were observed in spectra from chicken pectoralis muscle, and a peak from the H8 proton in the heterocyclic ring of ATP was seen in frog-muscle spectra. Additional peaks were detected, one of which showed signs of splitting, and it was suggested that these were due to various protons in either PCr or Cr. Mainly because of the high field strength, excellent resolution and signal strength were obtained, allowing spectra to be acquired in a matter of seconds. Peaks due to lactate, PCr, and Cr were particularly well observed, and it was possible to perform sequential data collections over a 10-hr period so as to demonstrate progressive PCr depletion and lactate and Cr accumulation accompanying muscle aging.

More recently, in vivo studies have been made of rat muscle metabolites at 8.5 tesla,[37] producing the spectra shown in Figure 1.6. Water in muscle tissue has a very short spin–spin relaxation time (T_2 = about 30 msec), and adequate suppression of the water signal can be achieved by using suitably long spin echoes. In this study, spectra were obtained using a saddle-shaped coil, tuned for both ^{31}P and ^{1}H, into which the rat's thigh was inserted. This allowed for the production of uniform flip angles across the tissue, and it was possible to use the Carr–Purcell–Meiboom–Gill[38] spin–echo method in order to measure T_2s. (This technique reduces the effect of magnetic field inhomogeneity on the measured T_2 and also self-corrects for flip-angle inhomogeneity which would lead to signal degradation.) T_2s were measured for PCr, taurine, anserine, water, and fat. The signal from the methyl lactate protons was also detected using a spin–echo double-resonance technique which removes the strong resonances from adjacent fatty-acid chains.[39] (This technique will be described in detail in Chapter 5.) Dual ^{31}P and ^{1}H spectra were obtained sequentially during ischemia, and the build-up of lactate and Pi, the decline of PCr and ATP, and the accompanying acidosis, as determined from the Pi chemical shift, were followed.

^{13}C MRS was first applied to mammalian muscle in 1977.[40] Three prominent peaks were detected in proton-decoupled, 25.2 MHz spectra and assigned to aliphatic, aromatic, and carbonyl carbons. Subsequently, in high-resolution, proton-decoupled studies of intact chicken pectoralis muscle made in 1981 at 90.5 MHz,[41] the origins of the spectral components

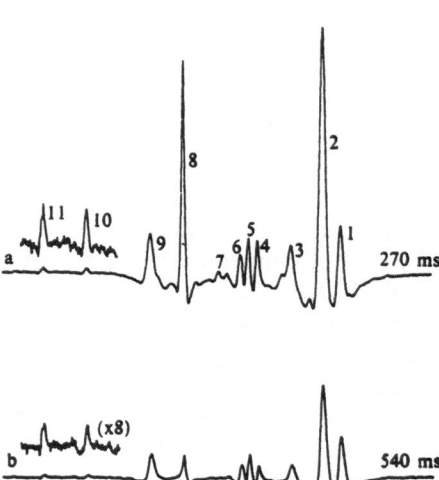

Figure 1.6. [1]H spectra obtained at 360 MHz from rat leg in vivo using Carr–Purcell–Meiboom–Gill (CPMG) spin–echo sequences of duration 270 msec [part (a)] and 540 msec [part (b)]. The thigh of the rat was inserted into a saddle-shaped coil of diameter 18 mm and height 17 mm and, for each spectrum, 16 scans were accumulated with a delay between consecutive pulse sequences of 1.7 sec. Signals are assigned as follows: 1, 2, 3, and 9, various fat protons; 4 and 7, the $N-CH_3$ and $N-CH_2$ protons, respectively, of PCr (with a smaller contribution from Cr); 5 and 6, predominantly taurine; 8, water; 10 and 11, anserine. Chemical shifts are referenced to the PCr signal from $N-CH_3$ protons at 3.03 ppm (from Williams et al.,[37] Academic Press Inc.; reprinted with permission).

were investigated by also obtaining spectra from microsome (containing phospholipids), soluble protein, and perchloric acid extract (low-molecular-weight metabolites) fractions. A major peak in the intact-muscle spectrum was that due to the CH_2 carbons in fatty acids. It was shown that this resonance was strong in the microsome fraction, very weak in spectra obtained from soluble proteins, and undetectable in extracts. This result was related to the phospholipid composition of the fractions. Similar logic was applied to the determination of the identity of many peaks, including those due to lactic acid, protein–amino-acid side-chain carbons, choline, PCr, Cr, residues from histidine, phenylalanine, tyrosine, arginine, tryptophan, and carbonyl–protein peptide-bond carbons. An important additional result of the study was the discovery that the strength of the prominent CH_2 fatty-acid peak was dependent on, and increased with, temperature. It was suggested that, at temperatures below about 10°C, the fatty-acid chains in membranes are largely immobilized but that, on increasing the temperature, molecular interactions are minimized, allowing rapid motion and thus increasing MRS detectability. In the same year, an in vivo study of the rat hind leg was reported.[42] However, due to the high concentration and close proximity to the coil of mobile fatty acids (mainly in subcutaneous adipose tissue), most of the signal detected came from superficial tissue.

1.4.2. Cardiac Muscle

The first MRS studies of whole hearts were reported by Gadian and colleagues in 1976.[43] Excised rat hearts were rapidly cooled to 0°C, arresting the heartbeat in the hope of slowing down biochemical processes involving high-energy phosphates. ^{31}P spectra obtained from the cooled tissue showed resonances assigned to sugar phosphates, Pi, PCr, and magnesium-bound ATP. On warming the unperfused tissue to 30°C, PCr and ATP decreased and the pH_i dropped from 7.0 to 6.0. This work was soon followed by investigations of the intact, perfused, beating heart,[44, 45] which revealed, among other things, the presence of a significant phosphodiester assigned to GPC.

High-resolution ^1H MRS at 361 MHz has also been used to study metabolism in the intact, perfused heart.[46] Resonances, including those due to taurine, carnitine, lactate, and glycerides, were detected. Spectra were collected using a spin–echo technique in which the methyl lactate doublet is inverted; during ischemia, it was possible to detect this metabolite, which resonates very close to the triglyceride CH_2 protons. Perchloric acid extracts from freeze-clamped tissue (normoxic and ischemic) showed additional peaks due to PCr, Cr, succinate, glutamate, acetate, alanine, and glucose.

In 1980, the first in vivo investigation of the whole heart using ^{31}P MRS was reported;[47] a solenoidal receiver coil was invasively placed around the beating organ. In the spectra (of which an example is shown in Figure 1.7), 2,3-diphosphoglycerate was only a minor component, implying that the high-energy phosphates detected were situated in cardiac muscle while signal from whole blood was minimal. Spectra were also obtained from perfused hearts, and the similarity of these to spectra collected in vivo supported the viability of the perfused heart as a model for the study of high-energy phosphate metabolism. During respiratory arrest over a 30-min period, PCr and ATP declined below the level of detectability and Pi was greatly increased. These changes were accompanied by a drop in pH$_i$ from 7.35 to 6.40. It is possible that a significant fraction of the observed Pi originated in whole blood. (In the perfused, normoxic, blood-free heart, Pi gave a pH$_i$ of 7.11.)

The response of compromised cardiac metabolism to pharmaceutical intervention has been investigated in perfused hearts.[48] Localized ventricular ischemia was produced by ligation of the left main coronary artery. Surface coils of diameter 8 to 11 mm were used, and the positioning of

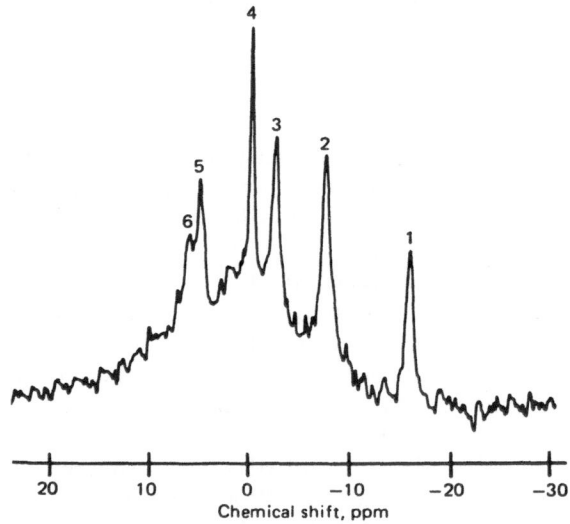

Figure 1.7. ^{31}P spectrum obtained from the in vivo, beating rat heart at 73.83 MHz. The data were acquired using an insulated four-turn solenoidal coil (1.8-cm length × 1.3-cm diameter) placed invasively around the heart. The spectrum represents the Fourier transform average of 1600 scans at 1-sec intervals using a 45° pulse. Peak assignments are as follows: 1, βATP; 2, αATP and αADP; 3, γATP and βADP; 4, PCr; 5, Pi and the 2-phosphate of 2,3-DPG; and 6, the 3-phosphate of 2,3-DPG (from Grove et al.,[47] National Academy of Sciences; reprinted with permission).

these with respect to the ischemic region was checked by staining the tissue with methylene blue dye. It was found that surface coils of suitable dimensions provided very good localization of the ischemic tissue and showed much greater metabolite changes than studies using conventional coils containing the entire organ. The effects of verapamil and chlorpromazine on ischemic metabolism were studied, and both were found to reduce Pi and allow the recovery of control PCr and ATP levels. The ^{31}P spectroscopy results were consistent with the hypothesis that coronary vasodilation can be induced by verapamil, and this may cause increased collateral blood flow in the ischemic region. There was also some indication that hearts treated with verapamil before ligation acquired a degree of mitochondrial protection from ischemia. Chlorpromazine had no apparent vascular effects but did provide sustained phosphorylated substrate levels during anoxia.

Chronically implanted solenoidal coils also have been used to study in vivo cardiac metabolism.[49] These were surgically positioned prior to study by MRS, and various investigations could be performed for up to 24 hr after implantation. Fructose was shown to have no effect on cardiac metabolite levels and, by triggering data collection with the aortic blood pressure, it was possible to show that spectra acquired at systole and diastole were comparable.

Recent work has studied the metabolic changes in cardiac muscle in response to ischemia with a time resolution of 40 sec for pH_i and 5 min for phosphate levels.[50] Within 40 sec of the onset of ischemia, PCr had been greatly reduced, with concomitant elevation of Pi, and the large Pi signal allowed changes in pH_i to be followed in detail. The pH_i decreased from 7.39 to 7.13 in the first 40 sec, to 6.71 after 4 min, and to 6.07 after 30 min. ATP decreased much less rapidly than PCr and had dropped only to 37% of its control level after 25 min of ischemia.

1.4.3. Smooth Muscle

Detailed investigations of guinea-pig taenia coli[51] and taenia caecum[52] using ^{31}P MRS were reported in 1983, although a preliminary report on the application of this technique to the study of the energetics of uterine muscle had already appeared.[53] In the taenia coli study, ^{31}P MRS at 103 MHz was used to investigate superfused, isometrically mounted guinea-pig smooth muscle under contraction with inhibition of respiration. Spectra obtained from fresh, relaxed tissue were very different from those obtained from skeletal muscle. PCr was much lower with respect to ATP, and a conspicuous peak, assigned to sugar phosphates (SP), was detected. Pi and PD were very weak. During contractions, initiated by superfusing with a phosphate-free physiological saline solution, a fully reversible reduction of PCr only was reported. Respiration was inhibited by addition of

NaCN to the superfusate. At 1 mM NaCN, PCr was reduced and a new peak was detected and tentatively assigned to Pi. This change was reversible. With the NaCN concentration at 2.5 mM, PCr was further reduced and ATP also started to decline. These alterations were found to be irreversible. Due to the very low Pi signal, it was very difficult to estimate the pH_i except in spectra acquired during inhibition of respiration. These gave a mean pH_i of 7.0 ± 0.1, and because there appeared to be no contraction- or anoxia-related shift in position of the phosphomonoester peak, it was concluded that the associated metabolic changes were not accompanied by significant alterations in pH_i. (Further work by the same group has confirmed the pH_i by treatment with 2-deoxyglucose-6-phosphate.[54]) Spectra were also shown from samples of superfused urinary bladder and portal vein. The study of taenia caecum[52] produced somewhat different results in that, under conditions of maximal contraction, PCr was reduced until it could no longer be detected, ATP was seen to decrease, and a strong acidosis to pH 6.5 was reported.

The uterus has been studied in some detail, including investigation of the effects of estrogen[55] and of pregnancy and parturition[56] on phosphate metabolism. Figure 1.8 shows a ^{31}P spectrum obtained from isolated uteri at 81 MHz.[53] The PCr content was reported to decrease within the first 1.5 to 3 hr after administration of 17β-estradiol, returning to the control level after between 6 and 12 hr and reaching a maximum at 24 hr. Similar changes were reported for ATP and the phosphomonoesters. An increase in the free Mg^{2+} concentration was detected 1.5 hr after estrogen injection.

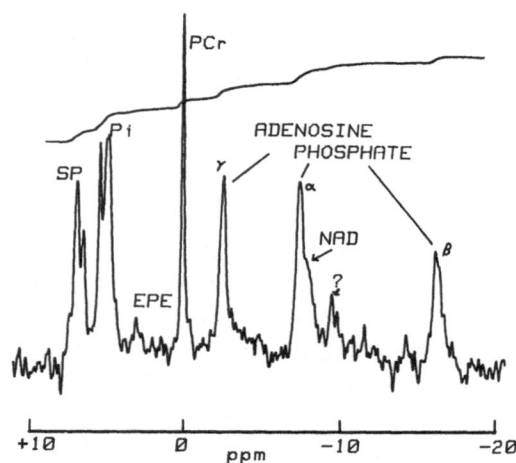

Figure 1.8. [31]P spectrum of four isolated rat uteri (nonpregnant). The spectrum was obtained at 81 MHz using 1198 pulses, each of 30-μsec duration, at 2-sec intervals. The tissue was superfused at 18°C with 25 mMole bicarbonate Krebs solution bubbled with 95% O_2 and 5% CO_2 (from Dawson and Wray,[53] the Physiological Society; reprinted with permission).

This was determined from changes in the chemical shifts of the ATP peaks. In late pregnancy, PCr has been observed to be 1.3 times larger than in the nonpregnant state and, following parturition, PCr fell rapidly to less than the nonpregnant level and remained low for at least 21 days. Pi and PME were seen to almost double their late-pregnancy levels within two days postpartum, and subsequently they declined but still had not regained their nonpregnant levels within 21 days. No significant changes in pH_i were noticed between the late-pregnant and nonpregnant uterus, but a large acid shift to pH_i 6.83 ± 0.11 was reported one day postpartum. The regulation of pH_i in uterine muscle has been investigated by varying the bicarbonate concentration in the superfusate,[57] and it was found that an increase or decrease of the extracellular pH by 1 unit produced a change in pH_i of about 0.3 pH units in the same direction. The acid change was reversible, whereas increases in pH_i were not and were accompanied by large decreases in the PCr concentration. More recent work on bladder and uterine muscle[58] has identified the peak in the monoester region of the spectrum as phosphoethanolamine (PE), and from the chemical shift of the βATP resonance it has been deduced that $[Mg^{2+}]$ is significantly lower than that encountered in striated muscle.

Arterial smooth muscle has received some attention, and multinuclear studies of the normo- and hypertensive aorta using ^{31}P, ^{13}C, ^{23}Na, and ^{1}H spectroscopy have been made.[59] The ^{31}P spectra from perfused tissue showed a PCr level much smaller than that encountered in skeletal muscle and smaller than that revealed by another arterial study.[60] Under conditions of hypertension, high phosphagen content was often found when compared with the normotensive state. Intracellular Pi was very difficult to detect, but other work has indicated a normal pH_i of 7.19 ± 0.03.[60] ^{13}C MRS of aorta fragments produced spectra which showed signals that were assigned mainly to CH_3-, CH_2-, CH-, and CHO-groups of lipids. Dysprosium tripolyphosphate (DysTPP) was used as a chemical shift reagent in the ^{23}Na studies. On addition of DysTPP to the perfusate, the ^{23}Na resonance was split into two components: a large peak, shifted by about 4 ppm, which originated from extracellular and perfusate sodium, and a much smaller, unshifted, peak due to intracellular Na. ^{1}H MRS results were reported from both tissue fragments and perfused tissue. Several molecules of biological interest were identified in spectra from fragments, including lactate, creatine, taurine, and choline. Signals from triglycerides in the perfused aortas made the spectra mainly uninterpretable.

1.4.4. Brain

In 1978, 4 years after the pioneering work of Hoult and colleagues on muscle tissue, the first studies of excised and in vivo mammalian brain tissue were made by Chance and co-workers at the University of Pennsylvania.[61] For studies of excised tissue it was necessary to freeze-trap the samples in order to maintain aerobic levels of energy-related phosphorus metabolites during examination. Rat brain tissue was freeze-trapped in situ using liquid nitrogen, and the sample was then transferred to the sample tube. Subsequently the material was allowed to warm to approximately $-15°C$, at which temperature there was sufficient molecular mobility to obtain adequate signal but not enough to permit rapid enzymatic activity. Spectra were obtained at various intervals after the tissue had reached the specified temperature range. Two strong resonances due to PCr and phosphomonoesters were detected in spectra obtained as soon as possible after freeze-trapping and warming. Other peaks were identified from Pi, PD, and mainly ATP. Within 30 min of reaching $-10°C$, PCr had been significantly reduced with an accompanying increase in Pi. The same group also obtained spectra from the whole heads of living mice which, under anerobic conditions, revealed depletion of PCr and ATP and greatly elevated Pi levels. All the spectra showed a broad feature, about 40 ppm wide, underlying the mobile phosphate peaks which has been identified as mainly due to membrane-bound phospholipids,[62] although spectra obtained in vivo also have a component resulting from phosphorus in bone[63] unless the coil is placed directly on the surface of the brain. Figures 1.9a and 1.9b show ^{31}P spectra obtained from living mammalian brain under normoxic and anoxic conditions, and Figure 1.9c also shows the broad underlying signal due to relatively immobile phosphorus nuclei in phospholipids and bone. The first convincing in vivo ^{31}P spectra were obtained from rat brain in 1980 by the use of a surface coil placed on the cranium.[8] By judicious positioning of the coil, signal from intervening muscle tissue was reduced to a minimum. The PCr/ATP ratio was found to be 1.93 ± 0.12, which was higher than that found by rapid freezing and using other assay techniques.[64] It was also estimated that the concentrations of both Pi and ADP were lower than those suggested by destructive analytical methods.

In two separate studies on the rabbit[65] and gerbil,[66] cerebral ischemia, induced by carotid-artery occlusion, has also been investigated using surface coils placed on the cranium. The study of rabbit brain was primarily motivated by the need to understand the metabolic changes occurring in the brain tissue of newborn infants suffering damage due to cerebral hemorrhage or ischemia with consequent infarction. Studies were done on mature rabbits, and control spectra showed prominent peaks from PCr, ATP, Pi, and phosphodiesters and a strong resonance in the phosphomonoester region. The control spectra were very similar to those obtained from rat brain by Ackerman et al.[8] In the animals with normally oxygenated brain tissue, the ratio of PCr to ATP was found to be

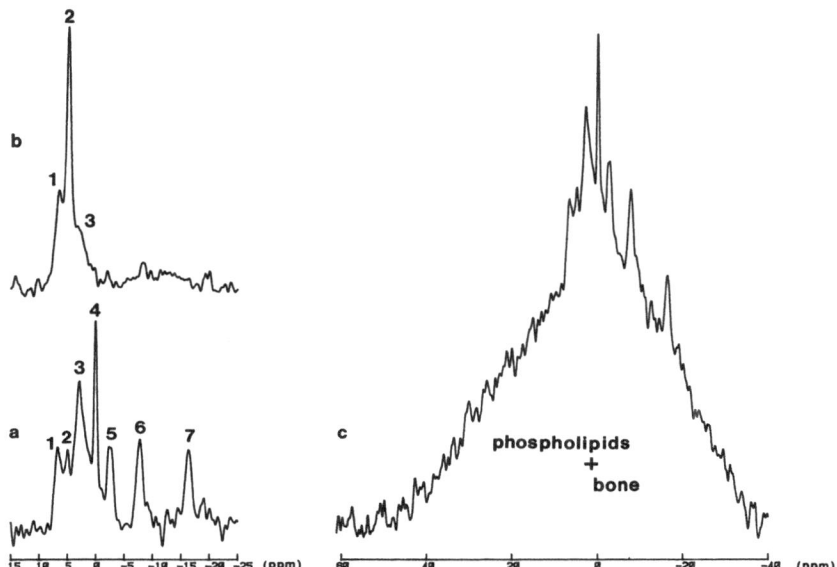

Figure 1.9. [31]P spectra from adult rat brain obtained at 32.5 MHz using a 12-mm, two-turn surface coil placed directly on the exposed skull. The spectra were collected using a 90° pulse at the coil center and a pulse interval of 2.256 sec: (a) Normoxic (768 scans); (b) Anoxic (512 scans); (c) The normoxic brain, as in (a), but including the broad underlying feature due to relatively immobile ^{31}P nuclei mainly in phospholipids and bone. The peak assignments are: 1, phosphomonoesters (mainly phosphoethanolamine and phosphocholine); 2, Pi; 3, phosphodiesters (mainly GPC and GPE); 4, PCr; 5, γATP and βADP; 6, αATP, αADP, and NAD/NADH; 7, βATP. Peaks 5, 6, and 7 also contain small but significant contributions from other nucleoside triphosphates. (Spectra obtained in collaboration with K. Ives.)

2.30 ± 0.13, and pH$_i$ was 7.33 ± 0.13. Under cerebral ischemia, PCr decreased rapidly but was accompanied by a reduction in ATP and an increase in Pi concentration. During ischemia, the brain tissue became acidotic, with a mean minimum pH of 6.7. Following reperfusion, metabolites returned to control levels except in one study in which the lowest pH$_i$ was found (6.54) and the spectra persistently showed a raised Pi.

The study of the effects of ischemia on gerbil brain metabolism was accomplished by occlusion of the right carotid artery only, and thus it was possible, by careful positioning of the surface coil, to try to investigate biochemical changes in each cerebral hemisphere and the cerebellum. In control spectra, pH$_i$ was estimated to be 7.2 ± 0.1 and PCr/ATP was found to be 1.5, although the data were not corrected for the effect of inadequate nuclear relaxation and the ratio served only as a comparison for values obtained during ischemia in the same study. After ligation of the right carotid artery, PCr and ATP were absent from spectra collected from the right hemisphere, which showed a large Pi peak. Spectra obtained from the left hemisphere also exhibited a raised Pi level but retained significant PCr and ATP peaks. In a separate set of experiments, ^{31}P spectra were monitored for an hour after ligation, and it was found that edema (assessed by grey-matter specific gravity measurements) correlated strongly with spectral abnormality. Several similar studies have been conducted to investigate cerebral energy metabolism in hypoxia, hypoglycemia, and status epilepticus[67, 68] and during acute potassium cyanide intoxication.[69] Serial studies giving time-resolved spectra have also been successful,[70] and it has been demonstrated that ^{31}P MRS can be used to investigate cerebral metabolism under hypoxic and hypoglycemic conditions using superfused slices of tissue.[71]

The identification of the ^{31}P resonances present in brain spectra has been greatly helped by work on tissue extracts and by investigation of the pH dependences of chemical shifts. Work carried out on perchloric acid extracts of guinea-pig brain using a 4.7-tesla system has produced a listing (see Table 1.2) of phosphorylated constituents, including PCh, glycerol 3-phosphorylethanolamine (GPE), GPC, Pi, PCr, ATP, ADP, NAD, NADH, and uridine diphosphates.[72] Unfortunately a prominent peak in the monoester region of the spectrum was originally misidentified as ribose 5-phosphate, but subsequent investigation by thin-layer chromatography and high-resolution MRS (including pH titration) has given strong indications that this peak is due to PE.[73, 74]

The application of ^{1}H MRS to studies of brain metabolism closely followed the use of ^{31}P, although an investigation at 220 MHz of intact rabbit sciatic nerve was reported as early as 1972.[75] Resonances due to choline, cholesterol, and phospholipids were identified, as were several unassigned signals. The first in vivo ^{1}H spectra were obtained from normoxic, hypoxic, and recovering rat brain in 1983 at 360 MHz using a surface coil placed on the exposed cranium.[76] Data were collected using a simple single-pulse technique with swept frequency saturation to suppress the signal from tissue water. The identification of resonances was assisted by reference to spectra obtained from perchloric acid extracts of brain tissue frozen in situ. Signals from PCr, Cr, PCh, aspartate, N-acasp, glutamate, GABA, alanine, and lactate were recognized both in the extracts and in intact excised samples, which also showed lipid resonances. The strongest peaks in the extracts and intact samples were identified as PCh, PCr and Cr, N-acasp, and the methyl doublet of lactate. The N-acasp resonance, the most intense, was used as an internal reference for chemical shift measurements. (This metabolite appears to be found only in brain

Table 1.2. Chemical Shifts and Relative Concentrations of ^{31}P Metabolites in the Perchloric Acid Extract Spectrum of Adult Guinea-Pig Brain[a,b]

^{31}P metabolite	Chemical shift relative to PCr (ppm; mean ± SEM)	Fraction of total extractable phosphate (%; mean ± SEM)
U[c]	13.81 ± 0.001	0.21 ± 0.027
U	13.09 ± 0.006	0.08 ± 0.012
U	7.93 ± 0.014	0.14 ± 0.024
U	7.77 ± 0.004	0.14 ± 0.025
U	7.74 ± 0.001	0.31 ± 0.035
Hexose-6-phosphates	7.59 ± 0.002	0.80 ± 0.041
U	7.45 ± 0.002	0.35 ± 0.038
α-glycerylphosphate	7.41 ± 0.001	0.62 ± 0.048
Fructose 1,6-diphosphate	{ 7.23 ± 0.004 / 7.17 ± 0.004	0.88 ± 0.059
2,3-diphosphoglycerate[d]	7.06 ± 0.003	1.68 ± 0.286
Ethanolamine phosphate	6.98 ± 0.002	9.57 ± 0.241
Inosine 5'-monophosphate	6.90 ± 0.001	1.25 ± 0.211
Adenosine 5'-monophosphate	6.87 ± 0.002	0.84 ± 0.080
NADP 2'-phosphate	6.68 ± 0.002	0.31 ± 0.016
U	6.56 ± 0.002	0.75 ± 0.084
Choline phosphate	6.45 ± 0.007	2.14 ± 0.160
U	6.05 ± 0.010	0.14 ± 0.039
Inorganic orthophosphate	5.75 ± 0.004	9.76 ± 0.403
Glucose 1-phosphate	5.42 ± 0.003	0.25 ± 0.016
U	4.97 ± 0.002	0.18 ± 0.011
U	4.04 ± 0.001	0.17 ± 0.012
Acid-labile phosphate	3.97 ± 0.006	0.79 ± 0.111
Glycerol 3-phosphorylethanolamine	3.93 ± 0.006	1.79 ± 0.109
Glycerol 3-phosphorylcholine	2.99[e]	3.78 ± 0.076
Phosphocreatine	−0.00 ± 0.002	11.21 ± 0.357
U	−1.15 ± 0.068	0.13 ± 0.034
αATP	−7.80 ± 0.003	39.28 ± 0.447
βATP	−18.33 ± 0.005	
γATP	−2.68 ± 0.004	
αADP	−7.49 ± 0.006	6.31 ± 0.216
βADP	−2.99 ± 0.004	
Nicotinamide adenine dinucleotides[f]	−8.25	6.85 ± 0.196
Uridine diphosphogalactose[g]	−9.71 ± 0.006	0.73 ± 0.096
Uridine diphosphoglucose[g]	−9.87 ± 0.010	0.91 ± 0.127
Uridine diphosphomannose[g]	−10.58 ± 0.010	0.52 ± 0.065

[a] From Glonek, T., Kopp, S. J., Kot, E., et al.: P-31 nuclear magnetic resonance analysis of brain: The perchloric acid extract spectrum, *J. Neurochem* 39:1210–1219.

[b] Data were collected at 81 MHz using a 45° flip angle, a pulse repetition time of 1.64 sec and with the extract's pH adjusted to 10. Chemical shifts are reported relative to 85% H_3PO_4.

[c] Unassigned.

[d] This is a complex resonance signal with several other phosphates contributing apart from 2,3-diphosphoglycerate.

[e] Internal reference.

[f] The chemical shift given is the approximate location of the center of the NAD and NADP multiplets.

[g] The chemical shifts denote the centers of the hexose end-group phosphate doublets.

tissue, and its function is as yet not clear.[77] Its concentration is nearly 10 mmole, but this varies in different parts of the brain, and there is evidence of rapid increase with development in neonatal tissue. The amide-proton resonance of this metabolite has recently been found to have a chemical shift which depends on temperature.[16]) In vivo spectra obtained under normoxic conditions showed a lactate signal that was small when compared with the adjacent N-acasp peak. However, during hypoxia this resonance increased dramatically until it was substantially larger than that of N-acasp. Upon returning to 25% O_2, the lactate peak was seen to fall almost to its initial level. Several successive hypoxic insults were applied, and lactate production appeared to be related to the severity of the insult. The same group subsequently demonstrated the feasibility of ^1H MRS at the low field strengths currently used for clinical work.[78] They looked at lactate levels in hypoxemic rabbit brain using a 1.9-tesla spectrometer (80-MHz proton frequency) with a surface coil placed directly on the exposed dura mater. A 120 msec spin–echo sequence was used which was sufficient to selectively suppress the signal from tissue water and to reduce spectral features due to lipid and other broad resonances. The possibility of collecting alternate, sequential ^1H and ^{31}P spectra so that lactate production during insulin-induced hypoglycemia can be related to pH_i has also been shown.[79] During a period of hypoglycemia, the ^1H spectra showed decreases in resonances assigned to glutamate and glutamine and an increase in that of aspartate, whereas ^{31}P spectra eventually revealed almost complete depletion of PCr and nucleoside triphosphates along with concomitant accumulation of Pi. On glucose administration, the glutamate and aspartate resonances reverted to their control levels while only incomplete recovery was observed for glutamine; PCr attained its control level, whereas Pi and the nucleoside triphosphates did not. During the recovery period, an increase in lactate was seen, but this did not appear to correlate with any alteration in pH_i, thus demonstrating some degree of buffering.

One problem with the application of ^1H MRS to many tissues is the presence of strong signals from fatty acids that overlap several important resonances, including those from methyl protons in lactate and several amino acids. The previously mentioned studies of brain tissue were all invasive because surgical procedures were required to remove most of the adipose tissue which had caused the intense peaks. In order to eliminate the necessity for surgery, a double-resonance method has been developed to detect resonances due to alanine, GABA, glutamate, taurine, and lactate in the brains of intact rats.[80] (This technique has already been mentioned with respect to ^1H studies of skeletal muscle.) Threefold increases in the strengths of the α- and γ-CH_2 signals from GABA were detected within 30 min of death. It has also been possible to obtain spectra limited to those protons that are directly bonded to ^{13}C nuclei and hence to determine

Table 1.3. Assignments of Proton Resonances in Excised Rat Brain[a,b]

Chemical shift[c] (ppm)	Resonance assignment
0.87	$-CH_3$ in amino-acid side chains of cytosolic proteins, $-CH_3$ in fatty
0.91	acyl chains of membrane phospholipids
1.19	$-CH_3$ and $>CH_2$ in amino-acid side chains of cytosolic proteins
1.32	$-CH_3$ lactate
1.39	$-CH_3$ and $>CH_2$ in amino-acid side chains of cytosolic proteins
1.46	$-CH_3$ of alanine
1.67	$>CH_2$ in amino-acid side chains of cytosolic proteins
1.90	$\beta-CH_2$ of γ-aminobutyrate
	$-CH_3$ of acetate
2.01	$-CH_3$ of N-acetylaspartate
2.11	$\beta-CH_2$ of glutamate and glutamine
2.27	$\alpha-CH_2$ of γ-aminobutyrate
2.35	$\gamma-CH_2$ of glutamate
	CH_2 of succinate
2.45	$\gamma-CH_2$ of glutamine
2.49	
2.52	
2.66	$\beta-CH_2$ of N-acetylaspartate
2.69	
2.78	$\beta-CH_2$ of aspartate
2.82	
3.02	$\gamma-CH_2$ of γ-aminobutyrate
	$>N-CH_3$ of creatine
3.20	$-N(CH_3)_3$ in choline
	$-N(CH_3)_3$ in phosphorylcholine and α-glycero-phosphorylcholine
	$>\overset{+}{N}-CH_2$ of taurine
3.24	$-CHOH$ of inositol (H5)
	$>\overset{+}{N}-(CH_3)_3$ in phosphatidylcholine in the membrane
3.42	$-S-CH_2$ of taurine
3.54	$-CHOH$ of inositol (H1, H3)
	$\alpha-CH_2$ of glycine
3.62	$-CHOH$ of inositol (H4, H6)
3.75	$\alpha-CH$ of glutamate, glutamine, and alanine
3.89	$\alpha-CH$ of aspartate
3.91	$>N-CH_2$ of creatine
4.06	$-CHOH$ of inositol (H2)
4.10	$-CHOH$ of lactate
5.74	$-NH_2$ in urea[d]

[a] From Arus et al.,[82] Meridional Publications. Reprinted with permission.
[b] Compiled from data acquired at 361 MHz with a sample temperature of about 36.8 °C.
[c] The $N-CH_3$ resonance of creatine at 3.022 ppm was used as an internal chemical shift standard-referenced to sodium 3-trimethylsilyl [2,2,3,3-$^2H_4$] propionate (TSP) at 0.000 ppm.
[d] Temperature-dependent chemical shift.

(*continued*)

Table 1.3. *Continued*

Chemical shift[c] (ppm)	Resonance assignment
6.14	H1' of ribose in nucleotides and nucleosides
6.68	$H_2 \overset{+}{N} = C-$ in creatine[d]
6.80	amino protons in glutamine[d]
7.08	C4-H proton histidine of cytosolic proteins[d], exchangeable amino protons of cytosolic proteins
7.30 to 7.40	exchangeable amide protons of cytosolic proteins[d]
7.51	amino protons in glutamine[d]
7.83	amide proton of *N*-acetylaspartate[d]
8.17 to 8.59	H8 and H2 protons of adenine nucleotides and nucleosides, exchangeable protons of cytosolic proteins[d]

(For the 6.68 entry, the structure shown includes NH_2 attached above the carbon: the group $H_2\overset{+}{N}=C-$ with an NH_2 substituent.)

fractional enrichments of this isotope with eleven times the sensitivity of the straightforward ^{13}C experiment.[81]

High-resolution, 360-MHz ^{1}H spectra have been obtained from excised brain tissue in order to facilitate the interpretation of spectra and the identification of resonances.[82] The spectra revealed that there were contributions from cytosolic proteins, membrane phospholipids, and sixteen different metabolites identified by examination of perchloric acid extracts and subcellular fractions (see Table 1.3). Changes in the spectra due to anesthesia, ischemia, and stage of brain development were reported. These alterations included an increase in myo-inositol and a decrease in taurine levels in 12-day-old rats when compared with adults. Further work by another group has described in greater detail the developmental dependences of fifteen brain metabolites including *N*-acasp, GABA, glutamate, glutamine, aspartate, GPC, taurine, myo-inositol, Cr and PCr.[83] The ^{1}H resonances detected in part of the 360-MHz extract spectrum of mammalian brain tissue are shown in Figure 1.10.

Recently, ^{1}H MRS has been applied to the study of a defect in amino-acid metabolism in the mutant histidinemic mouse.[14] The C2 and C4 proton signals from histidine both occur in a resonance band well away from potentially overlapping fatty acids, and it was possible to demonstrate raised levels of this amino acid in the mutants at both 8.5 and 1.9 tesla.

The first in vivo ^{13}C study of the brain was reported in 1981.[84] A surface coil was placed on the intact head of an anesthetized rat, and proton decoupling was accomplished by transmitting only during the data-acquisition period so as to reduce any heating effects. Undoubtedly some of the resonances detected originated in adipose tissue near the coil, but in addition to the strong signal from CH_2 carbons, there were also peaks assigned

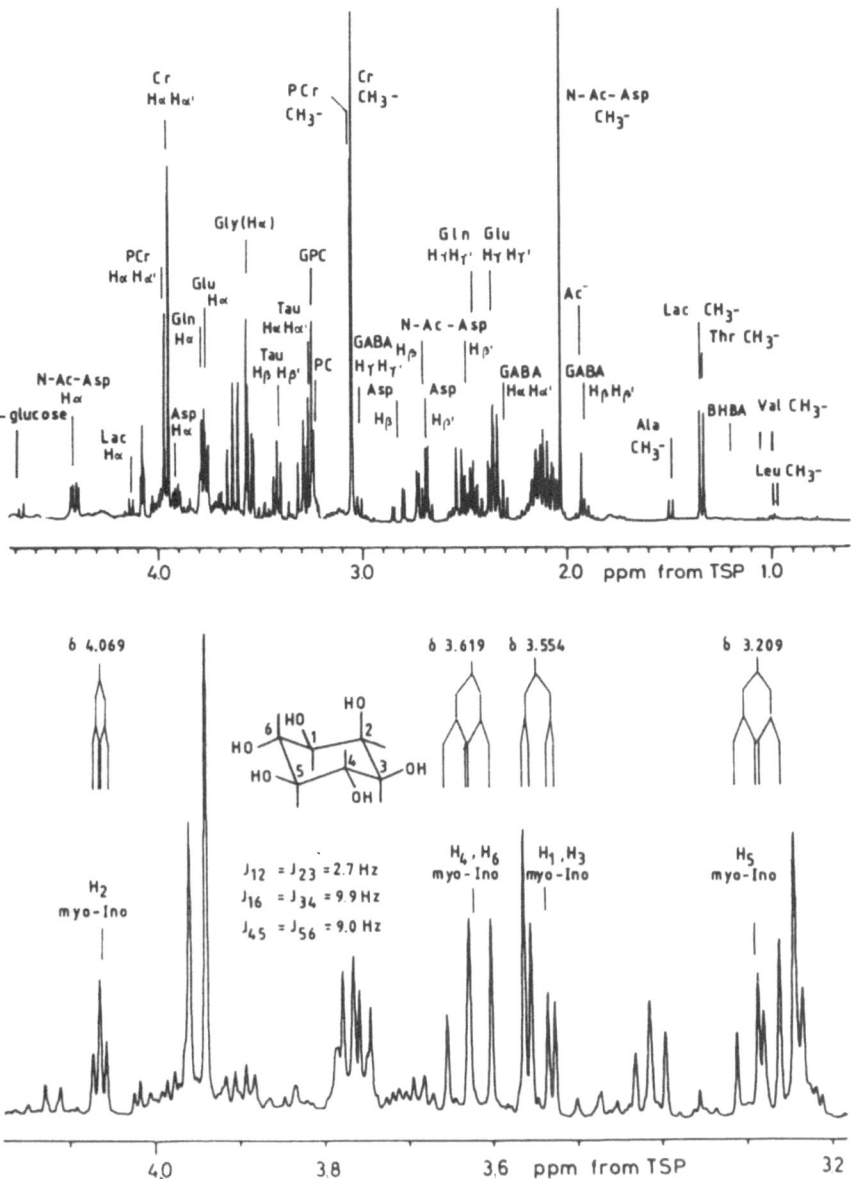

Figure 1.10. Part of the high-resolution ^{1}H spectrum obtained from a perchloric acid extract of normoxic, adult rat brain at 360 MHz. The spectrum was collected with a 90° pulse, a 3-sec pulse interval, and 256 scans: Lac, lactate; BHBA, β-hydroxybutyrate. The J values refer to line splittings, and the δ values are chemical shifts. The only resonances shown are those to be found at chemical shifts less than that of the water peak, at about 4.8 ppm, which in this spectrum was presaturated to reduce its intensity. Conspicuously absent from the extract spectrum are lipid resonances, which would be present in vivo (from Cerdan et al.,[83] Elsevier Science Publishers; reprinted with permission).

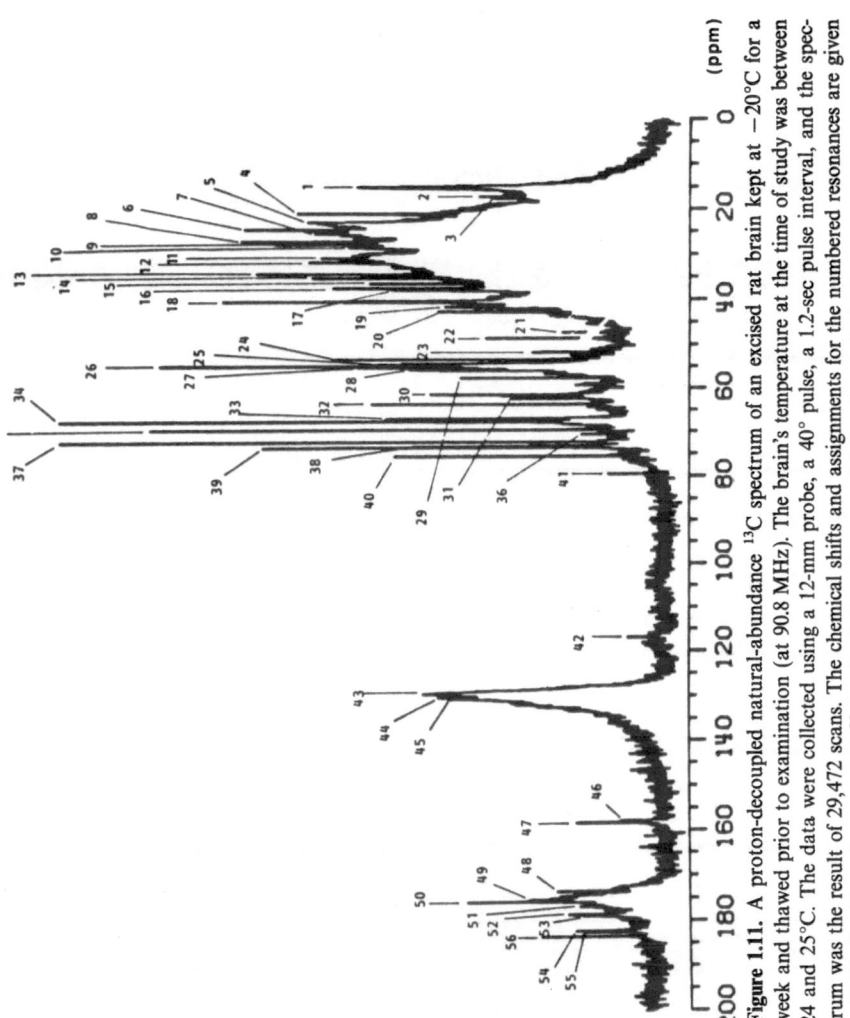

Figure 1.11. A proton-decoupled natural-abundance ^{13}C spectrum of an excised rat brain kept at $-20°C$ for a week and thawed prior to examination (at 90.8 MHz). The brain's temperature at the time of study was between 24 and 25°C. The data were collected using a 12-mm probe, a 40° pulse, a 1.2-sec pulse interval, and the spectrum was the result of 29,472 scans. The chemical shifts and assignments for the numbered resonances are given in Table 1.4 (from Barany et al.,[86] Academic Press Inc.; reprinted with permission).

to choline and arginine. It was suggested that the arginine is a constituent of the brain-myelin protein.

Most ^{13}C studies have so far been done with ^{1}H decoupling power from an additional, larger coil. This can present problems in terms of unnecessary heating of a large volume of tissue and detuning due to inductive coupling between the coils. A surface coil tunable at both the ^{13}C and ^{1}H frequencies has been designed which avoids these problems and has been successfully used to obtain high-resolution, decoupled ^{13}C spectra from excised mammalian brain tissue.[85] The coil was used to examine a sample

Table 1.4. Chemical Shifts of the Natural-Abundance [13]C Resonances of Excised Intact Rat Brain Correlated with Pure Brain Metabolites and Their Localization in Various Subfractions[a,b]

Carbon atom	δ^c (ppm)	Peak number[d]	δ^d (ppm)	Presence of the resonance[e]		
				in PCA extract	in Cytosol	in Microsomes
C1 glycine	173.30	48	173.21	+	+	
C2 glycine	42.48	20	42.43	+	+	
C1 PE	61.93	31	62.08	+	+	
C2 PE	41.55	19	41.40	+	+	
C1 taurine	48.41	22	48.37	+	+	
C2 taurine	36.17	15	36.35	+	+	
C1 alanine	176.58	51	176.61	+	+	
C2 alanine	51.50	23	51.47	+	+	
C3 alanine	17.11	2	17.09	+	+	
C1 lactate	183.28	56	183.40	+	+	
C2 lactate	69.24	35	69.33	+	+	
C3 lactate	20.95	4	20.98	+	+	
C1 aspartate	175.10	49	175.04	+	+	
C2 aspartate	52.96	24	53.14	+	+	
C3 aspartate	37.39	16	37.40	+	+	
C4 aspartate	178.33	52	178.48	+	+	
C1 creatine	175.57	50	175.34	+	+	
C2 creatine	37.78	17	37.78	+	+	
C3 creatine	158.00	47	157.95	+	+	
C4 creatine	54.66	26	54.67	+	+	
C1 GABA	182.28	55	182.26	+	+	
C2 GABA	35.17	14	35.18	+	+	

[a] From Barany et al.,[86] Academic Press Inc. Reprinted with permission.
[b] Chemical shifts are relative to that of dioxane at 67.40 ppm (peak number 34 in Fig. 1.11).
[c] Refers to the pure brain metabolite.
[d] Refers to the resonances in Fig. 1.11.
[e] + refers to a detectable peak.

(*continued*)

Table 1.4. *Continued*

Carbon atom	δ^c (ppm)	Peak number[d]	δ^d (ppm)	in PCA extract	in Cytosol	in Microsomes
				\multicolumn{3}{Presence of the resonance[e]}		
C3 GABA	24.54	6	24.55	+	+	
C4 GABA	40.11	18	40.29	+	+	
C1 + 3 inositol	73.29	39	73.31	+	+	
C2 inositol	73.10	38	73.10	+	+	
C4 + 6 inositol	71.91	37	72.02	+	+	
C5 inositol	75.13	40	75.25	+	+	
C1 glutamate	175.34	50	175.34	+	+	
C2 glutamate	55.43	28	55.57	+	+	
C3 glutamate	27.81	9	27.84	+	+	
C4 glutamate	34.19	13	34.27	+	+	
C5 glutamate	181.96	54	181.98	+	+	
C1 glutamine	174.79					
C2 glutamine	55.03	27	55.06	+	+	
C3 glutamine	27.08	8	27.13	+	+	
C4 glutamine	31.67	12	31.70	+	+	
C5 glutamine	178.42	52	178.45	+	+	
C1 + 3 GPC	62.89	32	63.40	+	+	
C2 GPC	71.47	36	70.43	+	+	
C3 GPC	66.81	33	66.77	+	+	
C4 GPC	60.26	30	61.11	+	+	
C5 GPC	54.78	26	54.67	+	+	
C1 *N*-acasp	179.65	53	178.89	+	+	
C2 *N*-acasp	53.99	25	54.08	+	+	
C3 *N*-acasp	40.44	18	40.29	+	+	
C4 *N*-acasp	179.52	53	178.89	+	+	
C5 *N*-acasp	174.27					
C6 *N*-acasp	22.81	−5	22.86	+	+	
U[f]		21	47.00	+	+	
U (protein side chains)		3	17.43		+	
		29	57.31		+	
		42	117.99		+	
U		1	15.07			+
U		7	25.45			+
U		10	28.45			+
U (membrane fatty acids)		11	30.75			+
U		43	129.01			+
U		44	129.86			+
U		45	130.16			+
U		41	79.33			
U		46	157.53			
Dioxane		34	67.40			

[f] U refers to an unassigned resonance.

of excised rabbit brain after enrichment of metabolic pools with ^{13}C. A high-resolution spectrum from 10 to 80 ppm was obtained and showed prominent peaks assigned to glutamate (C_4, C_3, and C_2), glutamine (C_4, C_3, and C_2), lactate (C_3), and GABA.

High-resolution ^{13}C spectra covering the complete range of biological chemical shifts (approximately 0 to 200 ppm) have recently been described.[86] The spectra, of which an example is shown in Figure 1.11, were obtained at 90.8 MHz from excised rat brain. The tissue had been stored at $-20°C$ for several days prior to study. Table 1.4 lists the fifty-five resonances that were resolved. These were identified by comparing the spectra from excised tissue with those obtained from perchloric acid extracts and from pure brain metabolites in solution. It is interesting to note that ^{13}C spectra obtained from brain samples acquired from other mammalian species (rabbit, pig, and human) were reported to be very similar.

1.4.5. Liver

In 1978, Cohen et al.[87] studied a suspension of rat hepatocytes at 4°C. The strongest signals detected were from Pi, GPE, GPC, ATP, and several overlapping resonances in the phosphomonoester region, including possibly PCh. PCr was not detected and, although this may have been an artifactual consequence of the assay technique, conventional biochemical methods also indicate that liver does not contain a significant amount of this metabolite. In addition, Cohen and colleagues described the presence of two Pi resonances in their spectra. These were assigned to mitochondrial and cytosolic Pi with each compartment at a slightly different pH. Studies have also been made of perfused livers, and the metabolic effects of hypoxia, glucose, insulin, and fructose metabolism have been investigated.[88-90] In 1980 the first serious attempt was made to obtain non-invasive in vivo spectra from a localized volume of rat liver.[91] Because liver tissue contains very little PCr, the efficacy of some localization methods (in particular, those designed to remove signals from surface tissues) can be tested by their ability to eliminate abdominal-muscle PCr signal from spectra. In the study under consideration, a surface coil was used in conjunction with profiling of the main magnetic field.[25] By arranging that only the tissue volume of interest experienced a homogeneous field (giving sharp resonances), while outside that volume large field gradients were present (producing greatly broadened spectral features), it was possible to separate out the spectral contributions from the liver and adjacent muscle. (This localization technique will also be described in Chapter 5.) In addition, liver metabolites have relatively short spin–lattice relaxation times, and it is thought that this is due mainly to the presence of paramagnetic ions in the tissue. For this reason, rapid pulsing could be used to saturate and hence to reduce the signal from the resonances of superficial muscle while still obtaining a good response from liver tissue. These approaches to the study of internal organs have also proved very useful in subsequent clinical studies on humans (described in Chapter 3). Invasive studies of the liver using only a surface coil have been reported as well,[92] and Figure 1.12 shows a ^{31}P spectrum obtained in this way from a live, anesthetized rat at 81 MHz.

A 470-MHz ^{1}H spectrum of intact, excised liver has been reported.[36] The main resonances detected were assigned to lactic acid, alanine, glutamic acid, glutamine, valine, leucine, betaine, carnitine, glucose, and glycogen. The metabolism of the analgesic drug acetaminophen has also been studied in intact hepatocytes and extracts, using single-pulse and spin–echo ^{1}H MRS at 500 MHz.[93] The main resonances detected in spectra from intact cells were due to glucose, fatty acids, and $-N(CH_3)_3$ groups, although data from extracts showed resonances from many more low-mol. wt. metabolites including amino acids, ketone bodies, lactate, and acetate.

[13]C studies have also been made, and Figure 1.13 shows a localized

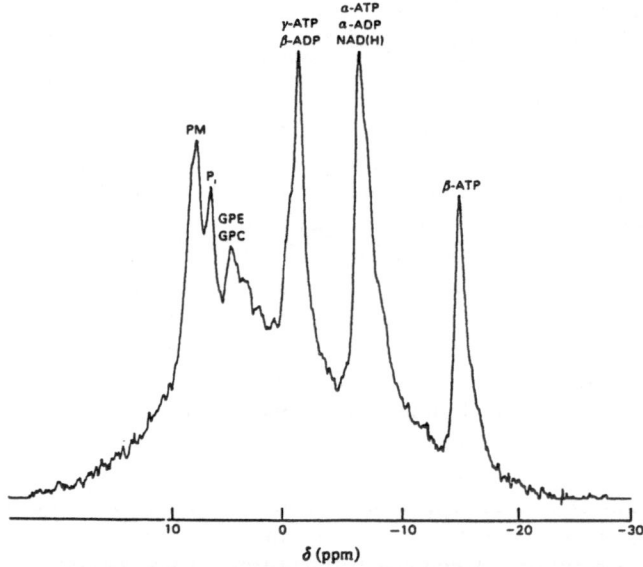

Figure 1.12. The ^{31}P spectrum of the liver of a fed, anesthetized rat (recorded at 81 MHz). The liver was exposed by an abdominal incision and covered in plastic film so that a two-turn, 15-mm-diameter surface coil could be placed on its surface. The spectrum was the sum of 1000 scans accumulated over a period of 34 min using a 30° pulse: PM, phosphomonoesters, mainly phosphocholine with small amounts of phosphorylated monosaccharides, e.g. glycerol-1-phosphate (from Iles et al.,[92] The Biochemical Society; reprinted with permission).

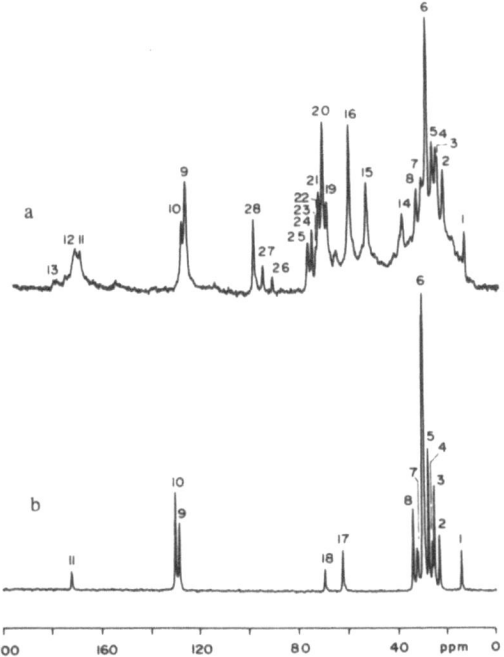

Figure 1.13. High-resolution proton-decoupled ¹³C spectra of excised liver (a) and excised adipose tissue (b) from a fed rat (recorded at 90.5 MHz). The samples were examined at 24°C using a 45° pulse and a 0.5-sec pulse interval, and the spectra are the result of 1024 scans. The assignments and chemical shifts of the numbered resonances are given in Table 1.5 (from Canioni et al.,[94] American Chemical Society; reprinted with permission).

spectrum obtained noninvasively from mammalian liver at 90.5 MHz using magnetic field profiling.[94] The resonances in the spectrum are identified in Table 1.5.

1.4.6. Kidney

The problem of kidney-tissue viability has been investigated using ³¹P MRS in a transplant model.[95] An excised rat kidney was perfused with blood from a second animal, and it was concluded that the maintenance of an almost normal pH indicated that irreversible damage had been avoided. The normal spectra showed prominent peaks from ATP, Pi, AMP, and possibly GPC. Very little PCr was detected. Acute renal acidosis has been examined using excised perfused rat kidneys challenged by the addition of HCl to the perfusion medium.[96, 97] The drop in perfusate pH was matched by a similar decline in pH$_i$ as measured from the chemical shift of the Pi peak.

Table 1.5. Assignments and Chemical Shift Values of ^{13}C Resonances of Rat-Liver and Rat–Adipose-Tissue Spectra[a]

Resonance	Assignment[b]	Chemical shift[c] (ppm)
1	$*CH_3-CH_2-CH_2-$, f.a.c.	14.20
2	$CH_3-*CH_2-CH_2-$, f.a.c.	23.05
3	$-*CH_2-CH_2-CO-$, f.a.c.	25.20
4	$-CH=CH-*CH_2-CH=CH-$, f.a.c.	25.45
5	$-CH=CH-*CH_2-CH^2-$, f.a.c.	27.60
6	$-(CH_2)_n-$, f.a.c.	30.05
7	$CH_3-CH_2-*CH_2-$, f.a.c.	32.30
8	$-CH_2-*CH_2-CO-$, f.a.c.	34.15
9	$-CH=*CH-CH_2-*CH=CH-$, f.a.c.	128.40
10	$-*CH=CH-CH_2-CH=*CH-$, f.a.c.	130.00
11	$-CH_2-CH_2-*CO-$, f.a.c.	172.10
12	$-CO-OR$, protein, phospholipid	173.80
13	$-COO-$, Glu, free fatty acid	182.20
14	$-CH_2-NH_2$, ethanolamine (protein)	40.00
15	$(CH_3)_3N-$, choline	54.60
16	C6, α,β-glucose, glycogen	61.40
17	C1, C3, glycerol (ester)	62.20
18	C2, glycerol (ester)	69.50
19	C4, α,β-glucose	70.40
20	C2, C5, α-glucose; C5, glycogen	72.10
21	C3, α-glucose	73.50
22	C3, glycogen	73.95
23	C2, β-glucose	74.90
24	C3, C5, β-glucose	76.50
25	C4, glycogen	78.05
26	C1, α-glucose	92.70
27	C1, β-glucose	96.60
28	C1, glycogen	100.50

[a] From Canioni et al.,[94] American Chemical Society. Reprinted with permission.
[b] * indicates the specifically assigned carbon; f.a.c. indicates fatty acyl chain.
[c] Chemical shifts are in ppm relative to tetramethylsilane. The methyl group of a fatty acyl chain is used as an internal reference at 14.2 ppm relative to tetramethylsilane. The resonances are numbered as indicated in Fig. 1.13.

This organ is situated at a greater depth than the liver and presents more difficulties for noninvasive in vivo study. However, ^{31}P studies of in vivo kidney, using magnetic field profiling to reduce signal from adjacent tissues, have also been carried out.[98]

1.4.7. Adipose Tissue

^{13}C MRS is usually handicapped greatly by the low natural abundance of the nucleus (1.11%, as shown in Table 1.1). However the high concentration of fatty acids in adipose tissue enables excellent signal strength to be obtained in reasonable time, and for this reason it has been possible to study effects such as those of dietary-fat modification on the composition of rat-epididymal adipose tissue.[94] Proton-decoupled spectra obtained from normally fed excised tissue showed strong resonances, which were assigned to carbon nuclei in many different fatty acyl-chain components, and two peaks were detected due to glycerol esters. A typical spectrum is shown in Figure 1.13, and resonance assignments are given in Table 1.5. Three dietary regimes were examined: control, fat-free, and high in polyunsaturated fat. The fat-free diet produced no change in the spectrum, but the regime high in polyunsaturated fat showed a large increase in a peak at 128.5 ppm corresponding to double-bond chain carbons. This was interpreted as indicating increased storage of polyunsaturated fat in adipose tissue.

1.4.8. Lung(s)

Very little MRS research has been done on this important tissue. The investigation of this organ can be hampered by alveolar gas–tissue interfaces, which generate magnetic field inhomogeneities and hence cause spectral-line broadening. A brief description of the ^{31}P spectrum has appeared in abstract form,[99] and the metabolic behavior of porcine lung tissue in response to anoxic-ischemic insult has recently been investigated.[100] In the latter investigation, the lungs were degassed in vivo by ventilating with 100% O_2 and allowing the circulation to remove the alveolar gas. This technique produced far better results than vacuum degassing. The lungs were perfused with autologous blood at 38°C, and blood gases, perfusion pressure, glucose, and pH were monitored and could be adjusted if required. Resonances due to Pi, ATP, ADP, NAD + NADH, and unassigned phosphodiesters and phosphomonoesters were reported in the

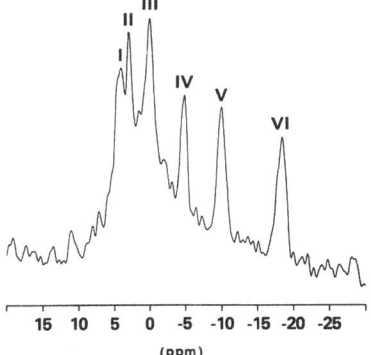

Figure 1.14. A ^{31}P spectrum obtained at 32.5 MHz from degassed, perfused pig lung. The left lobe of the organ was positioned over a 2-cm, two-turn surface coil, and 256 scans were collected using a 1-sec repetition time. The resonances are phosphomonoester (I); Pi (II); phosphodiester (III); γATP and βADP (IV); αATP, αADP, and NAD/NADH (V); and βATP (VI) (from Pillai et al.,[100] Academic Press Inc.; reprinted with permission).

normoxic perfused lung. PCr was below the level of detectability, as shown
in Figure 1.14, and this correlated with results from enzymatic assay of
freeze-clamped tissue. The normoxic pH_i was determined from the Pi
chemical shift and reported to be 7.47. After 15 min of anoxia, a 35%
decline in ATP, a 20% increase in Pi, a fall in pH to 7.11, and small
changes in phosphodiester and phosphomonoester levels were reported.
Complete recovery occurred during 25 min of normoxia. Ischemia
appeared to produce more extreme changes. After 15 min, ATP had fallen
by 90%, Pi had risen by 120%, and pH_i had dropped to 6.37, although
complete recovery was reported on reperfusion.

1.4.9. Testes

So far very little work has been done on this tissue, although the
normally exterior siting of the organ makes it ideal for examination with
surface coils. The clinical application of MRS to testes is therefore an
immediate possibility. Both ^{31}P and ^{1}H MRS have been used to study
whole, excised tissue and perchloric acid extracts, respectively.[101] Because,
in this investigation of rat tissue, little attempt was made to arrest
metabolic processes after dissection, no conclusive information was
obtained about the high-energy phosphates, PCr and ATP. However, the
^{31}P spectrum of a whole testis showed two prominent phosphomonoester
peaks which were assigned to PE (including some phosphoserine) and
PCh, a peak assigned to GPC, and large Pi and small PCr resonances. The
latter two features probably related strongly to the method of sample
preparation. Further dissection of the tissue allowed the elucidation of the
anatomical distribution of metabolites. The caput and cauda epididymis
produced similar spectra with large GPC resonances and smaller signals
identified as GPE. The seminal vesicles also showed a prominent GPC
resonance. High-resolution ^{1}H spectroscopy revealed resonances due to
carnitine, GPC, Cr, glycine, choline, citrate, glutamate, acetate, inositol,
leucine and isoleucine, lactate, alanine, succinate, and valine in varying
amounts, depending on the anatomical structure investigated. Some in vivo
^{31}P studies have also been made in order to elucidate the metabolic effects
of testicular torsion and the consequent ischemia.[102, 103] During an ischemic
insult induced by vascular occlusion, ATP decreased rapidly, with a concomitant increase in Pi, until, after 1 hr, ATP was undetectable. Spectra
obtained from testes which had experienced 30 minutes of ischemia showed
complete recovery of ATP 2 hr after reperfusion. At 10- to 16-day follow-
up, the spectra from this group were essentially normal. Testes which had
suffered 2 to 4 hr of vascular occlusion showed incomplete recovery of ATP
and Pi and at follow-up had significantly reduced overall signal when compared with an external reference. More recent research has revealed both a

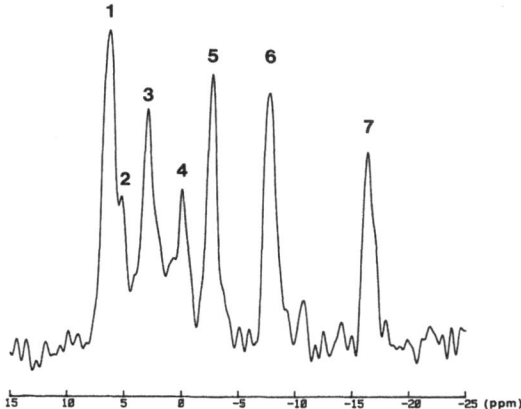

Figure 1.15. A ^{31}P spectrum obtained noninvasively from normal in vivo adult rat testes at 32.5 MHz. The testes were inserted into a three-turn solenoidal coil, 22 mm in diameter and 8 mm in length. The spectrum is the result of 1536 scans using 90° pulses at 2.256-sec intervals. Peak assignments are as follows: 1, mainly phosphoethanolamine and phosphocholine with some phosphoserine; 2, Pi; 3, phosphodiesters (mainly GPC but including some GPE); 4, PCr; 5, 6, and 7, mainly the γ, α, β resonances of ATP. (Spectrum obtained in collaboration with J. Nicholson.)

rapid loss of ATP and a decrease in the ratio of phosphomonoester to Pi following warm ischemia.[104] ATP regeneration was noticed during reperfusion. In the same study, suppressed spermatogenesis was related to decreased phosphomonoester–phosphodiester and phosphomonoester–ATP ratios. Figure 1.15 shows a ^{31}P spectrum obtained from normal mammalian testis in vivo.

1.4.10. Ovaries

The relative inaccessibility of these organs has limited their investigation by MRS. ^{31}P studies of perfused ovaries have been made, and Figure 1.16 shows spectra obtained from luteinized tissue.[105] The metabolic state of the tissue was monitored for several hours, and in four out of seven studies a gradual decrease in ATP to 70% of the original level was noticed. This was accompanied by a drop in pH_i from 7.4 to 6.7. Addition of 50 μmole adenosine to the perfusate was reported to return the ATP concentration and pH_i to their initial values. From the chemical shifts of the ATP peaks, it was claimed that, as in the case of uterine smooth muscle, the Mg^{2+} concentration was lower than that encountered in many other tissues.

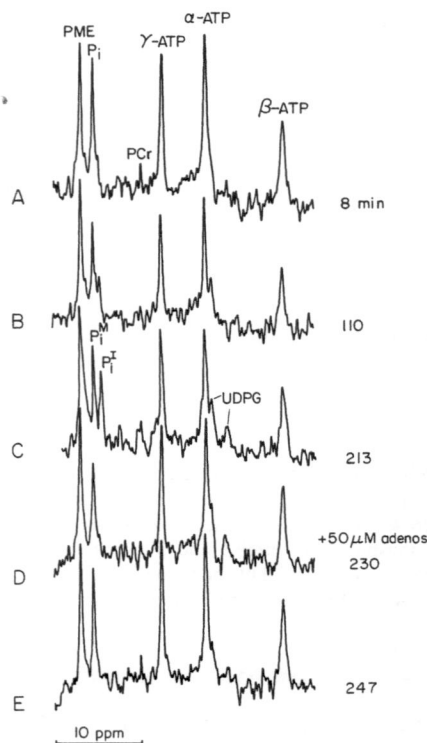

Figure 1.16. ^{31}P spectra of luteinized immature rat ovaries at 35°C perfused with Krebs–Ringer–Henseleit buffer at a rate of 1 ml/min. Each spectrum is the result of 1024 scans using a 60° pulse with a 1-sec pulse interval. The middle time points of the spectra are indicated in minutes. Spectra D and E were obtained after the addition of 50 μM adenosine to the medium: PME, phosphomonoesters; Pi_M, medium Pi; Pi_I, intracellular Pi; UDPG, uridine diphosphosugars (from Haseltine et al.,[105] Academic Press Inc.; reprinted with permission).

1.4.11. Skin

Skin should be one of the easiest tissues to which surface-coil MRS can be applied and yet, so far, it has not received much attention. The detection coil can be placed directly on the tissue and, by judicious selection of coil dimensions, pulse characteristics, and location, signal from underlying muscle and other tissues should be avoidable. However, only a few initial studies have been made.

Nunnally and colleagues[106] have used ^{31}P MRS at 300 MHz to study the effects of variations in Pco_2 and anoxia on frog epithelial sheets bathed in gassed phosphate-free Cl^- or SO_4^{2-} Ringer's solution. Control spectra obtained with 100% O_2 gassing showed peaks assigned to phosphomonoesters, Pi, ATP, ADP, NAD and NADH, and a small PCr resonance, as shown in Figure 1.17. The pH_i of the control spectra were reported as 7.19 and 7.42 for bathing in Cl^- and SO_4^{2-} Ringer's solution, respectively, compared to an extracellular pH of 7.52. Exposure to increasing extracellular Pco_2 revealed polynomial decreases in pH_i, reaching 6.40

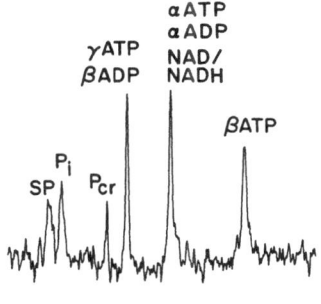

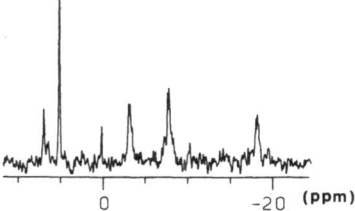

Figure 1.17. A 121.4-MHz ^{31}P spectrum of frog skin epithelial sheets (connective tissue separated) in SO_4^{2-} Ringer solution bubbled with 100% O_2 (top panel) compared with a similarly obtained spectrum of epithelial extracts (lower panel). The experimental conditions were as follows: 40° pulse, 1.13-sec pulse interval, 400 scans (top) and 1200 scans (bottom); SP, sugar phosphates (from Nunnally et al.,[106] The American Physiological Society; reprinted with permission).

(Cl^- Ringer's solution) and 6.61 (SO_4^{2-} Ringer's solution) at a Pco_2 of 15%. Changing the gassing from 100% O_2 to 100% N_2 produced, within 15 min, a large decrease in ATP, which was regenerated on re-oxygenation.

Preliminary work using ^{31}P MRS to study skin-graft metabolism has been reported.[107] Standard skin-graft flaps were employed in a mammalian model such that necrosis of the distal third could be predicted. Immediately after elevation of the flap, phosphometabolite levels were reported to be homogeneous. Within 30 min, PCr had dropped sharply in the distal third and somewhat less in the middle third of the flap. After 3 hr high-energy phosphates were undetectable from the distal third, and there was a large increase in Pi. Less severe changes were reported in the middle third.

More recent work used a Faraday-shielded solenoid coil to obtain ^{31}P spectra from adult rat skin.[108] Biochemical assays gave the following absolute quantitations for metabolites in the spectra: Pi, 1.63 ± 0.12 mM/kg wet wt. [mean $\pm$ standard error (S.E.)]; PCr, 1.40 ± 0.12 mM/kg wet wt.; ATP, 1.35 ± 0.22 mM/kg wet wt.

1.4.12. Neoplastic Tissues

It was claimed in the 1970s that the relaxation properties of tissue-water[109] and phosphate metabolites[110] are altered in malignant tissue. The changes have been attributed to increased hydration,[111, 112] alteration in the amounts of bound and free water,[24] and differences in the interaction of water molecules with macromolecular surfaces.[113] Observed increases in relaxation times for ^{31}P metabolites have been attributed to changes in their concentrations[110] and to the effects of tissue water.[24]

^{31}P MRS has been used to study cellular suspensions (e.g., Ehrlich ascites[114] and HeLa cells[115]), as has ^{1}H spectroscopy (e.g., neuroblastoma × glioma[116]). Excised tissues from various types of tumor have been studied, by, for example, the application of ^{19}F MRS[117, 118] to renal-cell carcinoma. Many in vivo studies have been carried out, and the reader is referred to the excellent review article by Evanochko and colleagues[119] published in 1984 for further details, including the use of MRS to study the effects of chemotherapy, X-radiation, and hyperthermia. Several examples of the application of MRS to the investigation of human tumors are given in Chapter 3.

REFERENCES

1. Bloch, F.: Nuclear induction, *Physical Review* 70:461–474, 1946.
2. Bloch, F., Hansen, W. W., and Packard, M.: The nuclear induction experiment, *Physical Review* 70:474–485, 1946.
3. Bracewell, R. N.: *The Fourier Transform and Its Applications*, New York, McGraw–Hill, 1965.
4. Purcell, E. M., Torrey, H. C., and Pound, R. V.: Resonance absorption by nuclear magnetic moments in a solid, *Physical Review* 69:37–38, 1946.
5. Bloch, F., Hansen, W. W., and Packard, M.: Nuclear induction, *Physical Review* 69:127, 1946.
6. Abragam, A.: *The Principles of Nuclear Magnetism*, Oxford, Clarendon Press, 1961.
7. Slichter, C. P.: *Principles of Magnetic Resonance*, 2nd edition, Berlin, Springer, 1978.
8. Ackerman, J. J. H., Grove, T. H., Wong, G. G., et al.: Mapping of metabolites in whole animals by ^{31}P NMR using surface coils, *Nature* 283:167–170, 1980.
9. Moon, R. B. and Richards, J. H.: Determination of intracellular pH by ^{31}P magnetic resonance, *J. Biol. Chem.* 248:7276–7278, 1973.
10. Cohn, M. and Hughes, T. R.: Phosphorus magnetic resonance spectra of adenosine di- and triphosphate. I. Effect of pH, *J. Biol. Chem.* 235:3250–3253, 1960.
11. Pettegrew, J. W., Withers, G., Panchalingam, K., et al.: Considerations for brain pH assessment by ^{31}P NMR, *Magn. Reson. Imaging* 6:135–142, 1988.
12. Cohn, M. and Hughes, T. R.: Nuclear magnetic resonance spectra of adenosine di- and triphosphate. II. Effect of complexing with divalent metal ions, *J. Biol. Chem.* 237:176–181, 1962.
13. Gupta, R. K. and Gupta, P.: A magnesium(II)ATP thermometer for ^{31}P NMR studies of biological systems, *Journal of Magnetic Resonance* 40:587–589, 1980.
14. Gadian, D. G., Proctor, E., Williams, S. R., et al.: Neurometabolic effects of an inborn error of amino acid metabolism demonstrated in vivo by ^{1}H NMR, *Mag. Reson. Med.* 3:150–156, 1986.
15. Wüthrich, K.: *NMR in Biological Research: Peptides and Proteins*, Amsterdam, North Holland, 1976, p. 86.
16. Arus, C., Chang, Y.–C., and Barany, M.: N-acetylaspartate as an intrinsic thermometer for ^{1}H NMR of brain slices, *Journal of Magnetic Resonance* 63:376–379, 1985.
17. Knowles, P. F., Marsh, D., and Rattle, H. W. E.: *Magnetic Resonance of Biomolecules*, London, Wiley, 1976.

18. Kanamori, K. and Roberts, J. D.: ^{15}N NMR studies of biological systems, *Accounts of Chemical Research* **16**:35–41, 1983.

19. Gupta, R. J. and Gupta, P.: Direct observation of resolved resonances from intra- and extracellular sodium-23 ions in NMR studies of intact cells and tissues using dysprosium (III) tripolyphosphate as paramagnetic shift reagent, *Journal of Magnetic Resonance* **47**:344–350, 1982.

20. Yeh, H. J. C., Brindley, F. J., and Becker, E. D.: Nuclear magnetic resonance studies on intra-cellular sodium in human erythrocytes and frog muscle, *Biophys. J.* **13**:56–71, 1973.

21. Pettegrew, J. W., Glonek, T., Minshew, N. J., et al.: Sodium-23 NMR of intact bovine lens and vitreous humor, *Journal of Magnetic Resonance* **63**:439–444, 1985.

22. Saunders, M., Wishnia, A., and Kirkwood, J.: The nuclear magnetic resonance spectrum of ribonuclease, *Journal of the American Chemical Society* **79**:3289–3290, 1957.

23. Gadian, D. G.: *Nuclear Magnetic Resonance and Its Applications to Living Systems*, Oxford, Clarendon Press, 1982.

24. Iles, R. A., Stevens, A. N., and Griffiths, J. R.: NMR studies of metabolites in living tissue, *Progress in Nuclear Magnetic Resonance Spectroscopy* **15**:49–200, 1982.

25. Gordon, R. E., Hanley, P. E., and Shaw, D.: Topical magnetic resonance, *Progress in Nuclear Magnetic Resonance Spectroscopy* **15**:1–47, 1982.

26. Petroff, O. A. C.: Biological ^{1}H NMR spectroscopy, *Comp. Biochem. Physiol.* [B] **90B**:249–260, 1988.

27. Hoult, D. I., Busby, S. J. W., Gadian, D. G., et al.: Observation of tissue metabolites using ^{31}P nuclear magnetic resonance, *Nature* **252**:285–287, 1974.

28. Dawson, M. J., Gadian, D. G., and Wilkie, D. R.: Contraction and recovery of living muscles studied by ^{31}P nuclear magnetic resonance, *J. Physiol.* (London) **267**:703–735, 1977.

29. Dawson, M. J., Gadian, D. G., and Wilkie, D. R.: Muscular fatigue investigated by phosphorus nuclear magnetic resonance, *Nature* **274**:861–866, 1978.

30. Chalovich, J. M., Tyler Burt, C., Cohen, S. M., et al.: Identification of an unknown ^{31}P nuclear magnetic resonance from dystrophic chicken as L-serine ethanolamine phosphodiester, *Arch. Biochem. Biophys.* **182**:683–689, 1977.

31. Lucy, J. A.: Is there a membrane defect in muscle and other cells? *Br. Med. Bull.* **36**:187–192, 1980.

32. Chalovich, J. M., Tyler Burt, C., Danon, M. J., et al.: Phosphodiesters in muscular dystrophies, *Ann. N.Y. Acad. Sci.* **317**:649–669, 1979.

33. Radda, G. K., Bore, P. J., and Rajagopalan, B.: Clinical aspects of ^{31}P NMR spectroscopy, *Br. Med. Bull.* **40**:155–159, 1983.

34. Yoshizaki, K., Seo, Y., and Nishikawa, H.: High-resolution proton magnetic resonance spectra of muscle, *Biochem. Biophys. Acta* **678**:283–291, 1981.

35. Arus, C., Barany, M., Westler, W. M., et al.: ^{1}H NMR of intact muscle at 11 T, *FEBS Lett.* **165**:231–237, 1984.

36. Arus, C. and Barany, M.: ^{1}H NMR of intact tissues at 11.1 T, *Journal of Magnetic Resonance* **57**:519–525, 1984.

37. Williams, S. R., Gadian, D. G., Proctor, E., et al.: Proton NMR studies of muscle metabolites in vivo, *Journal of Magnetic Resonance* **63**:406–412, 1985.

38. Meiboom, S. and Gill, D.: Modified spin–echo method for measuring nuclear relaxation times, *Review of Scientific Instruments* **29**:688–691, 1958.

39. Rothman, D. L., Mendoza, F. A., Shulman, G. I., et al.: A pulse sequence for simplifying hydrogen NMR spectra of biological tissues, *Journal of Magnetic Resonance* **60**:430–436, 1984.

40. Fung, B. M.: Carbon-13 and proton magnetic resonance of mouse muscle, *Biophys. J.* **19**:315–319, 1977.

41. Doyle, D. D., Chalovich, J. M., and Barany, M.: Natural abundance ^{13}C NMR spectra of intact muscle, *FEBS Lett.* 131:147–150, 1981.

42. Alger, J. R., Sillerud, L. O., Behar, K. L., et al.: In vivo carbon-13 nuclear magnetic resonance studies of mammals, *Science* 214:660–662, 1981.

43. Gadian, D. G., Hoult, D. I., Radda, G. K., et al.: Phosphorus nuclear magnetic resonance studies on normoxic and ischemic cardiac tissue, *Proc. Natl. Acad. Sci. U.S.A* 73:4446–4448, 1976.

44. Jacobus, W. E., Taylor, G. J., Hollis, D. P., et al.: Phosphorus nuclear magnetic resonance of perfused working rat hearts, *Nature* 265:756–758, 1977.

45. Garlick, P. B., Radda, G. K., Seeley, P. J., et al.: Phosphorus NMR studies on perfused heart, *Biochem. Biophys. Res. Commun.* 74:1256–1262, 1977.

46. Ugurbil, K., Petein, M., Maidan, R., et al.: High resolution proton NMR studies of perfused rat hearts, *FEBS Lett.* 167:73–78, 1984.

47. Grove, T. H., Ackerman, J. J. H., Radda, G. K., et al.: Analysis of rat heart in vivo by phosphorus nuclear magnetic resonance, *Proc. Natl. Acad. Sci. U.S.A* 77:299–302, 1980.

48. Nunnally, R. L. and Bottomley, P. A.: Assessment of pharmacological treatment of myocardial infarction by phosphorus-31 NMR with surface coils, *Science* 211:177–180, 1981.

49. Koretsky, A. P., Wang, S., Murphy–Boesch, J., et al.: ^{31}P NMR spectroscopy of rat organs, in situ, using chronically implanted radiofrequency coils, *Proc. Natl. Acad. Sci. U.S.A.* 80:7491–7495, 1983.

50. Stein, P. D., Goldstein, S., Sabbah, H. N., et al.: In vivo evaluation of intracellular pH and high-energy phosphate metabolites during regional ischemia in cats using ^{31}P nuclear magnetic resonance, *Magn. Reson. Med.* 3:262–269, 1986.

51. Vogel, H. J., Lilja, H., and Hellstrand, P.: Phosphorus-31 NMR studies of smooth muscle from guinea-pig taenia coli, *Biosci. Rep.* 3:863–870, 1983.

52. Vermue, N. A. and Nicolay, K.: Energetics of smooth muscle taenia caecum of guinea pig: A ^{31}P-NMR study, *FEBS Lett.* 156:293–297, 1983.

53. Dawson, M. J. and Wray, S.: [31]P nuclear magnetic resonance (NMR) studies of isolated rat uterus, *J. Physiol.* (London) 336:19–20P, 1983.

54. Hellstrand, P. and Vogel, H. J.: Phosphagens and intracellular pH in intact rabbit smooth muscle studied by ^{31}P-NMR, *American Journal of Physiology* 248:C320–C329, 1985.

55. Degani, H., Shaer, A., Victor, T. A., et al.: Estrogen-induced changes in high-energy phosphate metabolism in rat uterus: ^{31}P NMR studies, *Biochemistry* 23:2572–2577, 1984.

56. Dawson, M. J. and Wray, S.: The effects of pregnancy and parturition on phosphorus metabolites in rat uterus studied by ^{31}P nuclear magnetic resonance, *J. Physiol.* (London) 368:19–31, 1985.

57. Wray, S.: A [31]-phosphorus nuclear magnetic resonance study of the regulation of intracellular pH following changes in extracellular pH in rat uterine smooth muscle, *J. Physiol.* (London) 373:81P, 1986.

58. Kushmerick, M. J., Dillon, P. F., Meyer, R. A., et al.: ^{31}P NMR spectroscopy, chemical analysis and free Mg^{2+} of rabbit bladder and uterine smooth muscle, *J. Biol. Chem.* 261:14, 420–14, 429, 1986.

59. Carlier, P. G., Grandjean, J., Michel, P., et al.: Arterial metabolism as studied in vitro by NMR: Preliminary results in normotensive and hypertensive aortas, *Arch. Int. Physiol. Biochim.* 93:107–118, 1985.

60. Dawson, M. J., Spurway, N. C., and Wray, S.: A ^{31}P nuclear magnetic resonance (NMR) study of isolated rabbit arterial smooth muscle, *J. Physiol.* (London) 365:72P, 1985.

61. Chance, B., Nakase, Y., Bond, M., et al.: Detection of ^{31}P nuclear magnetic resonance signals in brain by in vivo and freeze-trapped assays, *Proc. Natl. Acad. Sci. U.S.A.* **75**:4925–4929, 1978.

62. Chang, Y.–C., Arus, C., and Barany, M.: Characterization of the broad resonance in ^{31}P NMR spectra of excised rat brain, *Physiol. Chem. Phys. Med. NMR* **17**:143–154, 1985.

63. Gonzalez–Mendez, R., Litt, L., Koretsky, A. P., et al.: Comparison of ^{31}P NMR spectra of in vivo rat brain using convolution difference and saturation with a surface coil. Source of the broad component in the brain spectrum, *Journal of Magnetic Resonance* **57**:526–533, 1984.

64. Veech, R. L., Harris, R. L., Veloso, D., et al.: Freeze blowing: A new technique for the study of brain in vivo, *J. Neurochem.* **20**:183–188, 1973.

65. Delpy, D. T., Gordon, R. E., Hope, P. L., et al.: Non-invasive investigation of cerebral ischemia by phosphorus nuclear magnetic resonance, *Pediatrics* **70**:310–313, 1982.

66. Thulborn, K. R., du Boulay, C. H., Duchen, L. W., et al.: A ^{31}P nuclear magnetic resonance in vivo study of cerebral ischemia in the gerbil, *J. Cereb. Blood Flow Metab.* **2**:299–306, 1982.

67. Prichard, J. W., Alger, J. R., Behar, K. L., et al.: Cerebral metabolic studies in vivo by ^{31}P NMR. *Proc. Natl. Acad. Sci. U.S.A* **80**:2748–2751, 1983.

68. Petroff, O. A. C., Prichard, J. W., Behar, K. L., et al.: In vivo phosphorus nuclear magnetic resonance spectroscopy in status epilepticus, *Ann. Neurol.* **16**:169–177, 1984.

69. Decorps, M., Lebas, J. F., Leviel, J. L., et al.: Analysis of brain metabolism changes induced by acute potassium cyanide intoxication by ^{31}P NMR in vivo using chronically implanted surface coils, *FEBS Lett.* **168**:1–6, 1984.

70. Hilberman, M., Subramanian, V. H., Haselgrove, J., et al.: In vivo time-resolved brain phosphorus nuclear magnetic resonance, *J. Cereb. Blood Flow Metab.* **4**:334–342, 1984.

71. Cox, D. W. G., Morris, P. G., Feeney, J., et al.: ^{31}P NMR studies on cerebral energy metabolism under conditions of hypoglycemia and hypoxia in vitro, *Biochem. J.* **212**:365–370, 1983.

72. Glonek, T., Kopp, S. J., Kot, E., et al.: P-31 nuclear magnetic resonance analysis of brain: The perchloric acid extract spectrum, *J. Neurochem.* **39**:1210–1219, 1982.

73. Brenton, D. P., Garrod, P. J., Krywawych, S., et al.: Phosphoethanolamine is major constituent of phosphomonoester peak detected by ^{31}P NMR in newborn brain, *Lancet* **1**:115, 1985.

74. Pettegrew, J. W., Kopp, S. J., Dadok, J., et al.: Chemical characterization of a prominent phosphomonoester resonance from mammalian brain. ^{31}P and ^{1}H NMR analysis at 4.7 and 14.1 tesla, *Journal of Magnetic Resonance* **67**:443–450, 1986.

75. Dea, P., Chan, S. I., and Dea, F. J.: High-resolution proton magnetic resonance spectra of a rabbit sciatic nerve, *Science* **175**:206–208, 1972.

76. Behar, K. L., den Hollander, J. A., Stromski, M. E., et al.: High-resolution ^{1}H nuclear magnetic resonance study of cerebral hypoxia in vivo, *Proc. Natl. Acad. Sci. U.S.A* **80**:4945–4948, 1983.

77. Tallan, H. H.: Studies on the distribution of N-acetyl-L-aspartic acid in brain, *J. Biol. Chem.* **223**:41–45, 1957.

78. Behar, K. L., Rothman, D. L., Shulman, R. G., et al.: Detection of cerebral lactate in vivo during hypoxemia by ^{1}H NMR at relatively low field strengths (1.9 T), *Proc. Natl. Acad. Sci. U.S.A* **81**:2517–2519, 1984.

79. Behar, K. L., den Hollander, J. A., Petroff, O. A. C., et al.: Effect of hypoglycemic encephalopathy upon amino acids, high energy phosphates and pH$_i$ in the rat brain

in vivo: Detection by sequential ^{1}H and ^{31}P NMR spectroscopy, *J. Neurochem.* **44**:1045–1055, 1985.

80. Rothman, D. L., Behar, K. L., Hetherington, H. P., et al.: Homonuclear ^{1}H double resonance difference spectroscopy of the rat brain in vivo, *Proc. Natl. Acad. Sci U.S.A* **81**:6330–6334, 1984.

81. Rothman, D. L., Behar, K. L., Hetherington, H. P., et al.: ^{1}H-observe ^{13}C-decouple spectroscopic measurements of lactate and glutamate in the rat brain in vivo, *Proc. Natl. Acad. Sci. U.S.A* **82**:1633–1637, 1985.

82. Arus, C., Chang, Y., and Barany, M.: Proton nuclear magnetic resonance spectra of excised rat brain. Assignment of resonances, *Physiol. Chem. Phys. Med. NMR* **17**:23–33, 1985.

83. Cerdan, S., Parrilla, R., Santoro, J., et al.: ^{1}H NMR detection of cerebral myo-inositol, *FEBS Lett.* **187**:167–172, 1985.

84. Alger, J. R., Sillerud, L. O., Behar, K. L., et al.: In vivo carbon-13 nuclear magnetic resonance studies of mammals, *Science* **214**:660–662, 1981.

85. Den Hollander, J. A., Behar, K. L., and Shulman, R. G.: Use of double tuned surface coils for the application of ^{13}C NMR to brain metabolism, *Journal of Magnetic Resonance* **57**:311–313, 1984.

86. Barany, M., Arus, C., and Chang, Y. C.: Natural abundance ^{13}C NMR of brain, *Magn. Reson. Med.* **2**:289–295, 1985.

87. Cohen, S. M., Ogawa, S., Rottenburg, H., et al.: ^{31}P nuclear magnetic resonance studies of isolated rat liver cells, *Nature* **273**:554–556, 1978.

88. McLaughlin, A. C., Takeda, H., and Chance, B.: Rapid ATP assays in perfused mouse liver by ^{31}P NMR, *Proc. Natl. Acad. Sci. U.S.A* **76**:5445–5449, 1979.

89. Salhany, J. M., Stohs, S. J., Reinke, L. A., et al.: ^{31}P nuclear magnetic resonance of metabolic changes associated with cyanide intoxication in the perfused rat liver, *Biochem. Biophys. Res. Commun.* **86**:1077–1083, 1979.

90. Iles, R. A., Griffiths, J. R., Stevens, A. N., et al.: Effects of fructose on the energy metabolism and acid-base status of the perfused starved rat liver. A ^{31} phosphorus nuclear magnetic resonance study, *Biochem. J.* **192**:191–202, 1980.

91. Gordon, R. E., Hanley, P. E., Shaw, D., et al.: Localization of metabolites in animals using ^{31}P topical magnetic resonance, *Nature* **287**:736–738, 1980.

92. Iles, R. A., Stevens, A. N., Griffiths, J. R., et al.: Phosphorylation status of liver by ^{31}P-n.m.r. spectroscopy, and its implications for metabolic control, *Biochem. J.* **229**:141–151, 1985.

93. Nicholson, J. K., Timbrell, J. A., Bales, J. R., et al.: A high resolution proton nuclear magnetic resonance approach to the study of hepatocyte and drug metabolism—application to acetaminophen, *Mol. Pharmacol.* **27**:634–643, 1985.

94. Canioni, P., Alger, J. R., and Shulman, R. G.: Natural abundance carbon-13 nuclear magnetic resonance spectroscopy of liver and adipose tissue of the living rat, *Biochemistry* **22**:4974–4980, 1983.

95. Sehr, P. A., Radda, G. K., Bore, P. J., et al.: A model kidney transplant studied by phosphorus nuclear magnetic resonance, *Biochem. Biophys. Res. Commun.* **77**:195–202, 1977.

96. Radda, G. K., Ackerman, J. J. H., Bore, P. J., et al.: ^{31}P NMR studies on kidney intracellular pH in acute renal acidosis, *Int. J. Biochem.* **12**:277–281, 1980.

97. Ackerman, J. J. H., Lowry, M., Radda, G. K., et al.: The role of intrarenal pH in regulation of ammoniagenesis: [^{31}P] NMR studies of the isolated perfused rat kidney, *J. Physiol.* (London) **319**:65–79, 1981.

98. Balaban, R. S., Gadian, D. G., and Radda, G. K.: Phosphorus nuclear magnetic resonance study of the rat kidney in vivo, *Kidney Int.* **20**:575–579, 1981.

99. Hexem, J. G., Marshall, C., Marshall, B. E., et al.: ^{31}P NMR studies of isolated perfused atelectatic rat lung, *Proceedings*, Third Annual Meeting, Society of Magnetic Resonance in Medicine, New York, 1984, p. 317.
100. Pillai, R. P., Buescher, P. C., Pearse, D. B., et al.: ^{31}P NMR spectroscopy of isolated perfused lungs, *Magn. Reson. Med.* 3:467–472, 1986.
101. Navon, G., Gogol, E., and Weissenberg, R.: Phosphorus-31 and proton NMR analysis of reproductive organs of male rats, *Arch. Androl.* 15:153–157, 1985.
102. See W. A., Richards, T., Mack, L. A., et al.: P-31 NMR spectroscopy as a predictor of testicular viability following ischemic injury, *Proceedings*, Third Annual Meeting, Society of Magnetic Resonance in Medicine, New York, 1986, p. 1013.
103. Vigneron, D. B., Bretan, P. N., Moseley, M. E., et al.: Assessment of testicular metabolic integrity by phosphorus-31 nuclear magnetic resonance, *Proceedings*, Fifth Annual Meeting, Society of Magnetic Resonance in Medicine, Montreal, 1986, p. 1023.
104. Bretan, P. N., Vigneron, D. B., Hricak, H., et al.: Assessment of testicular metabolic integrity with P-31 MR spectroscopy, *Radiology* 162:867–871, 1987.
105. Haseltine, F. P., Arias-Mendoza, F., Kaye, A. M., et al.: ^{31}P NMR studies of adenosine stimulated ATP synthesis in perfused luteinized ovaries, *Magn. Reson. Med.* 3:796–800, 1986.
106. Nunnally, R. L., Stoddard, J. S., Helman, S. I., et al.: Response of ^{31}P nuclear magnetic resonance spectra of frog skin to variations in Pco_2 and hypoxia, *Am. J. Physiol.* 245:F792–800, 1983.
107. Cuono, C. B. and Armitage, I. M.: Metabolic studies of skin grafts and flaps, *Proceedings*, Third Annual Meeting, Society of Magnetic Resonance in Medicine, New York, 1984, p. 175–176.
108. Stubbs, M., Vanstapel, F., Rodrigues, L. M., et al.: Phosphate metabolites in rat skin, *NMR in Biomedicine* 1:50–55, 1988.
109. Damadian, R.: Tumor detection by nuclear magnetic resonance, *Science* 171:1151–1153, 1971.
110. Zaner, K. S. and Damadian, R.: Phosphorus-31 as a nuclear probe for malignant tumors, *Science* 189:729–731, 1975.
111. Inch, W. R., McCredie, J. A., Knispel, R. R., et al.: Water content and proton spin relaxation time for neoplastic and non-neoplastic tissues from mice and humans, *J. Natl. Cancer Inst.* 52:353–356, 1974.
112. Bovee, W., Huisman, P., and Smidt, J.: Tumor detection and nuclear magnetic resonance, *J. Natl. Cancer Inst.* 52:595–597, 1974.
113. Beall, T., Asch, B. B., Chang, D. C., et al.: Distinction of normal, pre-neoplastic, and neoplastic mouse mammary primary cell cultures by water nuclear magnetic resonance relaxation times, *J. Natl. Cancer Inst.* 64:335–338, 1980.
114. Navon, G., Ogawa, S., Shulman, R. G., et al.: [31]P nuclear magnetic resonance studies of Ehrlich ascites tumor cells, *Proc. Natl. Acad. Sci. U.S.A* 74:87–91, 1977.
115. Evans, E. E. and Kaplan, N. O.: [31]P nuclear magnetic resonance studies of HeLa cells, *Proc. Natl. Acad. Sci. U.S.A* 74:4909–4913, 1977.
116. Navon, G., Burrows, H., and Cohen, J. S.: Differences in metabolite levels upon differentiation of intact neuroblastoma × glioma cells observed by proton NMR spectroscopy, *FEBS Lett.* 162:320–323, 1983.
117. Burt, C. T., Moore, R. R., and Roberts, M.: Fluorinated anesthetics as probes of lipophilic environments in tumors, *Journal of Magnetic Resonance* 53:163–166, 1983.
118. Stevens, A. N., Morris, P. G., Iles, R. A., et al.: 5-Fluorouracil metabolism monitored in vivo by ^{19}F NMR, *Br. J. Cancer* 50:113–117, 1984.
119. Evanochko, W. T., Ng, T. C., and Glickson, J. D.: Application of in vivo NMR spectroscopy to cancer, *Magn. Reson. Med.* 1:508–534, 1984.

Fundamentals of Clinical Magnetic Resonance

2.1. THE ORIGIN OF THE MAGNETIC RESONANCE SIGNAL

In the introduction to Chapter 1, a brief description was given of the principles upon which magnetic resonance spectroscopy (MRS) is based. This chapter will be devoted to a more-detailed discussion of the magnetic resonance (MR) phenomenon and, in particular, of aspects which are of great importance to clinical applications. Many texts[1-6] give complete and rigorous theoretical analyses of the underlying principles of MR, and the reader is referred to these for more advanced study.

2.1.1. The Behavior of Nuclear Spins in a Magnetic Field

All atomic nuclei have positive charge due to the protons that they possess, and many also exhibit properties which can be ascribed to spin. The spin is quantized (i.e., it takes discrete values) and is integral or half-integral (i.e., 1/2, 1, 3/2, etc.). Several isotopes which have spin are listed in Table 1.1. ^1H, ^{13}C, ^{19}F, and ^{31}P have been found to be of particular use in biomedical MR, and some of their applications are described in Chapters 1 and 3. Nuclei that have even numbers of both protons and neutrons (such as ^{12}C and ^{16}O) do not exhibit spin and hence are of no interest from an MR point of view. Nuclei which possess spin behave like small nuclear magnets and when placed in a strong magnetic field tend to align themselves parallel or antiparallel with it, as shown in Figure 1.1a. In fact, due to their spin, the nuclear magnets do not line up exactly but precess about the applied magnetic field with a frequency called the Larmor precession frequency v_0, given by

$$v_0 = \frac{\gamma}{2\pi} B_0 \tag{2.1}$$

where v_0 is the precession frequency in Hz, B_0 is the applied magnetic field, and γ is the gyromagnetic ratio for the nucleus concerned. For instance, for the hydrogen nucleus (a proton) and the ^{31}P nucleus, γ takes the values 26.75×10^7 and 10.80×10^7 rad/sec per tesla, respectively. This equation is one of the most important in MR because of the proportionality between v_0 and B_0. If B_0 is uniform, then all the nuclear magnets in a sample will precess at the same v_0. However, they will all be at different positions (phases) in their precession cycle and furthermore, some will be aligned with B_0 and some against B_0. As shown in Figure 1.1a, those nuclear magnets aligned with B_0 produce a parallel magnetic component, while those oriented in the opposite direction produce a resultant antiparallel magnetization. Due to the difference in the numbers of nuclei in the parallel and antiparallel states, a net magnetization is induced in the sample (represented by M_0 in Figure 1.1a). If the sample is in thermal equilibrium, this difference is given by the Boltzmann distribution, which for nuclei of spin $\frac{1}{2}$ is

$$N_2/N_1 = \exp(\gamma \hbar B_0/k_B T_s) \tag{2.2}$$

where N_2 and N_1 are the numbers of nuclei in the parallel and antiparallel states, respectively, $\hbar$ is Planck's constant/2π, k_B is Boltzmann's constant, and T_s is the temperature of the sample. Equation (2.2) reflects the order which the applied magnetic field B_0 imposes on a system of nuclei, which (due to random thermal motion) would prefer to be disordered. For a given temperature T_s, as B_0 increases, more order is produced and N_2/N_1 becomes much greater than unity, implying a larger number of nuclei in the parallel state. The greater difference between N_2 and N_1 produces a larger M_0. If B_0 is held constant and T_s is increased, thermal processes tend to equalize the numbers of nuclei in the parallel and antiparallel states, thereby reducing M_0. At 37°C in a field of 2 tesla (i.e., typical conditions for a clinical MR study) N_2/N_1 is very close to unity; this means that the induced magnetization is small, making MR studies relatively insensitive. In terms of energy, the parallel state is the lowest and tends to be preferred. Hence, N_2 will always be larger than N_1, though only slightly so under clinical conditions.

As a result of the application of an external field to a sample consisting of nuclei with nonzero spin, the induced magnetization is given by[2]

$$M_0 = N\gamma^2 \hbar^2 I(I+1) B_0/3k_B T_s \qquad (2.3)$$

where N is the number of nuclei per unit volume and I is the nuclear spin. From this equation it is quite clear that the induced magnetization is proportional to the density of nuclei (also known as the spin density) and to the applied magnetic field B_0.

In order to obtain useful information from the sample it is necessary to interrogate it in some manner. The pulse Fourier transform (FT) approach[3-6] is practically universal in clinical MR, although continuous-wave techniques[1,4,5] were employed for a long time in chemical laboratory studies. In its simplest form, the FT technique involves the transmission of a brief pulse of magnetic field B_1, varying at radio frequency (RF), from a coil as shown in Figure 1.1b. This causes the net magnetization M_0 to precess about B_1 and produces a rotation or flip whose angle is proportional to both the duration and the amplitude of the pulse. Thus, a component of the sample's magnetization will, in general, lie in the x, y plane and can induce RF current in a suitably tuned, adjacent detector coil. In the continuous-wave methods the sample is subjected to RF irradiation and either the RF frequency or the strength of the applied magnetic field (B_0) is swept (varied continuously) over a certain range. When the RF frequency or the magnetic field attains the resonance value, as given by equation (2.1), the RF flips M_0, and this change can then be measured to give the variation of signal with frequency or, in other words, a spectrum. For many applications, the FT approach has been found superior because, due to the broad band of frequencies present in the pulse, the whole spectrum can be interrogated at once, thereby introducing considerable reduction in data-collection time and increased signal-to-noise ratio (SNR). In addition, the Fourier technique requires digital methods for data acquisition, and this permits the implementation of a large number of computer-processing and analysis procedures which can, in turn, improve the quality of results. Because of the universal application of pulse FT methods in clinical MR, more importance will be attached to it in the following discussion.

2.1.2. The Rotating Frame of Reference

A great deal of the technique of MR depends on the detailed analysis of the way in which M_0 behaves when subjected to pulses which can be single or a complex sequence involving various pulse phases and interpulse delays as well as various pulse lengths, amplitudes, and shapes. In order to simplify the analysis, it has proven beneficial to invoke the idea of the rotating frame of reference. A laboratory frame of reference x, y, z is quite a familiar concept to many people. Using this reference system, a nuclear magnet will be observed to precess at v_0 when an external magnetic field is applied parallel to the z axis. A rotating reference frame can be imagined which also spins about the z axis at frequency v_0. In this reference system x', y', z', the nuclear magnetization is observed to be stationary, and B_0 now appears to have no effect. A constant magnetic field B_1 in the x', y' plane of the rotating frame will rotate at v_0 in the laboratory frame, and is produced by the magnetic field of an RF pulse of frequency v_0 transmitted into the sample by a tuned coil as shown in Figure 1.1b. For simplicity, the discussion which follows will assume that the pulse frequency and the nuclear resonant frequency are both v_0. This condition is often called "on resonance." In the frame rotating at v_0, the only field experienced by M_0 during the RF pulse will be B_1. As a result of this, in the rotating frame, M_0 precesses about B_1 alone. The difference between the two frames of reference is clearest if one considers the effect on the sample magnetization of a single brief RF pulse of duration and amplitude such as to produce a flip of 90°, as shown in Figs. 2.1a and b. In the laboratory frame (Figure 2.1a), the magnetization describes a spiral path on its way to the x, y plane, whereas in the rotating frame (Figure 2.1b) the path is direct. Because the precession in the rotating frame is due only to B_1, the angle θ through which M_0 is flipped is given in radians by an adaptation of equation (2.1):

$$\theta = \gamma B_1 t_p / 2 \qquad (2.4)$$

$B_1/2$ has been substituted for B_0, since in the rotating frame B_1 is resolved

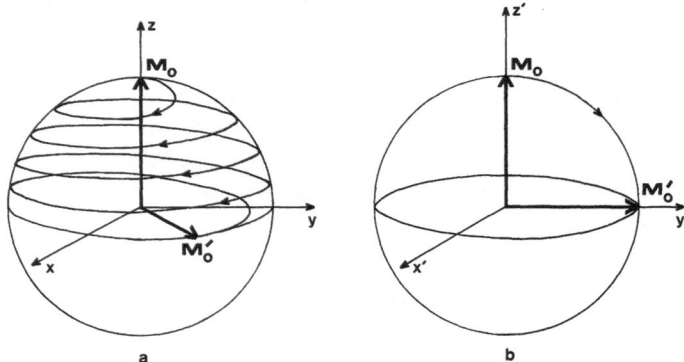

Figure 2.1. The path followed by the sample magnetization M_0 during the application of a brief RF pulse producing a 90° flip in (a) a laboratory frame of reference (x, y, z) and (b) a frame of reference (x', y', z') rotating about the z' axis with frequency v_0.

into a static component and a component rotating at $2\nu_0$, each of amplitude $B_1/2$, and the effect of the latter component can usually be ignored[7]; t_p is the duration of the pulse. At the end of the RF pulse, the magnetization M_0 in effect experiences no magnetic field in the rotating frame, and its subsequent evolution depends on relaxation phenomena (see Section 2.1.3). However, in the laboratory frame M_0 responds to B_0 and precesses about the z axis. An RF receiver coil observes a magnetic field varying at frequency ν_0. This is the free induction decay (FID). It induces a current in the coil, which produces an RF voltage at the coil terminals; this is measured by the RF receiver in the spectrometer.

2.1.3. The Bloch Equations and Relaxation Phenomena

In 1946, Bloch described the motion of M_0 in the presence of both a static magnetic field B_0 and the magnetic component B_1 of an RF field by a set of differential equations which carry his name.[8] The incomplete equations, taking no account of relaxation phenomena, are derived using the classical equation of motion of a magnetic moment in a static magnetic field.[1-6]

$$dM_x/dt = \gamma(M_y B_0 + M_z B_1 \sin \omega t) \qquad (2.5)$$

$$dM_y/dt = \gamma(M_z B_1 \cos \omega t - M_x B_0) \qquad (2.6)$$

$$dM_z/dt = -\gamma(M_x B_1 \sin \omega t + M_y B_1 \cos \omega t) \qquad (2.7)$$

M_x, M_y, and M_z are the components of M_0 along the x, y, and z axes, respectively, and ω is the frequency of B_1 ($\omega = 2\pi\nu$). In order to describe the motion of M_0 more exactly, it is necessary to include terms which take into account the ways in which nuclear relaxation occurs. These are spin–lattice relaxation, which is also known as longitudinal relaxation because by this means M_0 reverts to its original unperturbed (pre–RF pulse) state along the z axis, and spin–spin relaxation, which is sometimes called transverse relaxation due to its effect of causing decay of the component of M_0 in the x, y plane. These relaxation phenomena are quantified by the characteristic decay time constants T_1 and T_2, respectively. The following are the complete Bloch equations, including relaxation terms:

$$dM_x/dt = \gamma(M_y B_0 + M_z B_1 \sin \omega t) - M_x/T_2 \qquad (2.8)$$

$$dM_y/dt = \gamma(M_z B_1 \cos \omega t - M_x B_0) - M_y/T_2 \qquad (2.9)$$

$$dM_z/dt = -\gamma(M_x B_1 \sin \omega t + M_y B_1 \cos \omega t) - (M_z - M_0)/T_1 \qquad (2.10)$$

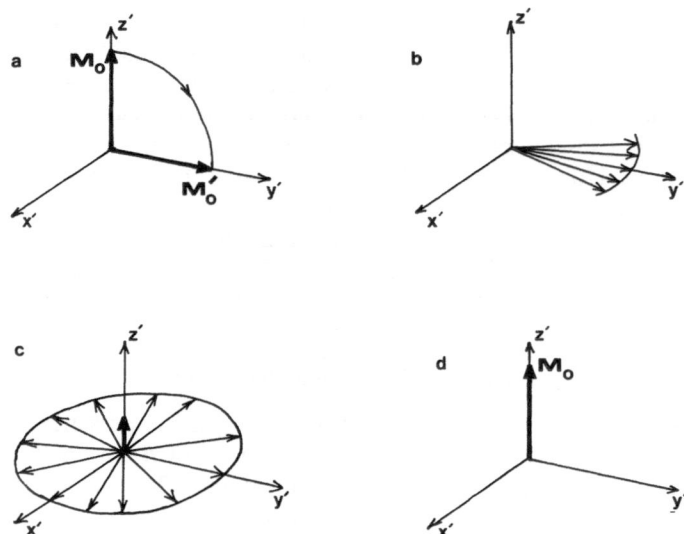

Figure 2.2. The evolution of the sample magnetization M_0 with time in the rotating frame (x', y', z') following the application of a 90° pulse including the effects of spin–lattice and spin–spin relaxation. (a) The magnetization M_0 is flipped into the $x' y'$ plane by a 90° pulse. (b) M_0' fans out in the $x' y'$ plane and the net magnetization decreases with characteristic time T_2^* due to spin–spin relaxation, magnetic field inhomogeneity, and molecular diffusion. (c) The magnetization in the $x' y'$ plane has now completely dephased, and no net $x' y'$ magnetization remains. Magnetization has begun to re-appear along the z' axis with characteristic time T_1 due to spin–lattice relaxation. (d) If sufficient time is allowed between successive pulses ($> 5T_1$), the z' magnetization recovers its equilibrium value M_0.

2.1.3.1. Spin–Lattice Relaxation

In practical terms, the effect of spin–lattice (longitudinal) relaxation is to gradually bring the perturbed M_0 back to its original equilibrium state, i.e., along the z axis. This process occurs as shown in Figure 2.2, where M_0 has been flipped into the x', y' plane and $M_{z'}$ (the component of M_0 parallel to the z' axis) recovers with a characteristic time T_1. There are many relaxation processes[3] which are responsible for this recovery of M_z, but they all result in the nuclear-spin system losing energy to the surrounding molecular environment. In water-proton studies of biological systems, T_1 varies from a few hundred milliseconds for most tissues to over 1 sec for cerebrospinal fluid, although there is a dependence on field strength. There is often a decline with subject's age during tissue maturation, and there are alterations due to pathological processes.

2.1.3.2. Spin–Spin Relaxation

This decay is due to a dephasing of the precessional motions of the nuclear spins in the x', y' plane, as shown in Figure 2.2. The dephasing continues until the nuclear spins are spread completely around the x', y' plane (also called transverse relaxation), and their resultant reduces exponentially to zero with the characteristic decay time T_2. Spin–spin relaxation leads to no energy loss from the spin system and results from an exchange of energy from one spin to another. This form of relaxation is caused by the same processes as spin–lattice relaxation, but additional effects also play a part. Hence T_2 is always less than T_1, although in pure liquids they can be nearly the same. Highly mobile nuclei such as protons in free water molecules relax slowly, whereas tightly bound nuclei (e.g., ^{31}P nuclei in mineralized bone) exhibit relatively short T_2s. In practice, it is often necessary to talk about T_2^*, which is the observed FID decay-time constant; this is less than T_2 due to enhanced dephasing caused by B_0 inhomogeneity and magnetic susceptibility variations in the tissue sample. If the local magnetic field differs in the various parts of the sample, then nuclear spins will precess at different rates in the x', y' plane, and this will then cause additional fanning out. The intrinsic T_2 effect is irreversible, whereas the dephasing due to inhomogeneity and susceptibility can be reversed by a "refocussing" pulse to produce a "spin echo," although diffusion, whereby a molecule can change its magnetic environment, has a significant effect. (This will be dealt with in greater detail in Chapter 5.) The water-proton T_2^*s determined from soft tissues tend to range from 40 to 100 msec and have a smaller dependence on field strength than the T_1s but also decline, in general, with maturation[7] and can be changed by disease. Changes in T_1 and T_2 in diseased tissue have proven useful in characterizing lesions on MR images, and knowledge of relaxation times has also allowed the design of optimum-contrast imaging pulse sequences.[9] Methods whereby T_1 and T_2 can be measured in biological samples will be described in chapter 5.

2.2. THE MAGNETIC RESONANCE SIGNAL

As shown in Figure 1.1b, if the equilibrium magnetization is perturbed so as to produce a component in the x, y plane, then a voltage will be induced in a suitably positioned RF coil. This voltage will vary sinusoidally at a frequency given by equation (2.1) because, in the laboratory frame of reference, the component of magnetization in the x, y plane will precess about the z axis at this same frequency. After processing by the receiver section of the spectrometer, the resultant voltage is digitized and displayed as a function of time, the FID.

2.2.1. The Free Induction Decay

As an example, an FID obtained from ^{31}P nuclei in living human muscle is shown in Figure 1.2. Two main components are present in the figure: a quasi-exponentially decaying signal due to the perturbed sample magnetization (the FID) and a superimposed, random background noise originating from various sources, including RF interference, the receiver, the coil, and the sample. The FID component contains contributions from ^{31}P nuclei in several muscle metabolites, each resonating at a different frequency and decaying exponentially with its own characteristic relaxation time T_2^*. Hence the envelope (shape) of the FID does not follow a single smoothly decaying exponential. Some components decay rapidly, whereas others take longer. In addition, components at different frequencies interfere, producing a complicated "beat" pattern whose shape depends on the relative frequencies and amplitudes of the components. Figure 2.3 shows this effect. An FID from a particular metabolite and its spectrum are shown in Figure 2.3a, while those from a metabolite resonating at another frequency are shown in Figure 2.3b. If the sample contains both these substances in solution, then the observed FID has the appearance of that in Figure 2.3c, and both peaks are found in the resultant spectrum. This effect is often seen in proton spectra obtained from human subjects. The

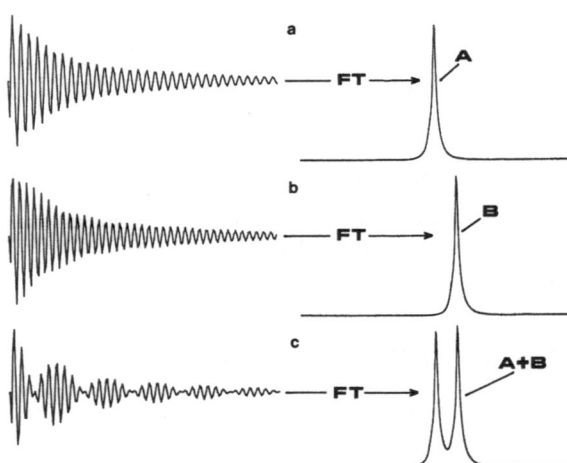

Figure 2.3. The effect of the presence of sample components, resonating at different frequencies, on the appearance of the FID and on the output of the Fourier transform (FT)—that is, the spectrum. Note the characteristic beats in (c), revealing the existence of two components resonating at almost the same frequency. (a) FID component A Fourier-transforms to give peak A. (b) FID component B Fourier-transforms to give peak B. (c) The FID of components A + B Fourier-transforms to give peaks A and B in the spectrum.

FID and spectrum from a lean subject will have characteristics similar to those shown in Figure 2.3a, consisting almost entirely of signal from tissue-water protons; if substantial subcutaneous fat is involved, the additional component of Figure 2.3b, mainly originating from the $(CH_2)_n$ fatty acyl-chain protons, is introduced into the signal. The FID and the spectrum are similar in appearance to those in Figure 2.3c.

The overall amplitude of the FID depends not only on the concentration of nuclear spins in the sample, but also on the interval T_R between interrogating RF pulses relative to T_1. The sequence of events required to accumulate data in a simple single-pulse experiment can be described by the following:

$$n[\theta\text{-Data Acquisition-Delay }(T_R)],$$

where n is the number of times the sequence is repeated and θ is the flip angle produced by the RF pulse. If sufficient time is allowed for spin–lattice relaxation to return the magnetization to its equilibrium state $(T_R > 5T_1)$, then almost the entire original magnetization is available for perturbation into the x, y plane by the next pulse. This results in the maximum FID amplitude. If T_R is shorter than about $5T_1$, then the nuclear-spin system will have recovered only partially, and a smaller z magnetization is available for flipping into the x, y plane. As a consequence, the FID amplitude is reduced and the sample is said to be partially saturated. This relaxation effect is demonstrated in Figure 2.4 where the relative amplitudes of FIDs obtained from brain water protons for T_Rs of approximately 10 sec and 0.1 sec are compared.

Because the metabolite concentrations encountered in biological

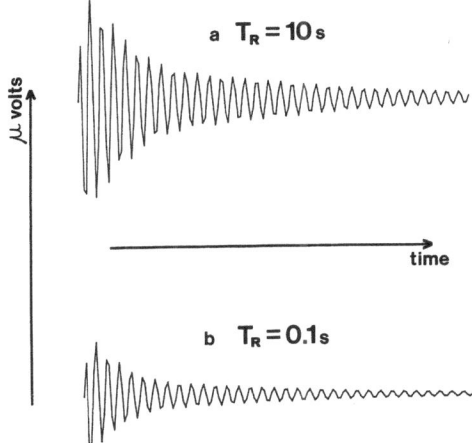

a $T_R = 10\,s$

b $T_R = 0.1\,s$

μ volts

time

Figure 2.4. The dependence of FID amplitude (μV) on the pulse repetition time T_R. The vertical and horizontal scales are identical for both FIDs. The single-scan ^1H FIDs were obtained from rat brain at 80.3 MHz using a 12-mm, two-turn circular surface coil placed directly on the exposed cranium. (a) $T_R = 10$ sec, which is many times more than the T_1 for the intracellular brain-water protons. (b) $T_R = 0.1$ sec, which is much less than T_1. Note the reduced signal amplitude in this scan.

systems are usually very small, it is often necessary to add together the FIDs from a sequence of n pulses in order to obtain a satisfactory SNR. The FID signal increases as n, whereas the random background noise is proportional to $\sqrt{n}$, and hence the SNR is raised by a factor of $\sqrt{n}$. In clinical studies, one of the prime considerations must be to optimize the SNR per unit of patient time. As already explained, the maximum signal per FID will be obtained when T_R is greater than $5T_1$; as T_R is reduced, the FID strength declines. However, by pulsing at $T_R = T_1$ one can accumulate five times as many FIDs per unit time with improved SNR, according to the $\sqrt{n}$ dependence. Because of the trade-off between reduced FID strength due to saturation and enhanced SNR from the increased number of pulses per unit time, in biological studies it is often found that SNR is optimized by pulsing with T_R less than $5T_1$. This will be dealt with in more detail later on in this chapter. It must be remembered that most spectra encountered in biological MR contain resonances from several metabolites, each of which has its own peculiar T_1. Hence, in general, it is not possible to optimize SNR for all the metabolites at once, and a compromise has to be found. An additional consequence is that the various resonances are saturated to different degrees and, in order to relate their strengths to metabolite concentrations, each has to be corrected by its own saturation factor. These factors are usually determined by comparing the resonance strengths of the metabolites using the selected T_R for optimal SNR and using a $T_R > 5T_1$.

The overall shape of the FID also depends on the T_2^*s of the sample components, as is demonstrated in Figure 2.5. A resonance with a very short T_2^* decays rapidly in the FID and on transformation produces a broad spectral feature as shown in Figure 2.5a. The ^{31}P resonances from mineralized bone phosphate and membrane phospholipids exhibit this characteristic. Highly mobile nuclei (e.g., those in free solution) have longer T_2^*s, the FID decays much more slowly, and the resultant spectral peaks

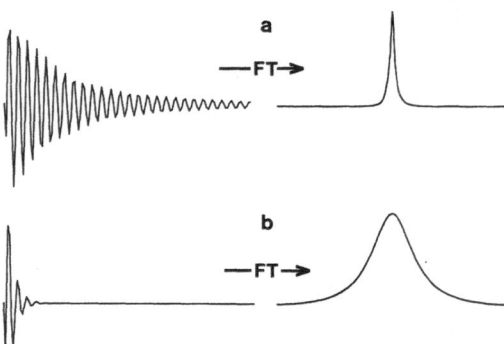

Figure 2.5. The dependence of the shape of the FID on the T_2^*s of the sample components and the resultant peak profiles after Fourier transformation (FT). (a) Large T_2^*, slow FID decay, narrow peak. (b) Small T_2^*, rapid FID decay, broad peak.

are considerably narrower, as shown in Figure 2.5b. It is useful to remember that signals of long duration produce sharp spectral peaks, whereas transient signals result in broad features.

2.2.2. The Spin Echo

Although the simple, single-pulse approach to spectroscopy is quite commonly adopted, many spectra, in particular those obtained using ^{1}H MRS, are cluttered with a multitude of overlapping peaks, making interpretation difficult. The spin–echo technique has been used to simplify spectra by editing resonances according to their spin–spin relaxation rates and can be used as a way of measuring T_2s. Additionally, spin echoes are commonly used in many imaging methods. The formation of a simple spin echo is shown in Figure 2.6. The magnetization is initially flipped through an angle θ (in this case, close to 90°), and then the magnetic vector fans out in the x, y plane during a dephasing period τ due to spin–spin relaxation and magnetic field inhomogeneity. Another pulse is now applied, of twice

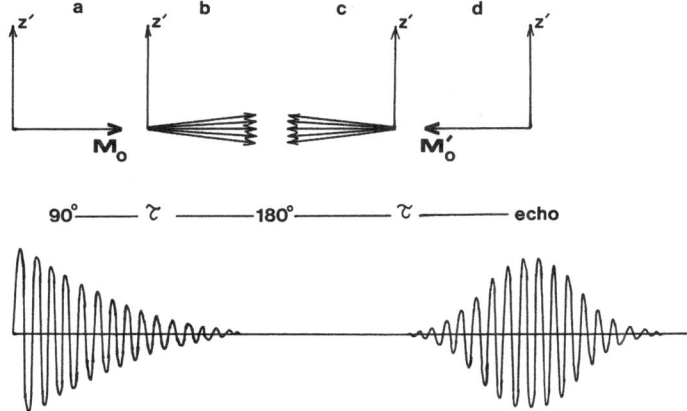

Figure 2.6. The necessary events leading to the generation of a spin echo in a rotating frame of reference $x'y'z'$. (a) The magnetization M_0 is flipped into the $x'y'$ plane by a 90° pulse, thereby initiating a FID. (b) During the period τ, M_0 dephases and fans out in the $x'y'$ plane due to field inhomogeneity, spin–spin relaxation and molecular diffusion. As a result the FID decays quasi-exponentially. (c) A 180° pulse is now applied, which flips the magnetization into the $x'y'$ plane onto the other side of the z' axis. The dephasing is reversed because the vectors which precess more slowly (and hence were phase retarded) are now phase advanced, whereas the more rapidly precessing vectors (which were phase advanced) are now phase retarded. Due to this, refocusing commences accompanied by the buildup of a spin echo. (d) After a further time τ, the xy magnetization refocuses to give M_0', and the spin echo has its maximum amplitude. M_0' is less than M_0 because only the dephasing due to field inhomogeneity is reversible. The spin–echo envelope is symmetrical because dephasing continues after M_0' is refocused.

the duration of the first, flipping the magnetization fan through 2θ (approximately 180°). This reverses some of the dephasing process; magnetization components which were retarded now precess faster, and those which were advanced precess more slowly, causing a refocusing during a further time interval τ. As the components of the magnetization vector refocus, they produce a detectable echo. The process is analogous to the situation of runners in a race. They all have different speeds, and after the start (the θ pulse) they spread out around the track. If their direction is now reversed (by the 2θ pulse), they run back to the starting line where they should all arrive at the same time (coinciding with the maximum amplitude of the spin echo). A simple spin–echo pulse sequence can be described in the following manner:

$$\theta\text{-}\tau\text{-}2\theta\text{-}\tau\text{-Data Acquisition-Delay } T_R$$

The usefulness of the spin echo lies in the fact that the refocusing is not complete; i.e., dephasing due to field inhomogeneity is reversed, but that caused by spin–spin relaxation is not. By increasing the delay τ between the θ and 2θ pulses, the amplitudes of resonances with longer T_2s are successively reduced, thereby greatly simplifying spectra and allowing more accurate measurement of the areas and positions of the remaining peaks. The pulse sequences used to produce spin echoes can be far more complicated than the simple, two-pulse sequence illustrated here, and there are many more applications, some of which will be dealt with in Chapter 5.

2.2.3. Fourier Transformation

The FT[3,5,10,11] is one of the most useful computational methods in spectral analysis. In MR spectroscopy, the FT is applied to the digitized FID using computer techniques, thereby producing a spectrum which is also in digital form. The function of this process is analogous to the role of the human brain in analyzing a complicated chord played on a musical instrument and distinguishing the constituent notes. The basis of Fourier analysis is the hypothesis that any waveform can be synthesized by adding together an ensemble of sine waves, provided that enough amplitudes and frequencies are considered. The FT is a method which accomplishes a computational search of the input waveform (the FID) for the frequencies and amplitudes present in the signal. By this process an input which is an amplitude as a function of time is converted to an amplitude as a function of frequency. The mathematical definition of the FT as the function of frequency $F(\omega)$ is related to the signal as a function of time $f(t)$, by

$$F(\omega) = \int_{-\infty}^{+\infty} f(t)e^{-i\omega t}\, dt, \tag{2.11}$$

where $\omega = 2\pi v$. For computational purposes this has to be interpreted as a discrete sum involving the N digitized values f_t for the FID.

$$F_v = \sum_{t=0}^{N-1} f_t e^{-2\pi i v t/N} \qquad (2.12)$$

Several methods for the computer implementation of equation (2.12) have been developed, and notable emong these are the fast Fourier transform (FFT) techniques. The evaluation of F_v by equation (2.12) involves approximately N^2 multiplications and, in order that the FT method may work efficiently, the size of the data set has to be 2^N, which is usually 1024, 2048, or 4096 ($N = 10$, 11, and 12, respectively). Hence N^2 may be rather a large number, and ordinary FT computer processing can be of some duration. The FFT techniques are based on the Cooley–Tukey algorithm[12] which requires only $2N \log_2 N$ arithmetic operations and hence saves considerable time, especially when N is large.

The output from the spectrometer's radio receiver consists of two channels of FID data (usually referred to as u and v, or "real" and "imaginary"). The FT is usually capable of handling two sets of input data and produces two output spectra. These are called the u and v or "real" and "imaginary" spectra, and both are available for analysis.

2.3. THE SPECTRUM

A spectrum obtained from a biological sample, such as that shown in Figure 2.7 (a ^{31}P spectrum obtained from the brain of a healthy human infant at full term), contains a wealth of potentially useful information. Correct interpretation of MR spectra can give a noninvasive insight into metabolite concentrations, pH_i, metal-ion concentrations, temperature and metabolic status of pathological tissue, and the degree of damage. A brief examination of Figure 2.7 reveals that two of the main characteristics of an MR spectrum are the frequency separations of the peaks and their amplitudes. Peak-profile shapes can be important also. Peaks do not always reveal finer structure when observed with the magnetic field homogeneities and strengths encountered clinically. Studies at higher fields can permit the detailed investigation of multiplets, leading to understanding of molecular structure. In addition, the width and shape of the Pi peak can give information concerning the range of pH_is present in the tissue. The peak separations are known as "chemical shifts," and we will now consider their origins and meanings in detail.

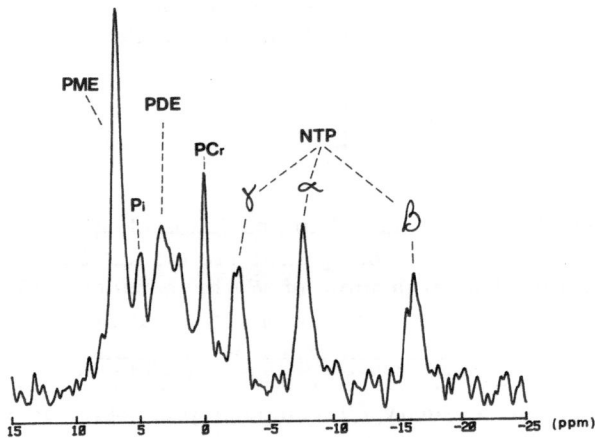

Figure 2.7. A ^{31}P spectrum obtained from the brain of a normal, full-term neonate. Peak identifications are PME, phosphomonoester, mainly phosphoethanolamine with some phosphocholine; Pi, inorganic phosphate; PDE, phosphodiesters, mainly glycerol-3-phosphorylethanolamine and glycerol-3-phosphorylcholine; PCr, phosphocreatine; NTP, nucleoside triphosphates, mainly ATP. The spectrum results from an accumulation of 128 scans with a 90° flip angle at the center of a 7-cm coil and a pulse interval of 20.256 sec. (Spectrum obtained in collaboration with E.O.R. Reynolds, D. Azzopardi, J. Wyatt, and D. T. Delpy.)

2.3.1. Chemical Shifts

Equation (2.1) shows that the resonant frequency of a nuclear-spin system is proportional to the applied magnetic field strength. Hence any effect which alters the magnetic field in the locality of the nucleus will also change the resonant frequency. The ^{31}P metabolites in human brain tissue produce peaks resonating at different frequencies, as shown in Figure 2.7. This is because the ^{31}P nuclei in these metabolites experience local magnetic field strengths which are slightly altered from that applied to the sample by the magnet of the spectrometer. The strength of the field in the vicinity of the nucleus depends not only on the metabolite species but also on the location of that particular nucleus in the molecule and the magnetic susceptibility of the tissue. Hence the ^{31}P nuclei in the α, β, and γ phosphate groups of the nucleoside triphosphates (NTP), shown in Figure 2.7, all resonate at different frequencies even though they are in the same molecule. The cause of chemical shift is the magnetic screening produced by electrons close to the nucleus. Because electrons have spin and charge, they are also magnetic particles and hence are influenced by the applied magnetic field. The screening effect produced by the electrons depends on the applied field strength and according to a fundamental law of physics (Lenz's law), opposes the main field. Thus, at the nucleus

$$B_0(\text{effective}) = B_0 - \sigma B_0 = B_0(1 - \sigma) \qquad (2.13)$$

The screening factor σ is very small, and this implies that the resonant frequencies of metabolites are clustered quite closely together. For protons, σ is $< 10^{-5}$, whereas for ^{31}P and ^{13}C, σ is $< 5 \times 10^{-3}$. ^{1}H resonances are much closer together than those found in spectra from most other nuclei, and therefore good resolution and field homogeneity are of great importance in proton spectroscopy.

In order that the applied magnetic field dependence can be taken into account and to make spectra obtained on different spectrometers comparable even though collected at varying field strengths, the chemical shift δ_r is defined by

$$\delta_r = \frac{(v_r - v_s)}{v_s} \times 10^6 \qquad (2.14)$$

in ppm, where v_r and v_s are the frequencies of a metabolite resonance and of a standard, respectively.

2.3.1.1. Chemical-Shift Reference Standards

Combining equations (2.1) and (2.13) gives

$$v_r = \frac{\gamma}{2\pi} B_0(1 - \sigma_r) \qquad (2.15)$$

It can be clearly seen that the entire spectrum will shift with B_0. It is therefore necessary to have some sort of reference which can be used to interrelate the metabolite chemical shifts brought about by the various values of σ_r present in the sample. In biological MR, both internal references (i.e., those contained within the tissue) and external references have been found useful. Internal references should be best on theoretical grounds because they are located in the same magnetic environment as the metabolites of interest, whereas external references are, by definition, outside the sample. Suitable internal reference standards should have resonant frequencies independent of changes in their local chemical and physical environment (i.e., insensitive to alterations in pH, metal-ion concentration, temperature, etc.). They must also be present in the tissue at a concentration which enables adequate SNR to be obtained from them in the study time available. In ^{31}P MRS, both phosphocreatine (PCr)[13] and the ^{1}H signal from H_2O[14,15] have been found to be very useful internal references. PCr is present in many mammalian tissues at concentrations which are usually high enough (e.g., about 30 mM in fresh, resting skeletal muscle) to give useful signals within reasonable data-collection times. (Some tissues such as liver and kidney contain very little PCr. For studies of these organs an alternative chemical-shift reference must be used.) The PCr chemical shift is pH-insensitive for pH values in the range 8.0 down to about 6.0. If the pH gets lower than this, the PCr moves significantly. The change in PCr chemical shift is approximately -0.5 ppm at pH 5.5.[16,17] Hence care must be exercised when studying tissues which are very acidotic, since PCr has often been depleted in such circumstances, and an alternative reference may need to be selected. A slight dependence of the PCr chemical shift on the concentration of Mg^{2+} has been found[17] which produces a shift of about -0.15 ppm for a metal-ion concentration of 30 mM.[13]

The ^1H H_2O chemical shift is insensitive to both pH and ionic strength changes within the normally encountered physiological ranges but has a dependence on temperature of $+0.015$ ppm/°C.[14] In many tissues, water has the added attraction of its high proton concentration (approximately 80 to 90 M), and hence an exceedingly good SNR can be obtained in minimal time. It is probable that water is better as a reference than PCr if the latter resonance has an SNR less than about 10 where the SNR is defined as the peak height divided by twice the root mean square (RMS) noise.[15] ^{31}P spectra can be bracketed by ^1H H_2O spectra in order to take into account any magnetic field drift that may have occurred during the ^{31}P collection. This can be particularly important in studies of tissues that under normal conditions contain adequate PCr for referencing purposes but can become depleted in PCr because of pathological or metabolic processes. Such tissues include fatigued muscle[19] and the brain during hypoxic-ischemic insult.[18]

For in vivo ^1H investigations, several metabolites have been employed as internal chemical-shift references. The resonance from tissue-water protons can be used, although in many studies the signal is greatly reduced by the techniques required to obtain adequate digitization of the resonances of interest which originate from low-concentration metabolites. Such data collections can, of course, be bracketed by ordinary proton-water spectra. In ^1H studies of brain tissue, both the combined $N-CH_3$ resonances of PCr and Cr[20] at 3.022 ppm and the $N=CH_3$ resonance of N-acasp[21] at 2.032 ppm have been used. The proximity of the PCr and Cr $N=CH_3$ signal to that of water may make it less useful than N-acasp for clinical investigations unless extremely good magnetic field homogeneity is obtained. It should also be noted that both N-acasp[22] and PCr[23] have been reported to be at much lower concentrations in immature rat brain than in the adult, and therefore it is possible that neither PCr and Cr nor N-acasp may be ideal internal reference standards in neonatal brain studies. The $N-CH_3$ resonances of PCr and Cr have been used as a chemical shift reference for ^1H studies of muscle.[24]

External reference standards have been employed mainly in extracted and excised-tissue studies, and many different compounds have been used. One of the main criteria for the suitability of an external reference is that it resonates some distance away from the spectrum of interest and can give an adequate signal from a filled and adjacently placed capillary. For ^{31}P studies both phosphoric and methylene biphosphonic acids have been used in some circumstances. In ^1H spectroscopy, sodium 3-trimethylsilyl [2,2,3,3-^2H] propionate (TSP) and 2,2-dimethylsilapentane 5-sulfonic acid, among others, have been used. Dioxane in ^2H_2O has been employed often in ^{13}C spectroscopy.

2.3.1.2. The Effect of the Molecular Environment

As has been noted in the previous discussion of chemical-shift reference standards, several factors can effect the position of peaks in the magnetic resonance spectrum. By accurate determination of spectral-peak positions, changes in chemical shift can be used to obtain physiologically useful information including pH_i, metal-ion concentration, and temperature.

2.3.1.2a. pH Effects. Most of the metabolites observed by ^{31}P spectroscopy have chemical shifts which depend to some extent on the pH of their environment. In the pH range 6 to 7.5, the resonances of γATP, Pi, 2,3-diphosphoglycerate (2,3-DPG), and many phosphomonoesters and sugar phosphates shift substantially.[16,17,25] The γATP resonance shows a very large shift, and at pH 5.5 has moved almost to the same position as that of αATP. However, because this peak is a doublet, and considering the resolution and SNR obtainable in vivo, it is not easy to measure its chemical shift accurately in practice. The chemical-shift dependences of various resonances can be calibrated to provide useful means of estimating pH_i. 2,3-DPG has been used for this purpose in studies of erythrocytes[16] because of its high intracellular concentration and its pH sensitivity. In many studies, however, Pi has been found the most useful metabolite for pH estimation because of its high sensitivity to pH changes, low sensitivity to metal-ion concentration, adequate tissue concentration in a variety of metabolic states, and intrinsic single-peak profile. In addition, during hypoxia–ischemia, Pi is elevated, thereby increasing its usefulness for pH estimation. The observed Pi chemical shift changes with pH because, under normal physiological conditions, it exists as the two ions HPO_4^{2-} and $H_2PO_4^-$. The MR signals from these are separated by about 2.4 ppm but, because there is rapid exchange between the two forms, they are observed as a single peak which has a chemical shift reflecting the relative proportions of the two ions. This rapid exchange between protonated and unprotonated species is common to all resonances exhibiting pH-dependent chemical shifts. The relationship between pH and Pi chemical shift fits the Henderson–Hasselbach equation

$$pH = pK_{Pi} + \log_{10}[(\delta_{Pi} - \delta_A)/(\delta_B - \delta_{Pi})] \tag{2.16}$$

where pK_{Pi} is approximately 6.75 and δ_A and δ_B are the acid and alkaline limits, respectively, for the Pi chemical shift δ_{Pi}. Figure 2.8 shows the effect of pH by comparing the Pi chemical shifts in human muscle before and after anerobic exercise. In detail, the dependence of the Pi chemical shift varies slightly with other parameters such as metal-ion concentration and temperature. Hence the relationship between pH_i and Pi chemical shift should be considered to be tissue specific. The dependence of Pi chemical shift on pH for physiological conditions similar to those found in mammalian brain cytosol[26] is shown in Figure 2.9, and Table 2.1 gives values of pH applicable to human brain and muscle tissue for Pi chemical shifts measured relative to PCr. In studies of implanted melanoma[27] and neonatal-brain[28] spectra, the use of the prominent PE peak as a probe of pH_i has been investigated. This resonance is very strong in spectra acquired from neonatal brain and many tumors but, unfortunately, the titration

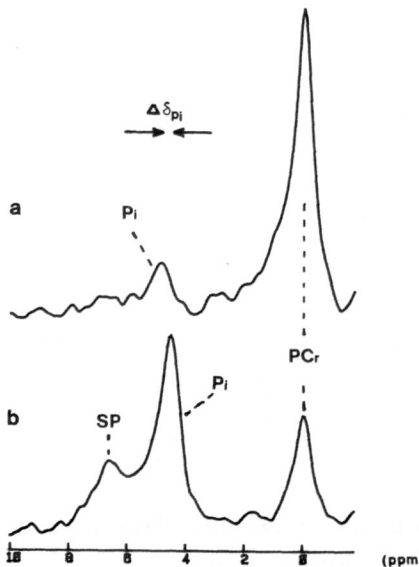

Figure 2.8. Part of the ^{31}P spectra obtained from in situ human skeletal muscle [the first dorsal interosseous muscle of the hand (FDI)] before (a) and after (b) anerobic exercise consisting of a 15-sec 100%-maximum voluntary contraction, showing the elevated, pH-shifted Pi peak. The spectra were collected using a 2.4-cm surface coil positioned over the FDI and a 90° pulse at the coil's center. Each spectrum is the result of accumulating 96 scans with a 2.256-sec pulse repetition interval: SP, sugar phosphates; Pi, inorganic phosphate; PCr, phosphocreatine; $\Delta\delta_{Pi}$, the chemical shift difference between the Pi peak positions in the two spectra. (Spectra obtained in collaboration with J. Lynn and D. A. Jones.)

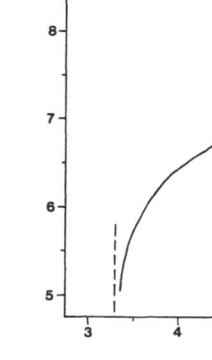

Figure 2.9. The relationship between intracellular pH and the Pi chemical shift δ_{Pi} for conditions resembling those found in mammalian brain tissue. The curve is a fit, using titration data obtained by Petroff et al.,[24] to the Henderson–Hasselbach equation: $pH = 6.77 + \log_{10}[(\delta_{Pi} - 3.29)/(5.68 - \delta_{Pi})]$. The dashed vertical lines define the limits of the range of Pi chemical shifts.

curve is such that, for pHs of about 7, slight changes in chemical shift, due perhaps to background noise or peak overlap, can produce large deviations in apparent pH. It is probable that the error in pH estimation using PE is roughly twice that for Pi. In addition, a systematic error due to peak overlap may affect results obtained by this method unless the PE is well resolved. It has recently been suggested that the difference in chemical shift between the γ and α resonances of ATP may give a good estimate of pH_i because the peaks are observed clearly over a wide range of physiological conditions.[29] The main advantage is that PCr does not have to be detectable if this method is adopted. However, ^1H water chemical shift referencing may be a better approach because of the high SNR attainable.

Many ^1H resonance chemical shifts also have been found to titrate against pH. In studies of muscle tissue, both carnosine[30,31] (β-alanylhistidine) and anserine[24] (β-alanyl-1-methylhistidine) have been employed as pH indicators. For carnosine, the C4-H, C2-H and N-H histidine ring proton resonances exhibit useful pH-dependent chemical shifts. The histidine pK is reported to be 7.06 ± 0.01, and δ_A and δ_B in equation (2.16) are 6.920 and 7.260 for C4-H, 7.657 and 8.576 for C2-H, and 7.980 and 8.224 for N-H, respectively.[30] Values for pH_i obtained from rat muscle in vivo using anserine resonances compare favorably with those determined using the Pi peak in ^{31}P spectra.[24]

2.3.1.2b. Metal-Ion Effects. Several metabolites are commonly bound to divalent metal ions to varying degrees. It is well known that this is a requirement for enzymatic reactions involving ATP and ADP. Binding with Mg^{2+}, Ca^{2+}, and Zn^{2+} has been observed to produce substantial shifts of both ^1H[32,33] and ^{31}P[17,33] resonances. The effects of the three ions are very similar, and all produce large shifts in the β and γATP resonances

Table 2.1. Human Brain and Muscle Intracellular pHs
for Pi Chemical Shifts $(\delta_{Pi})^a$

	Brain[b]	Muscle[c]
pK	6.77	6.73
δ_A (low δ_{Pi} limit)	3.29	3.28
δ_B (high δ_{Pi} limit)	5.68	5.69

δ_{Pi} (ppm)	Brain pH	Muscle pH
3.55	5.86	5.84
3.60	5.94	5.92
3.65	6.02	6.00
3.70	6.09	6.06
3.75	6.15	6.12
3.80	6.20	6.18
3.85	6.26	6.23
3.90	6.31	6.27
3.95	6.35	6.32
4.00	6.40	6.36
4.05	6.44	6.41
4.10	6.48	6.45
4.15	6.52	6.49
4.20	6.56	6.52
4.25	6.60	6.56
4.30	6.63	6.60
4.35	6.67	6.64
4.40	6.71	6.67
4.45	6.75	6.71
4.50	6.78	6.74
4.55	6.82	6.78
4.60	6.85	6.82
4.65	6.89	6.85
4.70	6.93	6.89
4.75	6.97	6.93
4.80	7.00	6.97
4.85	7.04	7.01
4.90	7.09	7.05
4.95	7.13	7.09
5.00	7.17	7.13
5.05	7.22	7.18
5.10	7.26	7.22
5.15	7.32	7.28
5.20	7.37	7.33
5.25	7.43	7.39
5.30	7.49	7.45

[a] Referenced to PCr; $pH = pK_{Pi} + \log_{10}[(\delta_{Pi} - \delta_A)/(\delta_B - \delta_{Pi})]$.
[b] Data taken from Petroff et al.[26]
[c] D. R. Wilkie, personal communication.

but practically no shift in that of αATP. For an Mg^{2+} concentration of 0.1 M, the observed shifts at pH 6.3 of the α, β, and γATP resonances are reported to be -0.4, -3.4, and -3.6 ppm, respectively.[33] Because of the small shift exhibited by the αATP peak, it is inferred that the metal ions bind to the β and γ phosphate groups only. In the case of ADP at pH 6.8 with a Mg^{2+} concentration of 0.1 M, the α and β peaks show shifts of -1.2 and -2.3 ppm, respectively, indicating that both the α and β phosphate groups are involved in the metal-ion complex.[33] More detailed work[17] has shown that the Mg^{2+} concentration can have a strong effect on the pH titration of the γATP resonance. Mg-bound ATP shows no shift of the γ peak until the pH has dropped to about 6.5, whereas the unbound ATP shifts markedly. At pH 7.0 with a Mg^{2+} concentration of 12 mM there is about 1.3 ppm difference in chemical shift between the bound and unbound states. In biological studies, the ^{31}P βATP-resonance chemical shift has been used to estimate the concentration of free Mg^{2+} in rabbit bladder and uterine smooth muscle[34] (0.5 mM), feline fast-twitch skeletal muscle[34] (1.5 mM). rat skeletal muscle[35] (dietically Mg-depleted and control; 0.83 mM and 0.95 mM, respectively), rat myocardium[34] (1.4 mM), rat brain[35] (dietically Mg-depleted and control; 0.48 mM and 0.71 mM respectively), and human liver[36] (0.3 mM). Table 2.2 gives the free Mg^{2+} concentrations for smooth muscle βATP chemical shifts (referenced to PCr) at pHs 6.6 and 7.0.

In ^1H studies of ATP[32,33] it has been found that only the H8 proton of the adenine ring exhibits a chemical shift dependent on metal-ion concentration, although the adenine ring H2 and ribose H'1 were also investigated. The effect was seen only for Zn^{2+}, and a metal-ion concentration of 100 mM produced a shift of approximately $+0.25$ ppm. All these changes in chemical shift are brought about by alterations in the nuclear shielding induced by the presence of the metal ion.

Paramagnetic ions such as Cu^{2+}, Mn^{2+}, and Co^{2+} are found to cause broadening of line profiles due to their effect on the nuclear relaxation rates and because of molecular exchange.[33] The increased nuclear relaxation is caused by fluctuating local magnetic fields generated by the Brownian (thermal) motion of the paramagnetic ions. In general, line profiles get broader and T_1 and T_2 are decreased. This latter effect has been exploited in magnetic resonance imaging in the development of agents for the enhancement of image contrast.[37] In detail, the effects on spectra are more complicated. Interactions between specific phosphate groups of ATP and Cu^{2+}, Mn^{2+}, and Co^{2+} have been investigated using ^{31}P spectroscopy for metal-ion concentrations of 10 and 100 μM.[33] Only the β and γ peaks are broadened by Cu^{2+}, implying that these phosphate groups can be complexed. However, all three ATP resonances are affected by Mn^{2+}, and this ion seems to form complexes with all the phosphate groups. The

Table 2.2. Concentration of Free Intracellular Mg^{2+} in Smooth Muscle for Given βATP Chemical Shift[a,b]

	pH	
	7.0	6.6
pK	4.11	4.04
δ_A (high $\delta_{\beta ATP}$ limit)	−15.86	−15.86
δ_B (low $\delta_{\beta ATP}$ limit)	−19.02	−19.19
$\delta_{\beta ATP}$ (ppm)	$[Mg^{2+}]$ (mM)	
−15.95	2.65	3.28
−16.00	1.67	2.08
−16.05	1.21	1.51
−16.10	0.94	1.17
−16.15	0.77	0.96
−16.20	0.64	0.80
−16.25	0.55	0.69
−16.30	0.48	0.60
−16.35	0.42	0.53
−16.40	0.38	0.47
−16.45	0.34	0.42
−16.50	0.31	0.38
−16.55	0.28	0.35
−16.60	0.25	0.32
−16.65	0.23	0.29
−16.70	0.21	0.27
−16.75	0.20	0.25
−16.80	0.18	0.23
−16.85	0.17	0.22
−16.90	0.16	0.20
−16.95	0.15	0.19
−17.00	0.14	0.18

[a] From Kushmerick et al.[34]
[b] $\delta_{\beta ATP}$ referenced to PCr at pH 7.0 and pH 6.6; $pMg^{2+} = pK + \log_{10}[(\delta_A - \delta_{\beta ATP})/(\delta_{\beta ATP} - \delta_B)]$.

broadening produces an increase in line width which varies approximately linearly with paramagnetic ion concentration. Co^{2+} has the same qualitative effect as Mn^{2+}.

In ^1H studies of ATP, the greatest broadening is observed for the H8 adenine-ring proton, which is obliterated by an Mn^{2+} concentration of 80 μM, although some degree of broadening is found for other peaks also.

2.3.1.2c. Temperature Effects. Many ^1H and ^{31}P resonances have chemical shifts which depend on temperature, and some of these have potential as probes of in situ tissue. A common spectroscopic approach to the determination of temperature has been to measure the chemical-shift difference between the methylene and alcohol protons of ethylene glycol contained in a capillary positioned within or close to the sample.[38] This has the obvious disadvantage, especially for in vivo work, that it is often not possible to place the temperature probe within the tissue itself. In order to avoid this problem, several internal temperature probes have been investigated. The amide proton resonance of N-acasp has been used in studies of rat brain,[39] and its chemical shift is found to vary linearly over the range 20°C to 40°C with the following dependence:

$$T = (8.127 - \delta_{N\text{-acasp}})/0.00789 \tag{2.17}$$

where T is the centigrade temperature and $\delta_{N\text{-acasp}}$ is the chemical shift of the N-acasp amide proton measured relative to the N-acasp CH_3 resonance at 2.015 ppm. In addition, the ^1H water resonance has been observed to have a shift of $+0.015$ ppm/°C, which (in combination with its high tissue concentration) may make this a potential candidate for temperature measurements in vivo.[14]

The use of the ^{31}P resonances of Mg-complexed ATP has been examined,[40] and the chemical-shift difference between the α and β resonances is found to be strongly temperature-dependent, changing by -0.012 ppm/°C at a frequency of 40.5 MHz. The decrease in chemical-shift difference as temperature increases is due mainly to change in the position of the βATP resonance, and this is thought to be caused by alteration in the conformation or metal-ion binding of the polyphosphate chain.

Fluorocarbons have been proposed also as thermometric fluids[41] which could be administered as artificial blood constituents. It has been claimed that 0.1°C accuracy can be obtained with ideal conditions.

2.3.2. The Shapes and Widths of Spectral Peaks

The resonances observed in biological tissue spectra vary not only in amplitude and position but also in width and shape. A large range of widths for spectral peaks is encountered. The widths depend on the mobility and hence the relaxation constants of the nuclei concerned which may be, for example, located in bone, cellular membranes, proteins, cytosol, or extracellular fluid. In addition many resonances exhibit a multiplet structure (doublet, triplet, or complex) which may not be resolved at the field strengths, tissue-susceptibility distributions, and field homogeneities found in clinical MR.

In order to be able to interpret spectra successfully, it is essential to be familiar with the underlying causes of the appearances of peaks.

2.3.2.1. The Theoretical Resonance Profile

Theoretical prediction based on the Bloch equations give a Lorentzian shape for spectral lines with a profile of the form[2]

$$g(\omega - \omega_0) = \frac{1}{\pi T_2} \frac{1}{(1/T_2)^2 + (\omega - \omega_0)^2} \tag{2.18}$$

where $\omega_0 = 2\pi\nu_0$ (ν_0 is the center frequency of the resonance) and T_2 is the spin–spin relaxation-time constant. Equation (2.18) has been normalized so that, for $\omega = \omega_0$, $g(0) = T_2/\pi$; at half this peak amplitude, the width of the profile is:

$$\nu_{1/2} = 1/(\pi T_2) \tag{2.19}$$

where $\nu_{1/2}$ is the half width (the width of the peak measured at half the peak amplitude). This relates the width of the peak to the inverse of the spin–spin relaxation-time constant. Hence, large values of T_2 give sharp spectral peaks, whereas rapidly relaxing nuclear species with short T_2s produce broad humps in the spectrum. The Lorentzian profile defined by equation (2.18) differs from other line shapes encountered in spectroscopic science (such as the Gaussian) in that it has very extensive wings. The tails of these are very long and are difficult to follow in most spectra because they rapidly merge into the background noise.

2.3.2.2. Spin–Spin Coupling and Multiplet Structure

Equation (2.18) gives the shape that one would expect to observe under ideal conditions from a sample, assuming that there are no inter-actions among the nuclei. However the phenomenon called spin–spin coupling occurs, in which the spins of adjacent nuclei (not necessarily of the same isotope) affect resonance-energy levels, splitting the peaks into various multiplet structures. Most resonances exhibit this splitting, but the chemical-shift differences between the components vary in magnitude and may not always be observable. The coupling is not direct but is mediated by electrons in chemical bonds between the nuclei and allows a given nucleus to sense the spin states of its neighbors. A suitable example to consider is the ^1H spectrum of lactate,[42] which consists of two multiplets (a doublet and a quartet) due to the $-CH_3$ and $-CH$ protons, respectively. The $-CH$ proton has two possible spin orientations, parallel or anti-parallel to the applied magnetic field, as shown in Figure 2.10. This proton is constantly changing from one spin state to the other. At any given time, half of the lactate $-CH_3$ protons will sense the $-CH$ proton in the

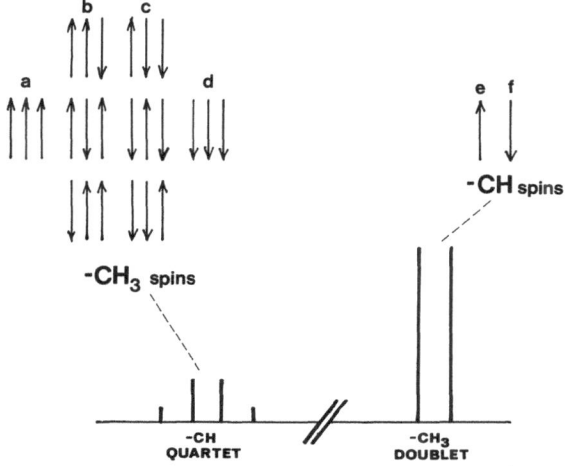

Figure 2.10. Spin–spin coupling in lactate. The $-CH_3$ doublet results from the two possible spin orientations (e and f) of the $-CH$ proton. The $-CH$ quartet is due to the statistical weighting of the eight possible spin-orientation combinations of the three equivalent methyl ($-CH_3$) protons: a, all up; b, two up, one down; c, one up, two down; and d, all down. Combinations b and c have three times the probability of a and d, and this ratio is reflected in the relative strengths of the quartet components.

parallel state; for the other half, this proton will be in the antiparallel state. The spin state of the $-CH$ proton causes a slight change in the local magnetic field experienced by the three equivalent $-CH_3$ protons, splitting the resonance into a doublet with components of equal amplitude. The situation is more complicated for the $-CH$ multiplet because there are eight possible spin combinations for the $-CH_3$ protons (see Figure 2.10). Not only does the $-CH$ resonance split up into a quartet because there are four possible spin states—all three spins parallel; all three spins antiparallel; two spins parallel and the third antiparallel; one spin parallel and the other two antiparallel—but the component amplitudes depend also on the relative probabilities for the possible states (1:3:3:1). The separation of multiplet components is known as the coupling constant J and if measured in Hz is independent of field strength. The coupling constants for the lactate $-CH_3$ doublet and the $-CH$ quartet are approximately 7 Hz. In high-resolution ^{31}P spectroscopy the α, β, and γATP resonances are observed to be a doublet, a triplet, and a doublet, respectively, with a coupling constant of about 20 Hz.[24]

2.3.2.2a. Decoupling. At the field strengths encountered in clinical spectroscopy, it is often difficult to resolve the ^1H and ^{31}P multiplet structure. However, for ^{13}C the coupling constants between directly bonded ^{13}C and ^{1}H nuclei are on the order of 100 Hz, and the splitting is relatively easy to detect. This creates a problem in ^{13}C studies in that the spectra become very complicated due to multiplet overlap, and the SNRs of the components are reduced; e.g., a methyl-carbon resonance is distributed over four peaks much as the ^{1}H-lactate $-CH$ resonance. In order to simplify ^{13}C spectra and improve the SNR, the spin–spin coupling can be eliminated by irradiating with RF at the ^{1}H frequency while collecting data. This process is called "heteronuclear decoupling" because the spins of another isotope are decoupled. The result is that the multiplets coalesce into single peaks. If the ^{1}H irradiation is continued during the interpulse period, then the ^{13}C sensitivity is further increased due to the nuclear Overhauser effect.[43] The decoupling irradiation is supplied by an additional transmitter channel of the spectrometer, which is designed to be capable of delivering continuous RF power. The decoupler unit should be able to produce either single-frequency irradiation (which can be used to decouple one specific resonance by transmitting at a particular resonant frequency, e.g., that of $-CH_3$ protons) or broad-band irradiation, which covers a large frequency range and decouples an entire spectrum. "Homonuclear decoupling" is also possible. This decouples spins of the same isotope. For instance, by irradiating at the lactate $^{1}H-CH$ frequency it is possible to collapse the $-CH_3$ doublet to a single peak. To be fully flexible, the decoupling transmitter should be capable of handling a very wide frequency band so that many different nuclei can be dealt with.

2.3.2.2b. Phase Modulation. An additional result of spin–spin coupling, called phase modulation, affects the amplitudes and phases of spectra obtained by spin–echo techniques.[44] Spin–spin coupling produces multiplets of varying degrees of complexity, each component of which resonates at a different frequency. During the delay between the pulses of a spin–echo sequence these components precess. For certain delays, their phases will be inverted and their amplitudes reduced with respect to other resonances (singlets) in the spectrum as shown for the case of the lactate $-CH_3$ doublet in Figure 2.11. At other spin–echo delays, the phase will be normal. The precession is not reversed by the 2θ pulse in the spin–echo sequence. In general, a doublet will be completely inverted when the spin–echo delay τ takes the values

$$\tau = 1/2J, \; 3/2J, \; 5/2J, \; \text{etc.} \tag{2.20}$$

and will be phase-normal with maximum amplitude when

$$\tau = 1/J, \; 2/J, \; 3/J, \; \text{etc.} \tag{2.21}$$

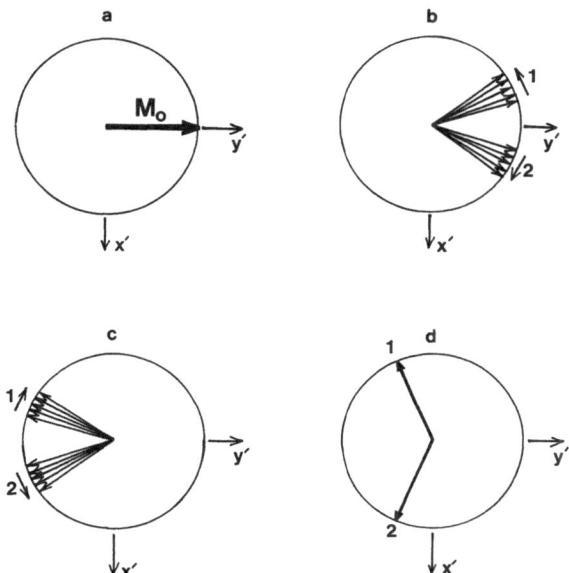

Figure 2.11. Phase modulation of the lactate methyl ($-CH_3$) doublet in a frame of reference $x'y'z'$ rotating at the mean frequency of the doublet components. (a) The magnetization M_0 in the $x'y'$ plane of the rotating frame immediately after a 90° pulse. (b) After a time τ, the doublet components (1 and 2) have separated because their frequencies differ slightly from that of the rotating frame. They have also begun to dephase in the $x'y'$ plane because of relaxation effects. (c) A 180° pulse has reflected 1 and 2 about the x' axis. Dephasing due to magnetic field inhomogeneity starts to refocus. However, the 180° pulse has also inverted the $-CH$ proton spin, causing the line splitting, and hence 1 now precesses at the previous frequency of 2 and vice versa. The spin–spin coupling is not refocused, and 1 and 2 get still further apart. (d) The refocused positions of 1 and 2 depend on τ. If the receiver is arranged to detect the x' magnetization, then no signal is picked up when 1 and 2 lie along the y' and $-y'$ axes, respectively; a 180° out-of-phase signal is obtained when 1 and 2 coincide with $-x'$; and an in-phase signal is collected when 1 and 2 are along the x' axis.

where J is the spin–coupling constant measured in Hz. Between these spin–echo times the amplitude varies with a cosine dependence and can become zero. As an example, the lactate $-CH_3$ doublet has a J spin-coupling constant of about 7 Hz and is phase inverted at 68 and 204 msec and phase normal at 136 and 272 msec. Phase modulation can provide a means whereby resonances with coupled spins can be individually selected by the use of editing sequences. These methods usually involve the decoupling of one of the coupled spins, thereby inhibiting the phase modulation; they will be described in more detail in Chapter 5.

2.3.3. Metabolite Concentrations

The tissue concentration of a particular metabolite is proportional to the area under its resonance profile in a fully relaxed spectrum. Serious errors can arise if peak heights are used to interpret spectra because not only may the constituent peaks have different intrinsic widths but some may also be multiplets that are unresolved with the fields available for clinical use. Thus peak area may not always be directly proportional to peak height. A brief examination of the heights and areas of the PCr and the βNTP (which has a triplet structure) in Figure 2.7 will suffice to demonstrate this. For most data-collection conditions, in order to maximise SNR per unit time, the FID will be accumulated with $T_R < 5T_1$. To make the measured areas relate directly to tissue concentrations, corrections have to be made for the partially saturated state of the nuclei. Correction factors are usually derived by making measurements with $T_R = 5T_1$ and comparing the results with those obtained under the normal pulsing conditions. Similar corrections have to be made to areas obtained from spin–echo studies, but in this case additional adjustments are necessary because the various resonances in the spectrum have different T_2s. In simple spin–echo spectra these corrections are made by measuring the various T_2s and multiplying each of the measured areas by $\exp(2\tau/T_2)$ where τ is the spin–echo delay.

2.4. THE SIGNAL-TO-NOISE RATIO

The SNR obtained per unit patient time is of fundamental importance in clinical studies. The operator would ideally like to collect data of sufficient quality in the minimum time. Studies which are performed quickly generally leave the subject comfortable and co-operative at their conclusion. It must always be remembered that being inserted in the bore of a large magnet can be quite stressful, especially for a patient who may have undergone many clinical tests already. Keeping the patient contented and comfortable is very important. Such subjects are more readily willing to remain still when requested to do so, and for this reason data quality is improved. This is of particular importance if the subjects are children.

In order to obtain a basic understanding of the influence of various factors which affect the SNR, it is best to begin by considering a sample in a solenoidal coil, i.e., one in which the B_1 field of the RF transmitter pulse is uniform over the whole sample and all the spins experience the same flip. The situation for surface coils is somewhat different because the flip angle produced by the RF pulse varies throughout the sample.

2.4.1. The SNR from a Single 90° Pulse

In order to obtain an understanding of the important factors, it is intuitive that one should investigate the SNR to be obtained by a single RF pulse giving a 90° flip to the sample spins. In this situation, the SNR has been studied in some detail, and usually it has been accepted that the various factors are involved according to the following equation[1,45]:

$$\frac{v_s}{v_n} = K \frac{\eta Q^{1/2} \omega_0^{1/2} V_c^{1/2} (\mu_0/4\pi)^{1/2}}{(F 4 k_B T_c \Delta_v)^{1/2}} M_0 \qquad (2.22)$$

where v_s and v_n are the signal and noise voltages, respectively; K is a factor (close to unity) depending on the coil's geometry; η is the filling factor (the fraction of the coil's volume filled by sample), Q is the quality factor (efficiency) of the coil (see Chapter 4); ω_0 is the frequency $(2\pi v_0)$; V_c is the coil's volume; μ_0 is the permeability of empty space; F is the "noise figure" of the spectrometer's receiver; k_B is Boltzmann's constant; T_c is the coil temperature; Δ_v is the bandwidth of the receiver; and M_0 is the nuclear magnetization given by equation (2.3). The numerator of the right-hand side of equation (2.22) contains factors which contribute to the signal strength, whereas the denominator contains factors pertaining to the noise voltage. As can be seen, the most important factors in this formulation are η, the filling factor, and M_0, the magnetization, which is directly proportional to the density of spins (metabolite concentration) and to the field strength (B_0). However, the interpretation of equation (2.22) is quite complicated because it contains several interacting factors. For instance, the Q of the coil depends on the resistance of its turns of conducting material, and this is a function of both temperature and frequency. The resistance of the coil's conductor is increased when carrying RF due to skin[46] and proximity[47] effects. At frequencies above about 5 MHz, the magnetic field associated with the RF current in a conductor ensures that the current flows only in a thin layer at the surface. Thus the cross-sectional area in which the RF current flows is greatly reduced. This is called the skin effect, and for a long, straight cylindrical conductor it can be shown that the resistance R is given by

$$R = (L/p)[\mu \mu_0 \omega_0 \rho(T_c)/2]^{1/2} \qquad (2.23)$$

where L is the conductor length, p is its circumference, μ is the permeability of the conductor, and $\rho(T_c)$ is its resistivity (which is temperature dependent). The resistance increases both with decreasing conductor diameter and with increasing frequency, and it is much greater than that for direct current. The proximity effect occurs when a conductor of RF is close to another conductor. The magnetic field of the current in one conductor has an influence on the flow of current in the other. The other conductor can be part of the coil itself or any other adjacent conducting object, including the sample. This disturbance of the current flow close to the conductor's surface further increases the resistance to RF. The coil's resistance may also depend on its volume by way of the length of conductor used for its construction. Coil resistance is of fundamental importance to an analysis of SNR because the coil's resistance is the source of thermal noise, according to

$$v_n = (4k_B T_c \Delta v \, R)^{1/2} \tag{2.24}$$

Additionally, the meaning of η (the filling factor) although obvious for a solenoidal or saddle type coil, which both have fairly uniform B_1 fields, is not so clear for a surface coil.

For the above reasons a more fundamental relationship between SNR and experimental conditions has been derived[48]:

$$\frac{v_s}{v_n} = \frac{K(B_1)_{xy} V_s N\gamma\hbar^2 I(I+1)}{7.12 k_B T_s} \left[\frac{P}{Fk_B T_c l\xi \, \Delta v} \right]^{1/2} \frac{\omega_0^{7/4}}{[\mu\mu_0 \rho(T_c)]^{1/4}}. \tag{2.25}$$

In this formulation, $K(B_1)_{xy}$ is the effective field throughout the sample produced by unit current flowing in the receiving coil, where K is an inhomogeneity factor; V_s is the sample volume; T_s is the sample's temperature; p and l are the perimeter and length of the coil's conductor, respectively; and ξ is a "proximity" factor. The strong dependence on V_s and N is still apparent but, in addition, the dependence on frequency is now $\omega_0^{7/4}$, showing the importance of magnetic field strength to SNR.

2.4.2. The Effect of T_1 and T_2 on SNR

The previous discussion has dealt only with the situation of an isolated 90° pulse. In most clinical applications, due to the low metabolite concentrations in tissue and in order to obtain the maximum signal-to-noise ratio per unit of patient time, MRS data is acquired as the sum of the FIDs resulting from a series of pulses or pulse sequences. Because the interval between pulses is often smaller than $5T_1$, the nuclear spins will not have had enough time to relax to their equilibrium state, and hence only a fraction of M_0 is available for the next pulse to flip into the xy plane. The FID will also decay with a spin–spin relaxation time constant T_2^*. It is found that the signal strength depends in a complicated manner on spin–lattice and spin–spin relaxation, the flip angle, and the resonance frequency. Due to the homogeneous B_1 field, the analysis for a sample in a solenoidal or saddle coil is much simpler than that for a surface coil. The homogeneous B_1 field situation has been studied in some detail[49] assuming that the effective B_1 field resulting from the RF pulse lies along the x-axis in the rotating frame. If the resonant frequency is ω_0, then B_0 has no apparent effect on M_0 in the rotating frame. During the RF pulse, precession occurs only about B_1 and, in this case, causes rotation about the x-axis. This simplifies the analysis of a system of isolated nuclear spins subject to brief RF pulses repeated at intervals T_R, which reveals that the resultant x and y magnetizations immediately after each pulse (i.e., at time $t = +0$) are given by

$$M_x(+0) = M_0 \frac{-(1-E_1)E_2 \sin\theta \sin\alpha}{(1-E_1\cos\alpha)(1-E_2\cos\theta) - (E_1-\cos\alpha)(E_2-\cos\theta)E_2}$$

(2.26)

$$M_y(+0) = M_0 \frac{(1-E_1)(1-E_2\cos\theta) \sin\alpha}{(1-E_1\cos\alpha)(1-E_2\cos\theta) - (E_1-\cos\alpha)(E_2-\cos\theta)E_2}$$

(2.27)

where $E_1 = \exp(-T_R/T_1)$, $E_2 = \exp(-T_R/T_2)$, α is the flip angle, and θ is the angle through which the magnetization precesses during time T_R and is equal to $T_R(\Omega_i - \omega_0)$ where ω_0 is the rotation frequency of the rotating frame and the frequency of the transmitter pulses and Ω_i is the resonant frequency of the nuclei. These two components are observed by the spectrometer's radio receiver (using phase-sensitive detection) as the magnetization $M_y(t)$.

$$M_y(t) = M_y(+0) \cos[(\Omega_i - \omega_0)t] + M_x(+0) \sin[(\Omega_i - \omega_0)t] \quad (2.28)$$

If $T_R \gg T_1$, then both E_1 and E_2 become insignificant and equations (2.26) and (2.27) reduce to

$$M_x(+0) = 0; \qquad M_y(+0) = \sin\alpha. \quad (2.29)$$

Equations (2.26), (2.27), and (2.28) can be further reduced if it is assumed that the transverse magnetization decays to zero during T_R, i.e., $E_2 = 0$. The detected signal is now given by[50]

$$M_y(t) = M_0 \frac{(1-E_1)\sin\alpha}{1-E_1\cos\alpha} \exp(-t/T_2) \cos[(\Omega_i - \omega_0)t] \quad (2.30)$$

Of obvious interest is the flip angle α, which gives maximum magnetization in the xy plane and is given by:

$$\cos \alpha_{max} = \frac{E_1 + E_2(\cos \theta - E_2)/(1 - E_2 \cos \theta)}{1 + E_1 E_2(\cos \theta - E_2)/(1 - E_2 \cos \theta)} \tag{2.31}$$

For $T_R \gg T_1$, $\cos \alpha_{max}$ gets close to 0 and hence, under this pulsing condition, maximum magnetization is given by flip angles approaching $90°$, $270°$, $450°$, etc. If the resonant frequency of the nucleus under investigation is very different from ω_0, then, in the rotating frame, the effective field will not lie in the $x'y'$ plane. The effective B_1 field now has a B_0 component and is out of the $x'y'$ plane. Precession during the RF pulse now takes place about the new effective B_1. The analysis of this more realistic situation is more complicated, and the reader is referred to the literature for more detailed study.[51]

As we have seen, the signal amplitude obtained in a repetitive-pulse experiment is a complicated function of flip angle α, T_1, T_2, and T_R. In a simplified case ($\alpha < 90°$), it is possible to show that the SNR can be derived from equation (2.2) to give[52]

$$\text{SNR} = \frac{M_0}{T_R^{1/2}} \frac{(1 - E_1) \sin \alpha}{1 - E_1 \cos \alpha} \tag{2.32}$$

Differentiation of equation (2.32) with respect to α gives:

$$\cos \alpha_{max} = E_1 \tag{2.33}$$

Plots of SNR versus flip angle for various values of T_1/T_R are shown in Figure 2.12. Fortunately, the SNR is fairly insensitive to flip angle, and a small amount of error in setting pulse lengths does not significantly degrade results. Figure 2.12 also shows the optimum flip angles for given T_1/T_R.

The above analysis has assumed a uniform B_1 field, a condition that holds good only for solenoidal, saddle, and some other coils. Surface coils, which are commonly used in clinical applications, have a very inhomogeneous B_1 distribution under RF-pulse excitation. The nuclear spins are flipped by angles which depend on spatial location. Thus, for regions in which the nuclei experience a large flip, more saturation will occur than in regions experiencing a small flip. The derivation of the expected signal strength is therefore more complicated than the previous discussion. In order to provide a solution to this problem, computer techniques are used to integrate the signal from incremental volumes throughout the space in

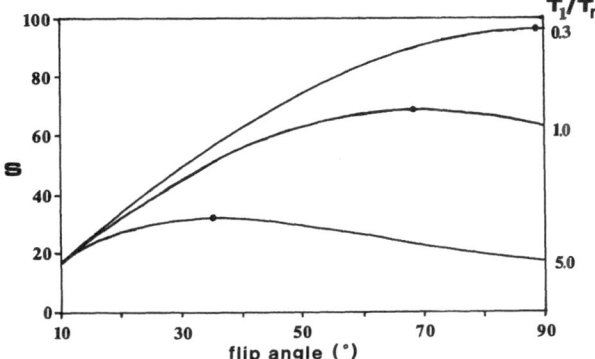

Figure 2.12. A plot of relative signal S as a function of flip angle, calculated according to equation (2.32) for three values of T_1/T_R, assuming a homogeneous B_1 field. Because of the spatial variation of flip angle, the relationship will be only approximately correct for a surface coil unless a point sample is being investigated. The closed circles on the curves give the optimum flip angles for a given T_1/T_R [from Becker et al.[49]].

front of the coil.[53,54] The actual signal obtained depends on the three-dimensional distribution of the sample and on fundamental parameters such as the flip angle at the center of the coil, T_R, and T_1. As pulsing gets more rapid (i.e., as T_R/T_1 gets smaller), one tends to get relatively more signal from parts of the sample further away from the coil. This is due to the nearby spins becoming more saturated due to the large flip they experience. Signal-amplitude maps derived using computerized integration techniques are given in the section on surface coils in Chapter 4.

REFERENCES

1. Abragam, A.: *The Principles of Nuclear Magnetism*, Oxford, Clarendon Press, 1961.
2. Kittel, C.: *Introduction to Solid State Physics*, 3rd ed., New York, John Wiley, 1968.
3. Farrar, T. C. and Becker, E. D.: *Pulse and Fourier Transform NMR*, New York, Academic Press, 1971.
4. Slichter, C. P.: *Principles of Magnetic Resonance*, 2nd ed., Berlin, Springer, 1978.
5. Becker, E. D.: *High Resolution NMR*, 2nd ed., New York, Academic Press, 1980.
6. Ernst, R. R., Bodenhausen, G., and Wokaun, A.: *Principles of Nuclear Magnetic Resonance in One and Two Dimensions*, Oxford, Clarendon Press, 1987.
7. Hutchinson, J. M. S. and Foster, M. A.: General principles, in Foster, M. A. and Hutchinson, J. M. S. (eds): *Practical NMR Imaging*, Oxford, IRL Press, 1987, p. 1.
8. Bloch, F.: Nuclear induction, *Physical Review* **70**:461–474, 1946.
9. Young, I. R.: Considerations affecting signal and contrast in NMR imaging, *Br. Med. Bull.* **40**:139–147, 1984.
10. Bracewell, R. N.: *The Fourier Transform and Its Applications*, New York, McGraw-Hill, 1965.
11. Lindon, J. C. and Ferrige, A. G.: Digitization and data processing in Fourier transform NMR, *Progress in Nuclear Magnetic Resonance Spectroscopy* **14**:27–66, 1980.

12. Cooley, J. W. and Tukey, J. W.: An algorithm for the machine calculation of complex Fourier series, *Mathematics Comput* **19**:297–301, 1965.
13. Dawson, M. J., Gadian, D. G., and Wilkie, D. R.: Contraction and recovery of living muscles studied by ^{31}P nuclear resonance, *J. Physiol.* (London) **267**:703–735, 1977.
14. Ackermann, J. J. H., Gadian, D. G., and Radda, G. K. et al.: Observation of ^{1}H NMR signals with receiver coils tuned for other nuclides, *J. Mag. Res.* **42**:498–500, 1981.
15. Cady, E. B. and Wilkie, D. R.: Estimation of cerebral intracellular pH by ^{31}P and ^{1}H nuclear magnetic resonance spectroscopy, in Rolfe, P. (ed.): *Neonatal Physiological Measurements*, London, Butterworth, 1986, p. 373.
16. Moon, R. B. and Richards, J. H.: Determination of intracellular pH by ^{31}P magnetic resonance, *J. Biol. Chem.* **248**:7276–7278, 1973.
17. Burt, C. T., Glonek, T., and Barany, M.: Analysis of phosphate metabolites, the intracellular pH and the state of adenosine triphosphate in intact muscle by phosphorus nuclear magnetic resonance, *J. Biol. Chem.* **251**:2584–2591, 1976.
18. Hope, P. L., Costello, A. M. de L., and Cady, E. B. et al.: Cerebral energy metabolism studied with phosphorus NMR spectroscopy in normal and birth asphyxiated infants, *Lancet* **2**:366–369, 1984.
19. Taylor, D. J., Styles, P., and Matthews, P. M. et al.: Energetics of human muscle: Exercise-induced ATP depletion, *Magn. Reson. Med.* **3**:44–54, 1986.
20. Arus, C., Chang, Y., and Barany, M.: Proton nuclear magnetic resonance spectra of excised rat brain. Assignment of resonances, *Physiol. Chem. Phys. Med. NMR* **17**:23–33, 1985.
21. Behar, K. L., Rothman, D. L., and Shulman, R. G. et al.: Detection of cerebral lactate in vivo during hypoxemia by ^{1}H NMR at relatively low field strengths (1.9 T), *Proc. Natl. Acad. Sci. U.S.A.* **81**:2517–2519, 1984.
22. Tallan, H. H.: Studies on the distribution of N-acetyl-L-aspartate acid in brain, *J. Biol. Chem.* **223**:41–45, 1957.
23. Tofts, P. S. and Wray, S.: Changes in brain phosphorus metabolites during the post-natal development of the rat, *J. Physiol.* (London) **359**:417–429, 1985.
24. Williams, S. R., Gadian, D. G., and Proctor, E. et al.: Proton NMR studies of muscle metabolites in vivo, *J. Mag. Res.* **63**:406–412, 1985.
25. Cohn, M. and Hughes, T. R.: Phosphorus magnetic resonance spectra of adenosine di- and triphosphate. I. Effect of pH, *J. Biol. Chem.* **235**:3250–3253, 1960.
26. Petroff, O. A. C., Prichard, M. D., and Behar, K. L. et al.: Cerebral intracellular pH by ^{31}P nuclear magnetic resonance spectroscopy, *Neurology* **35**:781–788, 1985.
27. Corbett, R. J. T., Nunnally, R. L., and Giovanella, B. C. et al.: Characterization of the ^{31}P nuclear magnetic resonance spectrum from human melanoma tumors implanted in nude mice, *Cancer Res.* **47**:5065–5069, 1987.
28. Corbett, R. J. T., Laptook, A. R., and Hassan, A. et al.: Quantitation of acidosis in neonatal brain tissue using the ^{31}P NMR resonance peak of phosphoethanolamine, *Magn. Reson. Med.* **6**:99–106, 1988.
29. Pettegrew, J. W., Withers, G., and Panchalingam, K. et al.: Considerations for brain pH assessment by ^{31}P NMR, *Magn. Reson. Imaging* **6**:135–142, 1988.
30. Arus, C., Barany, M., and Westler, W. M. et al.: ^{1}H NMR of intact muscle at 11 T, *FEBS Lett.* **165**:231–237, 1984.
31. Yoshizaki, K., Seo, Y., and Nishikawa, H.: High-resolution proton magnetic resonance spectra of muscle, *Biochim. Biophys. Acta* **678**:283–291, 1981.
32. Jardetzky, C. D. and Jardetzky, O.: Investigation of the structure of purines, pyrimidines, ribose nucleosides and nucleotides by proton magnetic resonance. II, *Journal of the American Chemical Society* **82**:222–229, 1960.
33. Cohn, M. and Hughes, T. R.: Nuclear magnetic resonance spectra of adenosine di- and

triphosphate. II. Effect of complexing with divalent metal ions, *J. Biol. Chem.* **237**:176–181, 1962.

34. Kushmerick, M. J., Dillon, P. F., and Meyer, R. A. et al.: ^{31}P NMR spectroscopy, chemical analysis and free Mg^{2+} of rabbit bladder and uterine smooth muscle, *J. Biol. Chem.* **261**:14,420–14,429, 1986.

35. Adam, W. R., Craik, D. J., and Hall, J. G. et al.: Problems in the assessment of magnesium depletion in the rat by in vivo ^{31}P NMR, *Magn. Reson. Med.* **7**:300–310, 1988.

36. Oberhaensli, R. D., Galloway, G. J., and Taylor, D. J. et al.: Assessment of human liver metabolism by phosphorus-31 magnetic resonance spectroscopy, *Br. J. Radiol.* **59**:695–699, 1986.

37. Gadian, D. G., Payne, J. A., and Bryant, D. J. et al.: Gadolinium-DTPA as a contrast agent in MR imaging—theoretical projections and practical observations, *J. Comput. Assist. Tomogr.* **9**:242–251, 1985.

38. Martin, M. L., Martin, G. J., and Delpuech, J. J.: *Practical NMR spectroscopy*, London, Heyden, 1980.

39. Arus, C., Chang, Y. C., and Barany, M.: *N*-acetylaspartate as an intrinsic thermometer for ^{1}H NMR of brain slices, *J. Mag. Res.* **63**:376–379, 1985.

40. Gupta, R. K. and Gupta, P.: A magnesium (II) ATP thermometer for ^{31}P NMR studies of biological systems, *J. Mag. Res.* **40**:587–589, 1980.

41. Ackerman, J. L., Clark, L. C., and Thomas, S. R. et al.: NMR thermal imaging, *Proceedings*, Third Annual Meeting, Society from Magnetic Resonance in Medicine, New York, 1984, p. 1.

42. Williams, S. R. and Cady, E. B.: In vivo spectroscopy, in Foster, M. A. and Hutchinson, J. (eds): *Practical NMR Imaging*, Oxford, IRL Press, 1987, pp. 249–274.

43. Freeman, R.: *A handbook of nuclear magnetic resonance*, New York, John Wiley, 1988.

44. Rabenstein, D. L. and Nakashima, T. T.: Spin-echo Fourier transform nuclear magnetic resonance spectroscopy, *Anal. Chem.* **51**:1465A–1474A, 1979.

45. Hill, H. D. W. and Richards, R. E.: Limits of measurement in magnetic resonance, *Journal of Scientific Instruments* (*Journal of Physics E*, Ser. 2) **1**:977–983, 1968.

46. Bleany, B. I. and Bleaney, B.: *Electricity and Magnetism*, 3rd ed., London, Oxford University Press, 1976.

47. Terman, F. E.: *Radio Engineer's Handbook*, 1st ed., New York, McGraw–Hill, 1943.

48. Hoult, D. I. and Richards, R. E.: The signal to noise ratio of the nuclear magnetic resonance experiment, *J. Mag. Res.* **24**:71–85, 1976.

49. Ernst, R. R. and Anderson, W. A.: Application of Fourier transform spectroscopy to magnetic resonance, *Review of Scientific Instruments* **37**:93–102, 1966.

50. Ernst, R. R. and Morgan, R. E.: Saturation effects in Fourier spectroscopy, *Molecular Physics* **26**:49–74, 1973.

51. Jones, D. E. and Sternlicht, H.: Fourier transform nuclear magnetic resonance. I. Repetitive pulses, *J. Mag. Res.* **6**:167–182, 1972.

52. Becker, E. D., Ferretti, J. A., and Gambhir, P. N.: Selection of optimum parameters for pulse Fourier transform nuclear magnetic resonance, *Anal. Chem.* **51**:1413–1420, 1979.

53. Evelhoch, J. L., Crowley, M. G., and Ackermann, J. J. H.: Signal to noise optimization and observed volume localization with circular surface coils, *J. Mag. Res.* **56**:110–124, 1984.

54. Haase, A., Hanicke, W., and Frahm, J.: The influence of experimental parameters in surface coil NMR, *J. Mag. Res.* **56**:401–412, 1984.

3

Clinical Studies

3.1. INTRODUCTION

After a very slow start, the clinical application of MRS is now beginning to expand. This expansion is in part related to the increased availability in hospital environments of suitable high-field imaging spectrometers with bores up to whole-body diameter. Other factors are also involved. Although the modality has been shown to be effective in certain specialized fields, the more general usefulness of MRS in health care has only recently become apparent. With the primitive methods of signal localization available using a surface coil alone, investigations were initially limited to superficial tissues (e.g., skeletal muscle) and a few special situations. Clinical users of MRS now have a selection of effective localization methods, which show great promise and can be applied to studies with many different nuclei (see Chapter 5). These techniques permit the selection of a volume of interest from a series of ^{1}H magnetic resonance images obtained from contiguous slices and the collection of spectroscopic data only from tissue within that selected volume. The power of these noninvasive biochemical assay methods is greatly enhanced by the fact that, in many tissues, magnetic susceptibility changes are small. This allows much better homogeneity of the field within the volume of interest and hence improved spectroscopic resolution.[1] As a result, it is becoming evident that ^{1}H and other nuclei will have a role in the future which will be as important as that of the ^{31}P nucleus. In addition, localization techniques are now available which allow some degree of flexibility not only in the position and size of the sensitive volume but also in its orientation and shape.

Although localization techniques are now proving to be very effective, a great deal of fundamental work has been done using either surface coils alone or early localization methods incapable of precisely defining a sensitive volume.[2] Because of the relatively recent availability of rigorous localization methods in a clinical setting, only a few results obtained using such techniques will be described in this chapter. The first in situ human tissue to be investigated was skeletal muscle.[3-6] This was because of its ready accessibility for study by surface coils and the possibility of positioning limbs within the magnets of small bore diameters (up to about 20 cm) available in the early 1980s. These studies were soon followed, in 1982, by the observation of phosphorus metabolites in the brains of newborn human infants.[7] The study of the adult human brain became possible only with the availability of larger-bore superconducting magnets permitting the insertion of most of the body.[8] A large proportion of subsequent investigations of in situ human tissues have focused on skeletal muscle and the brain. However, significant research has now also been carried out on many other tissues both in the normal physiological state and with various pathological conditions.

3.2. STUDIES OF THE HUMAN BRAIN

The use of MRS for the study of the human brain is now beginning to produce clinically important results. In certain applications, important prognostic information is obtainable. The main areas of research so far have been the investigation of the sequelae of hypoxic-ischemic episodes (e.g., birth asphyxia, intraventricular hemorrhage in neonates, and stroke in adults) or studies of intracerebral tumors. The ^{31}P nucleus has been used a great deal for this purpose, although recent advances in localization techniques have permitted the acquisition of well-resolved ^{1}H spectra. Other nuclei, such as ^{13}C, have not so far made a large impact in clinical studies of the brain.

Although the noninvasive measurement of biochemical data from the human brain has a lot of clinical potential, the application of MRS to the study of this organ was achieved only after studies of skeletal muscle were well under way. This is because detailed investigation of the brain requires techniques whereby precise localization is possible or, at least, the signals from superficial tissues (skin, fat, cranial muscle, bone, etc.) can be suppressed. Magnets of suitable dimensions are also necessary. (An important exception to this principle has proved to be the ^{31}P MRS study of the neonatal brain, in which many potentially contaminating signals are largely absent or can be removed adequately by computer processing.) Although the first ^{31}P human brain spectra were obtained from newborn infants, studies of normal adult and neonatal brain are complimentary and indicate age related biochemical changes which are consistent with brain development and maturation.

3.2.1. Studies of the Neonatal Brain

Persistent neurodevelopmental handicap or death in newborn infants is often because of cellular hypoxia caused either by ischemia related to cerebral hemorrhage[9] or by arterial hypoxemia.[10] Ultrasound, MRI, or CAT can bring about the visualization of changes due to hemorrhage. However, the occurrence of biochemical effects due to hypoxia is difficult to detect by these modalities until loss of brain tissue has taken place. MRS has the capability to monitor the levels of high-energy phosphates, such as PCr and ATP, as well as products of anerobic glycolysis, such as lactate. Thus MRS provides a method whereby the immediate biochemical response of brain tissue can be evaluated and sequentially monitored.

Clinical studies involving newborn infants have certain requirements peculiar to the nature of the subjects. The infant must be accomodated within the bore of the magnet and during transportation to the MRS suite in the same manner and with the same monitoring as in the neonatal intensive-care unit. Preferably, total data-collection time should be relatively short, and (unless sedation is permissible) the technique should allow data to be collected despite some degree of movement by the subject. MRS equipment should be safe for use with infants undergoing ventilation and within an atmosphere which may have a high O_2 concentration.

3.2.1.1. ^{31}P Spectroscopy of the Neonatal Brain

Early studies of the neonatal brain using ^{31}P MRS posed several interesting problems.[7,11,12] First, the ^{31}P spectra obtained from normal infants (see Figures 3.1a and b) were very different in appearance from those of skeletal muscle. Although the major mobile phosphorus metabolites were known in skeletal muscle, peak identification was not so easy for neonatal-brain spectra. In particular, the peak in the phosphomonoester (PME) region of the spectrum was found to be unexpectedy strong (peak 1 in Figure 3.1b). A moderately broad, unresolved peak was detected in the phosphodiester (PDE) region (peak 3 in Figure 3.1b), and the entire spectrum was superimposed on a broad hump (see Figure 3.1c). Hence one of the immediate tasks was to identify the peaks present in the spectrum. Second, the simple surface coil method by which spectra were obtained required detailed examination in order to assess the degree of spectral contamination by superficial tissues.

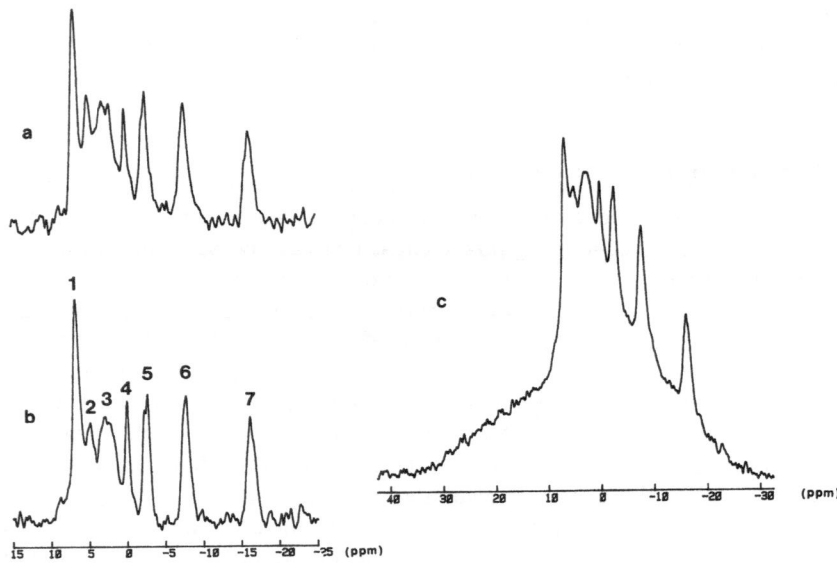

Figure 3.1. [31]P spectra obtained from the cerebral cortex of the normal human infant. The spectra were collected using a surface coil of diameter 7 cm, tuned to 32.5 MHz and placed directly on the scalp above the temporoparietal cortex. A flip angle of 90° at the coil's center and a pulse interval of 2.256 sec were used. (a) A spectrum obtained from a premature infant. (b) A spectrum collected from a full-term infant. (c) The same data as displayed in (b), but without computer processing, in order to show the broad spectral hump due to relatively immobile phosphorus nuclei in bone and membrane phospholipids. Peak identifications: 1, PME; 2, Pi; 3, PDE; 4, PCr; 5, 6, and 7, the γ, α, and β phosphorus nuclei of NTP (mainly ATP). Note the different relative sizes of the Pi and PCr peaks in spectra (a) and (b).

3.2.1.1a. Identification of the Resonances. The metabolites responsible for the peaks in Figures 3.1a and b were identified from the available literature on phosphorus-metabolite concentrations in the mammalian brain[13-16] and with reference to high-resolution ^{31}P spectra collected from guinea-pig brain extracts.[17] One erroneous resonance identification was made initially, but this was soon corrected. On the basis of the PME chemical shift, its strength in the normal neonatal-brain spectra, and comparison with guinea-pig extract data,[17] this resonance (peak 1 in Figure 3.1b) was initially identified as ribose-5-phosphate. However, it was soon realized that the high concentration implied by the relative area of peak 1 (when compared with NTP, for instance) would not be consistent with other measurements of brain-sugar phosphates. Further, the persistence of peak 1 in a postmortem study[11] gave strong evidence that the metabolite was not a sugar phosphate. Subsequent investigations of puppy-brain extracts by ^{31}P MRS,[18] of adult and neonatal rat-brain extracts by chromatography and ^{31}P MRS,[19] and of neonatal rabbit-brain extracts by chromatography and ^{31}P and ^{1}H MRS[20] identified phosphoethanolamine (PEt) as the major contributor to peak 1. The high concentrations of PEt measured in newborn mammalian[21] and fetal human[22] brain tissue provide a possible explanation for the relatively large size of this peak. Peaks 2, 3, and 4 were identified as Pi, PDE [including glycerol-3-phosphoryl-ethanolamine (GPE) and glycerol-3-phosphorylcholine (GPC)], and PCr, respectively. The remaining peaks (5, 6, and 7) resonated at the characteristic chemical shifts of magnesium-complexed ATP. It is known, however, that other nucleoside triphosphates (NTP; notably including GTP, UTP, and CTP) make up about 23% of the triphosphate pool in adult-rat cortical brain tissue.[23] It is probable, therefore, that these other NTPs contribute significantly to peaks 5, 6, and 7 in brain spectra, unless they are immobile. Peaks 5 and 6 have additional contributions from nucleoside diphosphates, and peak 6 contains contributions from nicotinamide adenine dinucleotides. The spectra in Figures 3.1a and b have been processed to remove the broad spectral feature due to signal from immobile phosphorus nuclei in bone and membrane phospholipids.[24] Figure 3.1c shows a spectrum without such processing, and the underlying broad signal from bound phosphorus nuclei is clearly visible. Many other resonances probably contribute to ^{31}P spectra from both neonatal and adult brain (see Chapter 1, Table 1.2).

3.2.1.1b. Spectral Contamination. The spectra obtained from initial studies of the neonatal brain were collected with a surface coil placed against the infant's scalp, adjacent to the temporoparietal cortex. This left open the possibility that signal from cranial skin, muscle, and bone situated within the sensitive volume of the coil could contribute to the observed spectrum. Skin contains phosphorus metabolites[25] but occupies only a very small fraction of the coil's sensitive volume. Phosphorus metabolites in skin also have relatively low concentrations (a few mM). In bone, the vast majority of phosphorus nuclei are relatively immobile and contribute only a broad hump to the spectrum (Figure 3.1c). This component can be separated easily from the mobile metabolite signal (Figures 3.1a and b). Skeletal muscle contains significant concentrations of PCr and ATP and is always a potential source of spectral contamination whenever a surface coil is the only available means of signal localization. Two preliminary analyses of the surface coil's spatial sensitivity,[26,27] in which the sensitive volume was assumed to approximate that of a right circular cone of radius and height each equal to the coil's radius, led to the conclusion that 46% to 66% of the sensitive volume is occupied by brain. More detailed analysis,[27] in which a more accurate description of the surface coil's spatial sensitivity was used,[28] indicated that, potentially, it was possible that 20% to 25% of the signal could have originated from extracerebral sources. However, postmortem examination of the cranial musculature at the coil's location revealed that very little muscle tissue was present. In addition, brain spectra obtained from an infant with propionic acidemia showed no detectable signals from PCr or NTP, whereas spectra obtained from the peripheral muscles showed large signals from these metabolites. These observations gave strong evidence that essentially all of the signals detected in the neonatal-brain spectra originated from brain metabolites. Thus the neonatal brain provides an almost unique opportunity to study in vivo metabolism using only a simple surface-coil localization technique.

3.2.1.1c. Results From Normal Infants. Initial measurements of phosphorus metabolite levels in the neonatal brain were published as relative concentrations. These have been quoted either as ratios (e.g., PCr/Pi, PCr/NTP) or in relation to the total mobile-phosphate pool (P_{total}). The brain of the human newborn infant is developing rapidly. ^{31}P spectra obtained from the human neonate differ greatly from those acquired from adult mammals. With reference to Figure 3.1a, in relative terms the PME resonance (peak 1) is much larger and the PDE and PCr resonances (peaks 3 and 4, respectively) are smaller in spectra from the human neonate when compared to spectra acquired from fully developed mammals, including adult humans.[29–35] It has also been observed that ^{31}P spectra obtained from a rat's brain gradually change between birth and maturity.[36] Hence it is important to take gestational + postnatal age into account when defining normal ranges for both relative and absolute metabolite concentrations in the neonatal brain. Figures 3.1a and b were collected from a normal premature infant and a normal full-term infant, respectively. Even though the difference in gestational plus postnatal age is only a few months, it is evident that the PCr peak (peak 4) is relatively larger in the full-term spectrum. Enough detailed work has now been carried out to attempt to define normal ranges and maturational changes for the neonatal brain.[37,38] Results for normal appropriate-for-gestational-age (AGA) infants are given in Table 3.1, and the changes in PCr/Pi, PME/P_{total}, and PDE/P_{total} with age are shown in Figures 3.2a, b, and c. Highly significant changes in PCr/Pi and PME/P_{total} and less significant, but still important, changes in PCr/P_{total} and PDE/P_{total} are seen over the period of gestational plus postnatal age from 28 to 42 weeks. Indications of possible trends are observed for Pi/P_{total}, PME/NTP, and the ratio of the broad bone- +-lipid resonance to NTP. The changes reveal a pattern of relatively decreasing PME and relatively increasing PCr, PDE, and bone-and-lipid concentrations. The increase in the signal from relatively immobile bone-and-lipid phosphorus nuclei probably reflects the advance of bone mineralization[39] and increased concentrations of membrane and

Table 3.1. Relative Phosphorus Metabolite Concentrations and Intracellular pH in the Brains of 30 Normal Infants Appropriate-for-Gestational Age[a]

	Age (Gestational plus Postnatal)[b]			
	28	(weeks)	42	p
PCr/Pi	0.85 ± 0.33		1.18 ± 0.33	<0.001
NTP/P$_{total}$	0.09 ± 0.03		0.10 ± 0.03	not sig.
PCr/P$_{total}$	0.09 ± 0.02		0.11 ± 0.02	<0.01
Pi/P$_{total}$	0.10 ± 0.02		0.09 ± 0.02	<0.05
PME/P$_{total}$	0.32 ± 0.04		0.26 ± 0.04	<0.001
PDE/P$_{total}$	0.19 ± 0.04		0.22 ± 0.04	<0.01
PCr/NTP	0.79 ± 0.42		1.09 ± 0.42	not sig.
Pi/NTP	1.13 ± 0.50		0.95 ± 0.50	not sig.
PME/NTP	3.49 ± 1.23		2.79 ± 1.25	<0.05
PDE/NTP	2.11 ± 1.03		2.34 ± 1.05	not sig.
Bone + Lipid/NTP[c]	20.5 ± 20.5		34.7 ± 21.6	<0.05
pH$_i$[d]	7.14 ± 0.28		7.09 ± 0.28	not sig.

[a] From Azzopardi et al.,[38] International Pediatric Research Foundation Inc. Reprinted with permission.
[b] Mean values $\pm 95\%$ confidence intervals from regression analysis.
[c] $n = 24$.
[d] $n = 23$.

myelin phospholipids.[40] Similar studies of otherwise normal, small-for-gestational-age (SGA) infants showed no significant differences in the phosphorus metabolites or pH$_i$ when compared with the AGA subjects.[37,38]

Recent work has also examined the dependence on head position of neonatal brain metabolism in normal infants.[41] Several reports have indicated that turning the head may compromise blood flow through the ipsilateral internal jugular vein, resulting in impeded outflow. It has been suggested that this can cause increased internal jugular-vein pressure and intracranial pressure. Brain-oxidative phosphorylation was investigated in 10 preterm and 8 full-term infants by adopting four distinct head positions: right and left neutral, supine (head turned 90° to the right), and prone (head turned 90° to the left). No significant change in PCr/Pi was detected for any of the head positions. This indicated that any changes in cerebral blood flow in a healthy infant as a result of altered head position can be compensated for by biochemical and physiological regulation.

Preliminary results have been produced from absolute quantitation studies, using a surface coil, of phosphorus metabolites in neonatal cortical brain.[42] This has been achieved by calibrating the spectrometer system using standard solutions and then comparing the ^{31}P signals from the brain with the ^{1}H signal from tissue water.[43] (Further details of this method are to be found in Chapter 5). The absolute concentrations obtained for normal infants are given in Table 3.2. The mean NTP concentration

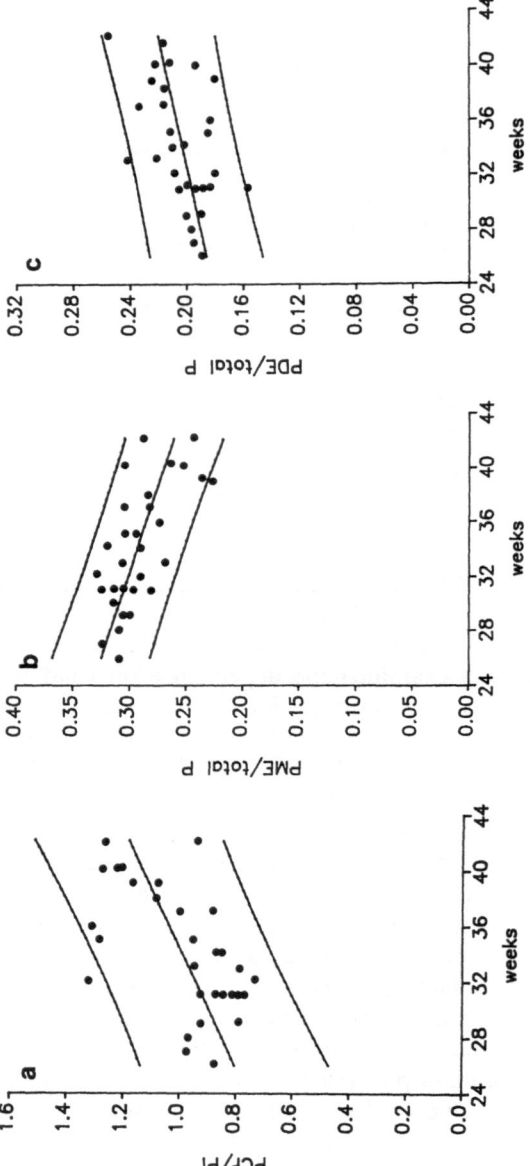

Figure 3.2. The relationships between phosphorus metabolite concentration ratios and gestational plus postnatal age in thirty normal infants of appropriate weight for gestational age. Hypothetical linear regressions and 95% confidence limits are shown (from Azzopardi et al.,[38] International Pediatric Research Foundation Inc.; reprinted with permission).

Table 3.2. Absolute Concentrations of Phosphorus Metabolites in the Cerebral Cortices of Normal Infants of Age (Gestational plus Postnatal) 35 to 37 weeks[a]

	Concentration (mMole/liter)					
	PME	PI	PDE	PCr	NTP	P_{total}
Mean ($n = 7$)	14.1	4.1	8.8	4.2	3.7	43.5
Standard deviation	2.2	0.7	1.3	0.6	0.6	6.2

[a] From Cady,[42] Academic Press Inc. Reprinted with permission.

[NTP] agrees well with high-pressure liquid–gas chromatographic measurements of the summed NTPs (adenosine, guanosine, cytidine, and uridine triphosphates) in adult rat-brain cortex extracts produced from in situ frozen tissue (3.69 ± 0.07 mM/kg wet wt.).[16] The PME peak from neonatal brain tissue is known to contain a significant contribution from PEt.[18–20] This has been reported to have a relatively high concentration in neonatal mammals.[21] For the rabbit this approaches 8 mM/kg wet wt.,[44] and a similar value has been obtained postmortem from the cerebral cortex of an 8-month human fetus.[22] However, it is known that several other metabolites, in combination, contribute significantly to this peak, and their presence may explain the high [PME] found in vivo.[17] The PDE peak contains resonances due to acid labile phosphates (ALP), GPC, and GPE. The [PDE] given in Table 3.2 is higher than the [ALP + GPC + GPE] estimated in adult guinea-pig brain,[17] and it has been suggested that the excess may originate from a mobile brain-phospholipid fraction.[45]

A further enhancement of the potential of MRS is provided by combined studies with near-infrared spectroscopy,[37,46] in which neonatal brain-metabolic status can be directly compared with cerebral oxygenation and hemodynamics.

3.2.1.1d. Clinical Results. ^{31}P MRS has been shown to be sensitive to abnormalities of oxidative phosphorylation in the neonatal brain.[7,11,12,37,46–48] Changes in relative metabolite concentrations, as determined from spectra, have been found following birth-asphyxia and also after the observation of abnormal cerebral echodensities on ultrasound scans. In particular, the ratios PCr/Pi and NTP/P_{total} have been shown to be useful prognostic indicators.[48] Infants who have suffered perinatal hypoxia–ischemia often, but not always, exhibit spectra similar to those obtained from normal infants on the first day of life,[12,48] as shown in Figure 3.3. Subsequent spectra obtained from severely affected infants show abnormalities in the spectra. PCr/Pi is observed to fall, mainly due to

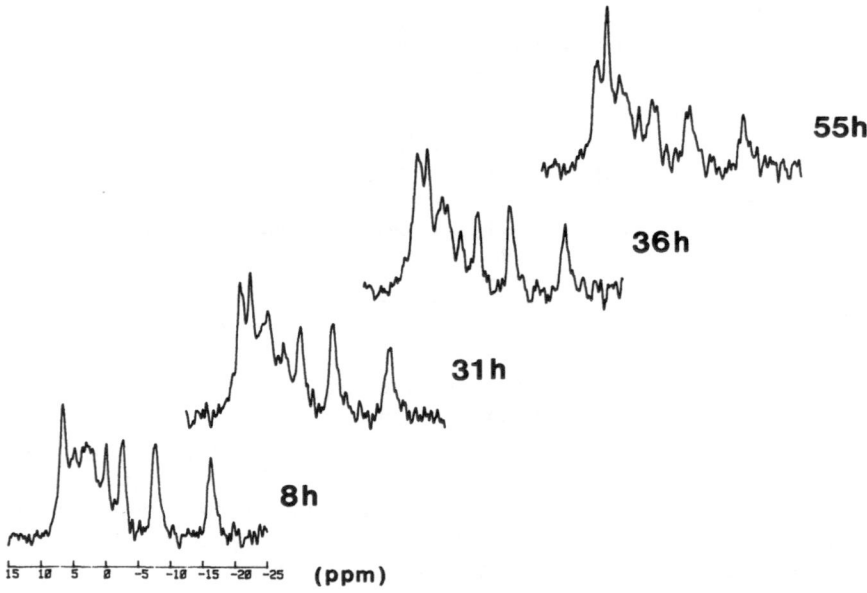

Figure 3.3. A series of ^{31}P spectra obtained from the cerebral cortex of a birth-asphyxiated infant born at 37 weeks gestation. The postnatal ages at the times of study are indicated. For peak assignments, see Figure 3.1. At 8 hr, PCr/Pi was 0.99, NTP/P_{total} 0.09, and pH_i 7.06; pH_i rose to a maximum of 7.28 at 36 hr. The minimum value for PCr/Pi was 0.32 at 55 hr, when NTP/P_{total} was 0.04 and pH_i was 6.99. The infant died aged 60 hours (from Azzopardi et al.,[48] International Pediatric Research Foundation Inc.; reprinted with permission).

increased Pi, and pH_i tends to indicate a degree of alkalosis. (The increase in pH_i provides evidence for the occurrence of different metabolic processes to those pertaining to acute hypoxic-ischemic insults in animal models during which acidosis is observed; see Chapter 1.) In severely affected infants NTP/P_{total} falls also. Eventually the Pi peak can become the predominant one in the spectrum and, in occasional extreme cases, PCr and NTP are undetectable. (Similar low or undetectable PCr and NTP levels have been found in spectra acquired from infants with inborn errors of metabolism such as propionic acidemia and arginosuccinic aciduria.) In surviving infants, the metabolite ratios measured from the spectra often return to normal values over a period of weeks, but the overall phosphorus signal strength may be reduced.[7,37,48] This decrease in signal is supported by results from absolute quantitation studies,[42] conducted several weeks after the initial insult, in which [NTP] and [P_{total}] have been found to be low in infants who, at the same time, had normal values for PCr/Pi and NTP/P_{total}. This probably indicates that permanent cell damage had taken place, causing some tissue destruction but leaving a certain amount of viable tissue in the sensitive volume of the coil. Enough data has now been acquired to relate metabolite ratios in the perinatal period with outcome. Figures 3.4a, b, and c show the relationships among the minimum measured values for PCr/Pi, survival, and neurodevelopmental outcome at one year of age.[48] Figures 3.5a, b, and c give the data for measurements of NTP/P_{total} determined from the same spectra when PCr/Pi was at a minimum. Values for NTP/P_{total} below the confidence limits for normal controls are almost always associated with a fatal outcome. Similar studies on infants with increased cerebral echodensities on ultrasound scans indicate that such infants were more likely to have low PCr/Pi values and a bad prognosis when the echodensities were diffuse or consistent with intraparenchymal hemorrhage than when they were indicative of periventricular leucomalacia. Independent neurodevelopmental assessment of survivors at age one year shows a correlation between the Griffiths general quotient[49] and the minimum PCr/Pi value, as shown in Figure 3.6.

The results herein described have all been obtained using simple surface-coil methods. The localization has, therefore, been relatively crude,

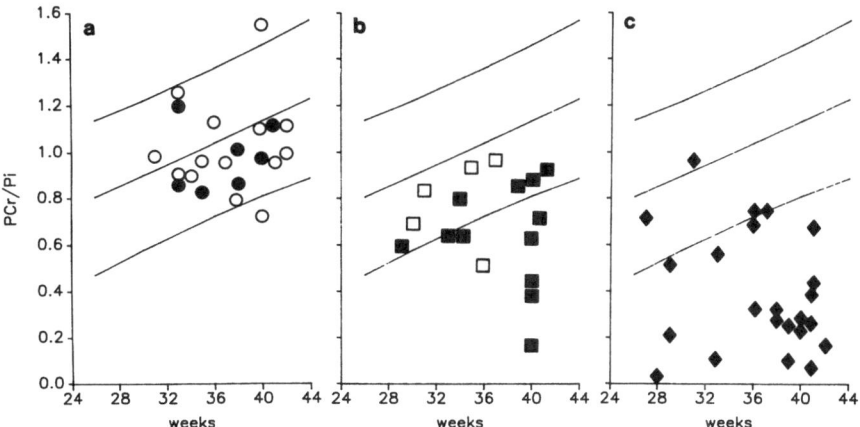

Figure 3.4. The relationship between minimum values for PCr/Pi in the neonatal cerebral cortex, survival, and neurodevelopmental outcome 1 year of age. The regression line and 95%-confidence limits for normal values versus gestational plus postnatal age are shown. (a) ○, normal progress [standard deviation scores as compared with normal infants (S.D.S.) (the standard deviation score is the number of standard deviations of the normal population distribution that a particular measurement is displaced from the normal population mean) -0.29 ± 1.21 (S.D.), not significant (ns)]; ●, minor impairments (S.D.S. -0.54 ± 0.52, ns). (b) □, major neuromotor impairment (S.D.S. -1.06 ± 1.22, $p < 0.05$); ■, multiple major impairments (S.D.S. -2.60 ± 1.44, $p < 0.001$). (c) ◆, died (S.D.S. -3.82 ± 1.89, $p < 0.001$) (from Azzopardi et al.,[48] International Pediatric Research Foundation Inc.; reprinted with permission).

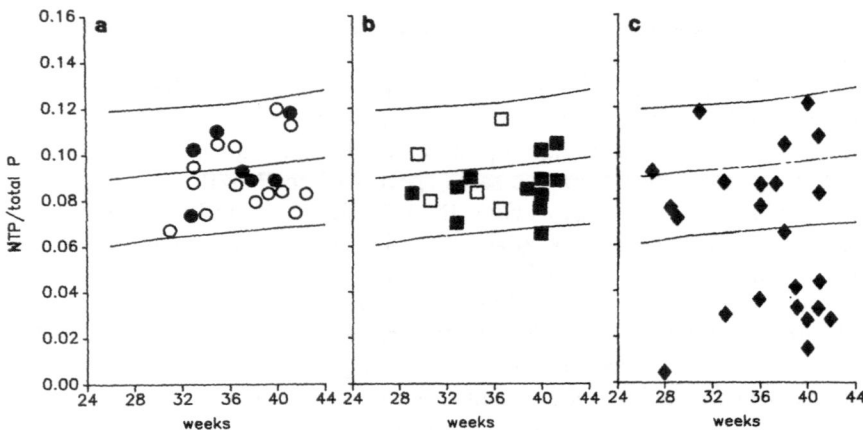

Figure 3.5. Relationships between the values for NTP/P$_{total}$ (measured from the same spectra as the minimum PCr/Pi), survival, and neurodevelopmental outcome as determined from ^{31}P spectra collected from the neonatal cerebral cortex. The linear regression and 95% confidence limits for normal values as compared with gestational plus postnatal age are shown. (a) ○, normal progress [standard deviation scores as compared with normal infants (S.D.S.) -0.22 ± 1.01 (S.D.), not significant (ns)]; ●, minor impairments (S.D.S. -0.03 ± 0.98, ns). (b) □, major neuromotor impairments (S.D.S. $+0.13 \pm 1.12$, ns); ■, multiple major impairments (S.D.S. -0.67 ± 0.74, $p < 0.05$). (c) ◆, died (S.D.S. -2.13 ± 2.45, $p < 0.002$) (from Azzopardi et al.,[48] International Pediatric Research Foundation Inc.; reprinted with permission).

and the sensitive volume from which spectroscopic information has been obtained is approximately the same as that of a hemisphere of radius equal to that of the coil. Hence, the data collected originates from a superficial layer of the brain, extending to only a few centimeters in depth. Thus the information obtained, including that from normal infants, relates mainly to cortical tissue. This provides a possible reason why lesions nearer to the midline need not necessarily have produced observable changes in relative.

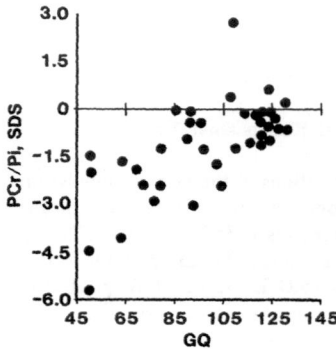

Figure 3.6. The relationship between minimum observed values for PCr/Pi, expressed as a standard deviation score (S.D.S.) as compared with normal infants, and Griffiths general quotient (GQ) assessed at age one year in 38 surviving infants. Scores of 50 or below are recorded as 50 ($r = 0.67$, $p < 0.001$) (from Azzopardi et al.,[48] International Pediatric Research Foundation Inc.; reprinted with permission).

metabolite levels. It is to be anticipated that the application of advanced localization techniques will allow the more detailed investigation not only of deeper lesions but also of the anatomical distribution of metabolites in the normally developing neonatal brain.

3.2.2. Spectroscopic Studies of the Adult Brain

The detailed study of the adult human brain has had to await the development of techniques for adequate localization or suppression of signals from surface tissues. Because of the presence of cranial muscle and a substantial layer of highly mineralized cranial bone there is no convenient "window" by which simple surface-coil methods can produce adequate, uncontaminated spectra. Advances in methods of localization, taking advantage of the small variations in magnetic susceptibility within the brain tissue, have now allowed the collection not only of high-quality, interpretable ^{31}P spectra, but also of well-resolved ^{1}H spectra.

3.2.2.1. ^{31}P Spectroscopy of the Adult Brain

Nonlocalized, spectra from the normal adult human brain[8,29] were described shortly after the publication of the first ^{31}P spectra from the neonatal brain. These were obtained with a 1.5-tesla, whole-body super-conducting magnet and a spectrometer capable of both ^{1}H imaging and multinuclear spectroscopy. The spectra were collected using a single 6.5-cm surface coil, pressed against the temple, which both transmitted the RF pulses and also detected the resulting FID. The length and amplitude of the transmitter pulse were such that it produced a 180° flip at the coil's center. Fifty FIDs collected at 16-sec intervals were averaged together for each spectrum (total collection time was about 13 min). The relatively long pulse-repetition interval reduced the possible effect of differential saturation (T_1 weighting) on the relative peak areas. Even though the only localization used was that provided by the surface coil in combination with a 180° pulse (which gave some suppression of signals from surface tissues), the spectra obtained were substantially different from those acquired from both skeletal muscle and the neonatal brain. The PCr/NTP ratio was higher relative to that obtained in neonatal brain spectra, although this result may have been partly due to spectral contamination from cranial-muscle PCr. Significant resonances were also seen in both the PME and PDE regions of the spectrum. However, the PME peak was not as dominant as that present in neonatal-brain spectra. T_1s were measured for PCr (3.5 sec), NTP (1 sec), and the PME and PDE peaks (1 sec). The metabolite ratios measured in this study were βNTP/PCr $= 0.86 \pm 0.2$, PDE/PCr $=$

0.87 ± 0.2, Pi/PCr = 0.26 ± 0.15, and PME/PCr = 0.53 ± 0.2. Similar studies of both forebrain and hindbrain have been made at 1.4 tesla, and no differences were found between the two anatomical locations.[33]

In order to ensure that the metabolite resonances observed in adult human spectra originate from the brain and not from superficial tissues, localization methods have to be applied, and several different techniques have been used. One of the first approaches adopted for localized studies of the adult brain was depth-resolved surface-coil spectroscopy (DRESS).[30] In this method a 90° selective pulse is used to define a slice, from which spectroscopic data are acquired and which is parallel to the surface coil (see Chapter 5). In the first published use of this technique, the surface coil was placed on the temple. Spectra were collected from parallel slices, each approximately 12 mm thick, at the surface and at a depth of 35 mm. The magnetic field homogeneity was optimized for each slice using ^{1}H DRESS. The spectra obtained from the two locations had different appearances. The surface spectrum had metabolite ratios similar to, but not exactly the same as, those obtained from skeletal muscle. The spectrum collected from a depth of 35 mm had characteristics indicating that a substantial amount of the signal had originated in brain tissue. Later work demonstrated the application of slice-interleaved (SLIT) DRESS[50] in which data was collected from a series of contiguous slices in the same total time as it would normally take to acquire a spectrum from one slice. This is achieved by applying selective pulses to the other slices during the relaxation delay between pulses applied to the first slice, thereby avoiding saturation. Careful sequencing of the slice selection ensures that adjacent slices do not receive successive pulses which could increase saturation due to some degree of slice overlap. The result of such a SLIT-DRESS study with the head lying sideways on a combination of coaxial, coplanar surface coils (6-cm-diameter receiver and 27-cm-diameter transmitter) situated above the ear is shown in Figure 3.7. The use of a large transmitter coil facilitated the generation of flip angles of greater uniformity within the interrogated slice. The surface spectrum (depth = 1 cm) differs from skeletal-muscle spectra in that it exhibits significant PME and PDE peaks and a reduced PCr signal. This indicates a major contamination fraction from deeper tissues. As the slice depth increases, PCr becomes less prominent and PDE shows a relative increase. There is an overall decrease in signal strength with depth, reflecting the spatial-sensitivity profile of the surface coil. The ratio PCr/βNTP in the brain was reported to be 0.6 ± 0.1 for a relaxation delay of 2 sec. In a similar study using conventional DRESS with a relaxation delay of 8 sec, PCr/βNTP was found to be 1.1 ± 0.1, and it was suggested that this difference could be completely explained by saturation effects.

In later work, further DRESS studies were undertaken on normal volunteers and on patients who had chronic cerebral infarction.[51] No significant

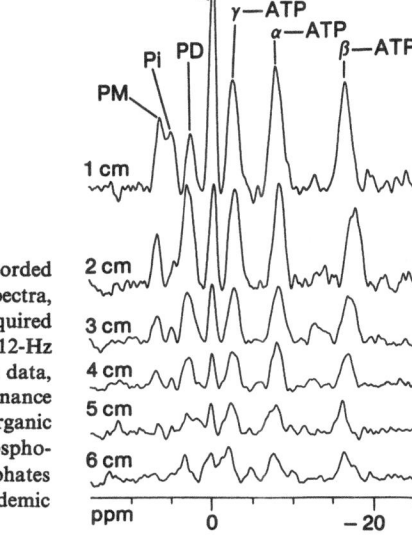

Figure 3.7. In vivo ^{31}P SLIT DRESS spectra recorded from the adult human head. The series of six spectra, collected at depths from 1 to 6 cm, were acquired within 10 min using a repetition time of 2 sec. 12-Hz exponential broadening has been applied to the data, and the spectra were baseline flattened. Resonance identifications: PM, phosphomonoester; Pi, inorganic orthophosphate; PD, phosphodiester; PCr, phosphocreatine; γ, α, and β ATP, nucleoside triphosphates (mainly ATP) (from Bottomley et al.,[50] Academic Press Inc.; reprinted with permission).

differences in phosphorus metabolites and pH$_i$ were found between the normal subjects and the patients, as shown in Table 3.3. Spectra obtained from infarcted tissue were reported to have signal strengths decreased by up to 40%, possibly indicating a reduction in the amount of viable brain tissue. As a conclusion, the study indicated that in chronic brain infarction in adults, phosphorus metabolite ratios are not sensitive indicators of abnormal metabolism. This correlates with studies on the brains of neonates carried out some time after the initial hypoxic-ischemic insult in which reduced [NTP] and [P$_{total}$] are observed while PCr/Pi and other ratios are close to normal values.[7,37,46,48]

Heteronuclear spectral editing of ^{31}P adult human brain spectra has been used mainly in order to improve the resolution of the PCr and Pi peaks and hence to obtain a better indication of metabolic status.[32] Spectra were obtained from the side of the subject's head using a 15-cm-diameter surface coil modified by the insertion of a slotted copper-foil disk, with a central aperture of 7 cm, which allowed (it was claimed) greater RF-field penetration into the subject. To obtain unedited spectra (see Figure 3.8a; for peak identifications, see Figure 3.7), a composite pulse giving a 180° flip at the coil's center was used, thereby providing some suppression of signal from superficial tissues. Although further localization methods were not used, the unedited spectrum obtained in this manner had very similar characteristics to those acquired within the brain using the DRESS[30] and SLIT-DRESS[50] techniques. Due to their short T_2s, the PME, PDE, and NTP

Table 3.3. ^{31}P DRESS Results Obtained from the Brains of Normal Adult Subjects and from Patients with Chronic Cerebral Infarction[a]

	Normals	Patients			
	(n = 9)	1	2	3	4
Age (years)	22–50	18	37	19	60
Time after infarct (weeks)	—	12	7	10	7
Total P (%)[b]	—	60 ± 15	63 ± 10	77 ± 20	94 ± 10
PCr/βNTP	1.2 ± 0.3	1.3 ± 0.2	1.2	1.5 ± 0.4	0.85 ± 0.15
PCr/Pi	7.7 ± 2.3	6 ± 2	>10	5.5 ± 1.5	>7
PDE/PCr	1.14 ± 0.14	1.18 ± 0.35	0.91	1.18 ± 0.21	0.9 ± 0.2
PME/PCr	0.56 ± 0.15	0.71 ± 0.05	0.52	0.74 ± 0.27	0.7 ± 0.4
pH$_i$	7.01 ± 0.05	7.07 ± 0.05	NA[c]	7.0 ± 0.1	7.0 ± 0.1

[a] From Bottomley et al.,[51] Radiological Society of North America. Reprinted with permission.
[b] Average ratio of the total integrated ^{31}P signal in a spectrum obtained from the infarct site to that from normal tissue in the same patient at a corresponding depth.
[c] Pi too small to allow pH$_i$ determination.

resonances were significantly suppressed using a 30-msec spin–echo sequence, as shown in Figure 3.8b. Furthermore, on application of a 180° ^{1}H pulse immediately before the 180° ^{31}P pulse in the spin–echo sequence, even better suppression of the PME and PDE resonances was obtained, leaving the PCr and Pi peaks well resolved, as shown in Figure 3.8c. The ^{1}H pulse permits phase modulation of the proton-coupled signals, which include the constituents of the PME and PDE peaks. For a spin–echo delay of 30 msec, these resonances are nulled.

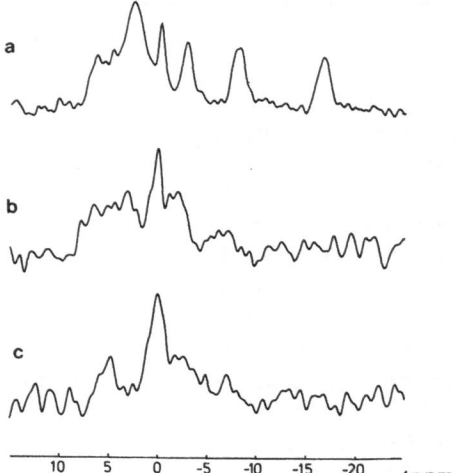

Figure 3.8. Unedited (a) and edited (b and c) ^{31}P spectra obtained from the adult human brain. The collection conditions are described in the text (from Brindle et al.,[32] Academic Press Inc.; reprinted with permission).

One of the earliest techniques suggested for spectral localization was topical magnetic resonance (TMR).[52,53] In this method high-order magnetic gradients are applied in order to produce a definable region in which the static magnetic field is homogeneous. Within this volume, highly resolved spectra could be collected. The field was designed so as to decrease rapidly outside this region, and resonances from metabolites in this exterior volume were therefore broadened. Using computer techniques, the high-resolution and broad components can be separated, thereby producing a spectrum localized to the homogeneous volume. This approach has been used to obtain localized ^{31}P spectra of the adult brain.[34] Data were obtained using two different types of coil. A whole-head coil (Dadok type) was used in conjunction with a 10-cm homogeneous TMR volume in order to obtain spectra representative of the entire brain. The homogeneous volume was symmetrically positioned about the midline, touching the vertex. In other studies, a surface coil of 8-cm diameter was employed so that particular regions of the brain could be studied. In order to achieve this, the surface coil was positioned at various locations on the skull. It was reported that significant differences in the ratio PCr/βNTP were found between the temporal and frontal regions in studies on six normal subjects (1.08 ± 0.03 and 0.79 ± 0.08, respectively; [mean ± standard error of the mean (S.E.M.)]. The difference may have been due to biological causes, but dissimilar saturation conditions at the two locations could have been partially or totally responsible. (In one of the subjects examined, the distance from the coil's center to the surface of the cortex was measured as 0.8 cm frontally and 1.4 cm temporally. Saturation effects may have been more severe in frontal collections due to the larger flip angle experienced by the brain tissue.) The spectra obtained had characteristics similar to those described previously, using localization methods (e.g., SLIT-DRESS) for data collection. The pH$_i$ was reported to be 7.03 ± 0.04 (mean ± S.D.). T_1s were measured in two subjects using a saturation recovery sequence ($90°_x$, $90°_y$-τ-$90°$-Acquire Data). The mean values so obtained were Pi, 2.5 sec; PME, 2.8 sec, PDE, 2.4 sec, PCr, 4.8 sec; and γNTP, 1.0 sec. Reproducibility of the measurement of metabolite ratios was assessed in eight subjects. PCr/βNTP, PDE/βNTP, PME/βNTP, and Pi/βNTP were found to be measurable with coefficients of variation of 10%, 17%, 20%, and 28%, respectively.

A further development of the DRESS technique[30] has recently been applied to obtain spectra from the adult human brain. This variant of DRESS is called point-resolved rotating-gradient surface-coil spectroscopy (PROGRESS)[54] and, like one-dimensional image-selected in vivo spectroscopy (ISIS),[55] involves performing two experiments. In the first experiment, a 180° pulse is applied, using a surface coil in conjunction with one magnetic gradient that varies sinusoidally with time along the x-axis and another magnetic gradient that varies along the z-axis with a cosine function at twice the x-gradient frequency. This is followed by a selective 90° pulse as in conventional DRESS. The second experiment consists of the conventional DRESS pulse only. The results of the second experiments are subtracted from those of the first. The procedure produces a sensitive volume similar to that of the DRESS method, but its extent is more limited in directions parallel to the surface coil. In fact, the sensitive disk has a radius approximately equal to that of the coil up to a localization depth of about three coil radii. The technique should therefore be capable of producing spectra from a better-defined region of tissue. The spectrum obtained by using this technique on the normal adult human brain has similar characteristics to that obtained at a depth of 2 to 3 cm by SLIT-DRESS.[50]

In combination with ^{1}H MRI, a modified version of ISIS has been applied to the study of the metabolism of both the normal adult brain and cerebral tumors.[56] A small-volume coil, tuned for ^{31}P, was inserted into a larger ^{1}H imaging coil, thereby allowing rapid switching from MRI to spectroscopy without changing coils or moving the subject. The ^{31}P volume coil permitted greater flexibility in the location of the ISIS-sensitive volume and also provided uniform spatial receptivity. In order to obtain an optimum sensitive volume profile and improved magnetization inversion, Hoult–Silver hyperbolic secant selective pulses[57] were used. In studies on normal subjects, using a $6 \times 6 \times 6$-cm volume of interest (VOI), high-quality spectra were obtained within about 13 min. These exhibited spectral characteristics similar to those obtained using SLIT-DRESS.[50] Image-guided spectra were collected from tumor tissue using smaller VOIs. The dimensions of the VOI were adjusted so as to minimize contamination of the spectra by tissue outside the tumor. As a consequence, data collection times were longer. The smallest VOI used was 40 cm^3, and this required 30 min total acquisition time in order to obtain a spectrum of acceptable quality. The main conclusions from the studies were that the spectra from cerebral tumors showed abnormally high PME and low PCr signals. The effects of chemotherapy and radiation therapy were monitored in three patients and, in each case, changes were detected in the spectra.

One- and two-dimensional phase encoding have also been used to obtain localized spectra of the adult human brain. In one-dimensional studies,[1,35] a receiver surface coil 6 cm in diameter was used in conjunction with a much larger saddle-coil transmitter (39 cm long by 34 cm wide). The transmitter coil completely enclosed the part of the body under study and gave a uniform flip angle over the entire sensitive volume of the receiver coil. This produced a larger sensitive volume than would have been obtained had the surface coil by itself been used in combined transmit-receive mode. The phase-encoding sequence consisted of a hard (broadband) 45° pulse with a 1-sec repetition time. The pulse was followed immediately by a 2-msec phase-encoding gradient pulse. The amplitude of the phase-encoding gradient was incremented through sixteen equal steps to give a two-dimensional data set. This was repeated up to 128 times in order to improve the SNR. Two-dimensional Fourier transformation of the data set produced spectra from contiguous parallel slices at 10-mm depth intervals. The spectra obtained from a normal subject at depths of 3 and 4 cm were similar in appearance to those collected using other localization methods (e.g., SLIT-DRESS) and described previously in this section. A spectrum obtained from a depth of 2 cm had elevated PCr, which probably originated from cranial skeletal muscle, and reduced PDE, indicating the smaller brain-tissue fraction. In this technique, a delay of about 3 msec is required to allow for the decay of eddy currents induced by the gradient pulse before commencing data collection. This produces a severely rolling baseline under the peaks in the spectrum and makes accurate quantitation difficult. Comparison between spectra obtained from normal and abnormal tissue is therefore not easy. When his method was used to study the brain of a patient with a superficial glioma, it was claimed that spectra obtained from the abnormal tissue had Pi elevated by 50% to 100% when compared with normal brain tissue. More recently, two-dimensional phase encoding has been applied to 1.5-tesla[58] and 4-tesla[59] whole-body systems. For the 1.5-tesla study, instead of a hard pulse, a narrow-band sinc-function-modulated pulse (see Section 5.2.2.2) was used in combination with a magnetic field gradient in order to select a slice. Phase encoding was then applied using two orthogonal stepped magnetic gradients, and the resulting FIDs were digitized. The data set was then subjected to three-dimensional Fourier transformation. Further processing consisted of inverse Fourier transformation of the individual spectra, application of 12-Hz line broadening to reduce the noise level, and zero filling to improve interpolation. The data were then Fourier transformed again to give improved spectra, which were phase-corrected and baseline-flattened by subtraction of a broad curve fitted to six points in the spectrum baseline. A typical example of the results obtained is shown in Figure 3.9. Each spectrum consists of 256 points, and the volume cells (voxels) are $3 \times 3 \times 3$ cm. The data thus give a checkerboard display of the variation of metabolite concentrations across the selected brain slice. The data shown in Figure 3.9 were obtained in 34 min. Direct, absolute quantitation was possible using this method. A vial containing a suitable external concentration reference (in this case, 2 ml of 1-M H_3PO_4) was positioned so that it was in the field of view as shown in Figure 3.9 (bottom left-hand voxels). Providing that corrections are applied for differential saturation, absolute concentrations of cerebral metabolites can be obtained by comparison with the signal from the reference. The mean NTP concentrations determined in this manner from several spectroscopic images (including that shown in Figure 3.9) were

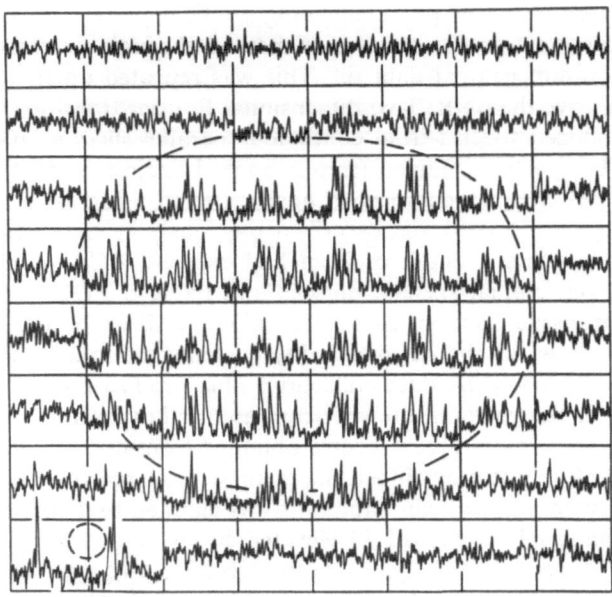

Figure 3.9. A transaxial, $8 \times 8 \times 256$-point three-dimensional Fourier transform ^{31}P spectroscopic image of the adult human brain, acquired in 34 min. The voxel was $3 \times 3 \times 3$-cm^3. The two voxels at the bottom left (containing the small, dashed circle) show a resonance from a vial containing 2 ml of 1M H_3PO_4, included in the field of view to facilitate absolute quantitation. The large, dashed ellipse delineates the periphery of the head (as determined from a ^1H image) with the forehead at the left (from Bottomley et al.,[58] Academic Press Inc.; reprinted with permission).

3.1 ± 0.6, 3.2 ± 1.1, 3.0 ± 1.2, and 3.1 ± 0.6 mmole of tissue (mean $\pm$ S.D.). In studies on three normal volunteers using a 10-sec repetition time, the ratio PCr/βNTP was 1.2 ± 0.1 (mean $\pm$ S.D.). On the 4-tesla system,[59] substantial reductions in data-collection time and improvements in spectral resolution were achieved. An 8×8-voxel image with 256 point spectra gave adequate SNR in just 4 min with a voxel size of $3 \times 3 \times 3$ cm. Further studies of the adult human brain using two-dimensional phase encoding have been published.[60,61]

Additional absolute quantitation data for phosphorus metabolites in the adult brain have been reported using other methods.[62,63] Mean concentrations from slices across whole adult brain have been obtained using two selective pulse methods, $180°_{sel} - 90°$ and $90°_{sel}$. In the former method, the $180°_{sel}$ pulse is used only on every other scan and data are alternately added to and subtracted from the integration array. This approach has the advantage that, due to the last pulse being a nonselective 90° (i.e., no gradients are used), there is no requirement for a delay to allow for eddy-current decay before commencing data acquisition. Spectroscopic data were collected from an entire slice across the head. The fractions of the slice volume containing brain and ventricles ($\sim 75\%$) and muscle ($\sim 5\%$) were measured from ^1H images obtained from the same slice. Ventricular contributions to the brain volume were then subtracted. Absolute concentrations were calculated from the peak areas in the spectrum, the brain volume in the slice, and the signal obtained from an adjacent external concentration reference. The concentrations obtained from spectra acquired from 15 normal subjects aged 31 to 62 years using the $180°_{sel} - 90°$ method were NTP, 3.03 ± 0.49 mmole/liter wet (mean $\pm$ S.D.); PCr, 5.18 ± 0.89 mmole/liter wet; and Pi, 1.5 ± 0.7 mmole/liter wet. A slight tendency for [NTP] to decrease with age was seen. Concentrations obtained with the $90°_{sel}$ method were significantly lower, and it was suggested that this was due to the delay required for gradient stabilization. Three-dimensional ISIS has also been used to obtain absolute concentrations of adult brain metabolites using $5 \times 5 \times 5$ cm^3 (volume coil) and $3 \times 5 \times 5$ cm^3 (surface coil) sensitive volumes.[63] Absolute quantitation was achieved by using the same sensitive volumes on a standard solution. Loading conditions were taken into account using measurements of signals from a common external reference (see Section 5.6.3). The concentrations measured for PME, Pi, PDE, PCr, and NTP were 4.1, 2.0, 11.6, 4.9, and 2.3 mM, respectively, using the $5 \times 5 \times 5$ cm^3 volume ($n = 5$), and 3.4, 1.7, 11.0, 5.1, and 2.9 mM, respectively, using the $3 \times 5 \times 5$ cm^3 volume ($n = 1$). Longitudinal relaxation times were also measured using inversion recovery. The T_1 results obtained were as follows: PME, 1.42 ± 0.43 sec

Table 3.4. Phospholipid Profiles of Normal Adult Human Brain[a]

Phospholipid	Proportion of total phospholipid content (Mean % $\pm$ S.D.)		
	Gray matter[b]	White matter[c]	Myelin[d]
Phosphatidylcholine	43.6 ± 2.7	28.1 ± 1.9	27.1 ± 3.0
Phosphatidylethanolamine	27.5 ± 2.3	26.8 ± 1.7	23.3 ± 4.0
Phosphatidylserine	14.3 ± 1.7	19.5 ± 1.5	19.2 ± 2.2
Phosphatidylinositol	2.8 ± 1.9	<1	<1
Sphingomyelin	12.5 ± 1.7	19.8 ± 3.0	22.3 ± 1.7
Ethanolamine plasmalogen	<1	5.7 ± 1.2	7.6 ± 4.2

[a] From Kwee and Nakada,[64] Academic Press Inc. Reprinted with permission.
[b] Cerebral cortices, thalamus, and basal ganglia.
[c] Cerebral and cerebellar subcortical white matter.
[d] Corpus callosum.

(mean $\pm$ S.D.); Pi, 1.45 $\pm$ 0.26 sec; PDE, 1.32 $\pm$ 0.14 sec; PCr, 3.14 $\pm$ 0.47 sec; γNTP, 0.65 $\pm$ 0.15 sec; αNTP, 0.85 $\pm$ 0.21 sec; and βNTP, 0.80 $\pm$ 0.16 sec.

In vitro studies of postmortem adult brain tissue have focussed on the PME and PDE resonances. In samples obtained from patients with Huntington's chorea and from patients with Alzheimer's disease, the PME signal has been observed to be elevated when compared with that of normal brain tissue.[20] The PME signal was elevated even in histologically normal areas of the brain. The increased signal was presumed to originate from a raised PEt concentration. The observations suggested that chemical alterations in phospholipid metabolism were detectable by ^{31}P MRS before changes in cellular structure were apparent from neuropathological studies. The ^{31}P spectra of the PDE resonances of brain tissue have been investigated using a chloroform–methanol lipid extraction method, and anatomically dependent phospholipid profiles have been obtained for normal tissue as shown in Table 3.4.[64] Preliminary results for tissue obtained from patients with Alzheimer's disease show a significant alteration in the PDE profile of the frontal and parietal cortices. [31]P PDE spectra obtained from different anatomical locations in the brain are shown in Figure 3.10.

3.2.2.2. [1]H Spectroscopy of the Adult Human Brain

In order to obtain useful ^{1}H spectra from the human brain in vivo, several problems must be overcome. The strong resonance due to tissue

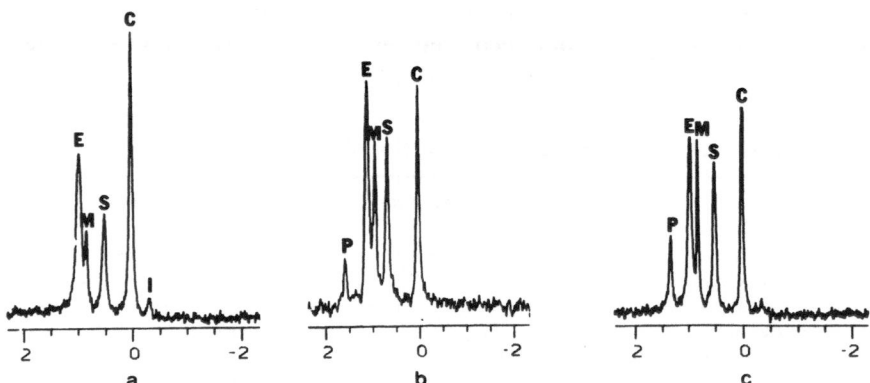

Figure 3.10. [31]P phospholipid spectra from chloroform–methanol extracts of three types of adult human brain tissue obtained postmortem: (a) Gray matter, (b) Frontal white matter, (c) Corpus callosum. Data were collected at 145.8 MHz using a proton decoupling sequence. Peak identifications: C, phosphatidylcholine; E, phosphatidylethanolamine; S, phosphatidylserine; I, phosphatidylinositol; M, sphingomyelin; P, ethanolamine plasmalogen (represented by lysophosphatidylethanolamine) (from Kwee and Nakada,[64] Academic Press Inc.; reprinted with permission).

water has to be suppressed so that the weak signals from cerebral metabolites can be adequately digitized. Also, because of the narrow bandwidth within which most proton resonances are found (about 8 ppm), spin–echo techniques are usually necessary in order to simplify the spectra by reducing the broad resonances from proteins and other relatively immobile molecules. Furthermore, some form of localization has to be used; otherwise, signals from superficial tissues (particularly subcutaneous fat) contaminate the spectra. These problems have been solved to a large extent, and it has been demonstrated that high-quality ^{1}H spectra can be obtained from the human brain. The problem of the narrow ^{1}H resonance bandwidth has been dealt with by attempting to optimize magnetic homogeneity on the volume of interest using "single shot" localization sequences (i.e., sequences which achieve localization without the need for adding and subtracting the results of separate data collections). The observation that the magnetic susceptibility is very uniform within the brain, allowing good field homogeneity, has also been of assistance.[1]

One of the earliest attempts to obtain ^{1}H spectra from the human brain utilized variations of the DRESS[30,65] technique in which water-signal suppression as well as slice selection was employed. The localized volume was a slice parallel to the surface coil similar to that obtained using ^{31}P DRESS. Four different pulse sequences were tried. Water suppression was achieved in two of these sequences by a 90° narrow-band sinc-function-modulated saturation pulse of 100 msec duration centered at the water frequency. In a third sequence, a similar pulse was tailored to excite, rather than saturate, the metabolite resonances of interest while leaving the water unexcited. Finally, a broad-band inversion pulse was used, which allows the water signal to be nulled by waiting 0.69 T_1 (the time required for the water magnetization to recover from $-M_z$ to zero) before performing the localized study. The last method depends on the metabolites of interest having substantially different T_1s from that of tissue water. Following water suppression, spectrum localization was obtained by slice selection using a combination of a magnetic gradient pulse and a selective RF pulse. In two of the sequences, the selective RF pulse was followed by a nonselective 180° pulse, thereby producing a spin–echo sequence. In one sequence the selective pulse itself was 180°. ^{1}H spectra were obtained from the human brain at 62.2 MHz with a 25-cm-diameter saddle coil for transmission (allowing a more uniform spatial distribution of flip angles) and a 3-cm-diameter surface coil as a receiver. Optimization of the magnetic field's homogeneity was accomplished by using the volume-selective pulse sequences without water suppression. This technique enabled the homogeneity to be optimized for the same tissue volume of interest. A combination of water saturation and a 20-msec slice-selective spin echo produced spectra from a depth of 5 cm which showed signals from

ing a DANTE[73] pulse (see Section 5.5.1) into a two-dimensional ISIS–DEPTH pulse sequence, homonuclear editing for lactate was possible.[74] In a normoxic brain, the measured lactate concentration was ~0.5 mM. (Quantitation was achieved by assuming $[PCr] + [Cr] = 10.5$ mM.) Furthermore, spectra acquired using a three-dimensional ISIS–DEPTH pulse sequence with binomial-pulse water-signal suppression have shown the N-acasp/ $(PCr + Cr)$ ratio to be lower in white matter than in gray matter. The T_2 relaxation times for N-acasp (at 2.02 ppm) and for $PCr + Cr$ (at 3.03 ppm) were also measured. Although the $PCr + Cr$ T_2 was similar in both white and gray matter, that determined for N-acasp was longer in white matter. Spectral editing methods have also allowed the isolation of the C4 and C3 resonances of glutamate and glutamine[75] and also of the signal from the methyl protons of imbibed alcohol.[76]

An alternative approach to localized ^1H spectroscopy in the human brain has been the application of the stimulated echo technique (STEAM (see Section 5.2.2.6)).[77] Three mutually orthogonal slice-selective pulses are applied. The resultant stimulated echo originates solely from the sensitive volume defined by the intersection of the three slices. Suppression techniques to reduce the water signal are also employed. ^1H spectra obtained from the insular area of normal adult brain are shown in Figures 3.12a and b.[78] The spectra show that a variety of metabolites are detectable in vivo.

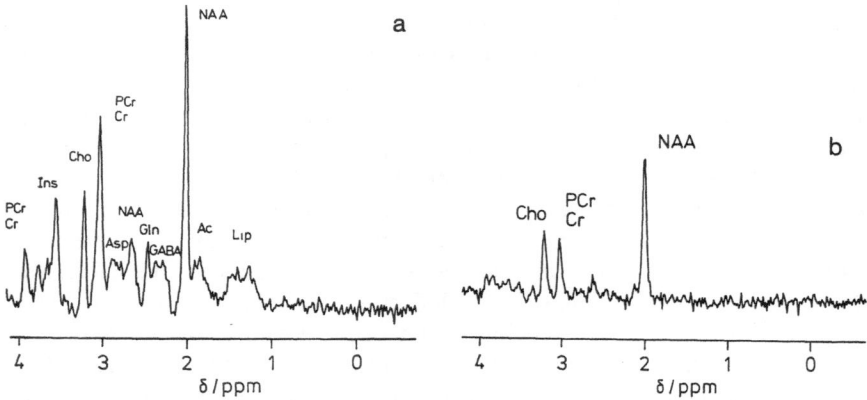

Figure 3.12. Localized ^1H spectra obtained from adult human brain in vivo using the STEAM technique. The spectra were collected at 64 MHz (1.5 tesla) with a $3 \times 3 \times 3$-cm sensitive volume located in the insular area and were acquired with a 1.5-sec repetition time and 512 scans. (a) Echo time = 50 msec. (b) Echo time = 270 msec. Intensities in spectra (a) and (b) are directly comparable. Resonance identifications: Lip, residual lipids; Ac, acetate; NAA, N-acetylaspartate; GABA, γ-aminobutyric acid; Gln, glutamine; Asp, aspartate; Cr/PCr, creatine + phosphocreatine; Cho, choline compounds; and Ins, inositols (from Bruhn et al.,[78] Academic Press Inc.; reprinted with permission).

Assuming a N-acasp concentration of 6 mM, a normoxic lactate concentration of 0.5 mM was estimated.[77] Similar studies on patients with acute cerebral infarction[78] have indicated N-acasp depletion in the affected region. A dramatic increase in lactate (up to 16 mM) was also seen. Signal strength from within the infarcted region was markedly reduced, indicating metabolite concentrations at about a third of the normal levels. It was suggested that edema was contributory to this observation.

3.3. STUDIES OF HUMAN SKELETAL MUSCLE

As mentioned in the introduction to this chapter, due to the ease with which ^{31}P spectroscopic data could be obtained using surface coil techniques only, skeletal muscle was the first human in vivo tissue to be studied.[3–6,79] It has also been possible to conduct studies of peripheral skeletal muscle with magnet systems having only a moderate bore diameter. Mainly for these reasons, a large amount of information has been gathered about phosphorus metabolites in both normal and abnormal human muscle at rest and while performing exercise. Because of the requirement for localization methods in order to reduce contaminant signals, mainly from subcutaneous fat, other nuclei have not been so intensively used, although recent developments in this field are beginning to show promise.

3.3.1. ^{31}P Spectroscopy of Human Muscle

Due to the relatively modest requirements of ^{31}P skeletal-muscle studies in terms of magnet bore, of simple surface coil localization, and consequently of elementary spectrometer capabilities, investigations have been carried out on humans by many MRS research groups. This has resulted not only in the gathering of a great deal of data but also in corroboration of results to a greater extent than has been possible for many other tissues. Of particular interest are exercise studies, in which spectra are collected at intervals of the order of 1 min and the metabolic response to muscular work can be successfully followed noninvasively. Several reviews are available in the literature concerning the application of ^{31}P MRS to the study of human skeletal muscle.[2,80–82]

3.3.1.1. Normal, Resting Human Muscle

The relative concentrations of phosphorus metabolites in normal, resting skeletal muscle have been well characterized. (For an example of a ^{31}P skeletal muscle spectrum, see Figure 3.13a, which was collected from

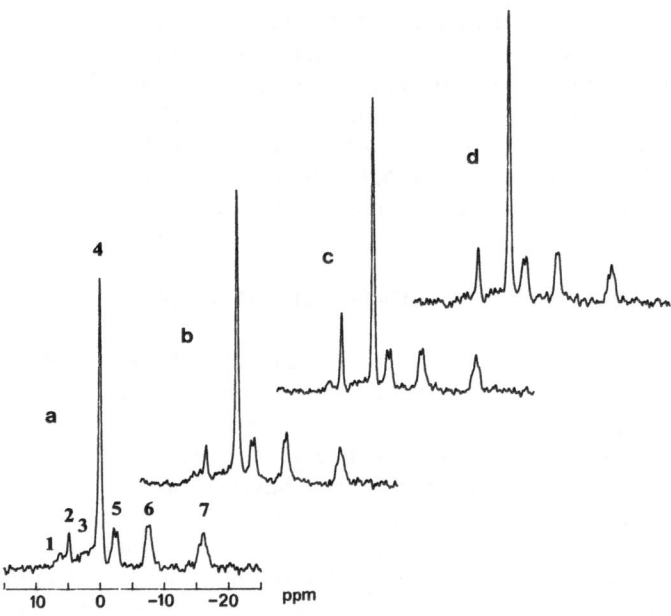

Figure 3.13. ^{31}P spectra of forearm muscle obtained from one subject using a 6-cm-diameter surface coil tuned to 32.5 MHz. Each spectrum results from 512 scans using a 90° flip angle at the coil's center and a pulse interval of 2.256 sec. (a) Fresh, resting muscle before exercise. (b) 1 hr after muscle-stretching exercises of 20 min duration (four contractions per min). (c) 24 hr after exercise. (d) 48 hr after exercise. Peak identifications: 1, PME (mainly sugar phosphates); 2, Pi; 3, PDE; 4, PCr; 5, γATP; 6, αATP; and 7, βATP. Resonance 6 includes a contribution from nicotinamide dinucleotides (NAD + NADH), and peaks 5 and 6 contain small amounts of ADP. Note the elevated Pi in spectra (c) and (d) (from Aldridge et al.,[84] Plenum Press; reprinted with permission).

normal, unexercised wrist flexors.) Much of this work has been carried out on the superficial muscles of the forearm, the hand, and the calf. Although most of the muscles studied so far have similar characteristics, it would be wrong to assume that all muscles in the human body will turn out to have precisely the same metabolite ratios, particularly when muscles with different fiber-type compositions are considered. Studies of cat skeletal muscles have demonstrated large differences in PCr/Pi ratios between muscles with predominantly fast-twitch glycolytic and predominantly slow-twitch oxidative fibers.[83] It has also been shown that the relative size of the Pi resonance in normal, fresh, resting muscle can vary. The amplitude of the Pi peak can depend both on whether a subject has performed strenuous exercise some time prior to the study[84] (as shown in Figures 3.13a to d) and on the recent intake of glucose.[85] Thus it is important that the

Table 3.5. Composition of Resting Human Forearm Muscle[a]

	Concentration (mMole/kg wet wt., mean $\pm$ S.E.)		
	Biopsy[b]	^{31}P MRS[c]	^{31}P MRS[d]
ATP	5.5 ± 0.1 (81)	5.5	5.1 ± 0.1
PCr	17.4 ± 0.2 (81)	29.0 ± 0.7	27.4 ± 0.2
Pi	10 (3)	4.4 ± 0.3	4.3 ± 0.3
PCr + Pi	31.6 ± 3.3 (11)	33.4 ± 0.8	31.7 ± 0.3
PDE	—	0.7 ± 0.1	0.8 ± 0.1

[a] From Dawson M. J.,[86] Plenum Publishing Corp. Data reprinted with permission.
[b] Concentration followed by (n).
[c] [ATP] = 5.5 mMole/kg wet wt., ($n = 7$).
[d] [Mobile ^{31}P] = 49.5 mMole/kg wet wt., ($n = 7$).

immediate exercise and dietary history of the subject be known when studies of resting human muscle are undertaken.

^{31}P MRS appears to give consistently higher relative concentrations for PCr and lower relative concentrations for Pi when compared with data obtained by needle biopsy, as shown in Table 3.5. It has been suggested that this is due to the hydrolysis of PCr and the consequent production of Pi during the interval (of the order of a few seconds) between the collection of the biopsy sample and its freeze-clamping by immersion in liquid N_2. Measurements of metabolite concentrations from in situ human skeletal muscle are now also available from MRS absolute quantitation studies in which either tissue water has been used as an internal concentration reference[42] (see Table 3.6) or isis localization has been used in combination with calibration using standard techniques.[63] These measurements nearly all agree closely with the MRS data given in Table 3.5, which was obtained using assumed concentrations for either the ATP or the total mobile phosphate. ([PDE] is somewhat high in Table 3.6 and may contain con-

Table 3.6. Absolute Concentration of Phosphorus Metabolites in Fresh, Resting Adult Forearm Muscle[a]

	Concentration (mMole/liter)							
	PME	Pi	PDE	PCr	ATP	PCr + Pi	P_{total}	NAD + NADH
Mean ($n = 6$)	2.2	4.7	4.2	32.0	6.3	36.6	63.9	0.5
S.D.	0.7	1.0	0.8	4.5	0.8	5.2	8.0	0.2

[a] From Cady,[42] Academic Press Inc. Reprinted with permission.

tamination from the neighboring PCr resonance.) It appears that, regardless of whether quantitation of phosphorus muscle metabolites is accomplished by comparison with assumed total mobile phosphate or ATP concentrations or by direct MRS absolute quantitation methods, the summed concentration of PCr + Pi is essentially the same as that obtained by biopsy.[79,86] Thus, in addition to being noninvasive and rapidly repeatable, it seems that, for determining certain metabolite concentrations, MRS may be a more accurate technique than conventional biopsy. Studies using localized spectroscopy (one-dimensional[35] and two-dimensional[60,61] phase encoding) have also been made on normal resting muscle.

3.3.1.2. Pathological Changes in Resting Muscle Spectra

Although the most spectacular results from studies of human muscle have been produced by periodically collecting data during exercise, some pathologies exhibit altered metabolite levels in the fresh, resting state. An early study on the forearm muscles of a 16-year-old boy with a mitochondrial myopathy[6] showed a reduced PCr/Pi ratio, and this result indicated the potential usefulness of ^{31}P MRS as a rapid, noninvasive technique for investigating muscle weakness.

3.3.1.2a. Duchenne Muscular Dystrophy. This was one of the first pathological conditions to be observed in vivo by ^{31}P MRS. Early results suggested substantially altered relative phosphorus metabolite levels when compared with data acquired from normal tissue.[79,87,88] Recent work[89] has provided a more detailed ^{31}P MRS characterization of Duchenne dystrophy and its development. Figure 3.14 shows ^{31}P spectra obtained from the gastrocnemius of a normal, age-matched boy and from a boy with Duchenne dystrophy. In Duchenne dystrophy, Pi and PDE are elevated as fractions of the total mobile phosphorus and PCr is lowered, although ATP is not significantly different from normal. (See Table 3.7.) pH_i may be slightly more alkaline in Duchenne muscle. The age dependences of phosphorus metabolite levels (expressed as a percentage of the total mobile phosphate) in normal muscle and Duchenne dystrophy muscle are shown in Figures 3.15a to d. For normal boys, PCr/P_{total}, ATP/P_{total}, and PCr/Pi all increase with age, whereas in cases of Duchenne dystrophy these quantities tend to decrease. PDE and Pi both show strong increases with age in boys with Duchenne dystrophy. These results imply that a progressive metabolic deterioration exists in Duchenne muscular dystrophy. There is also a tendency for ^1H spectra from normal boys to have higher H_2O/CH_2 ratios. This probably indicates that there is a higher proportion of healthy muscle tissue within the coil's sensitive volume in normal boys when compared with Duchenne dystrophy patients.

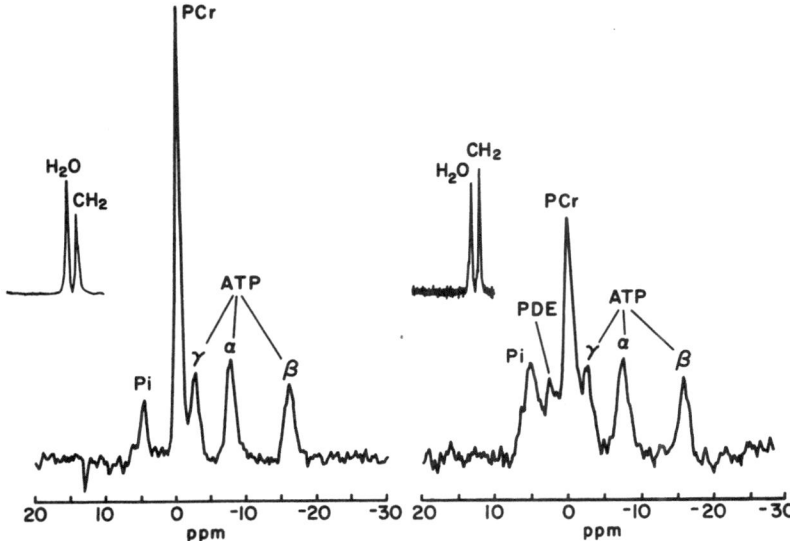

Figure 3.14. ^{31}P and ^{1}H spectra collected in vivo from the resting gastrocnemius muscle at 32.5 MHz. The spectrum on the left was collected from a normal eight-year-old boy, while that on the right was obtained from a boy of the same age with Duchenne dystrophy. Both ^{31}P spectra resulted from 240 scans with a 3-sec pulse interval using a 4-cm-diameter surface coil. 10-Hz line broadening was used in the processing. Relative metabolite concentrations are very different in the two ^{31}P spectra (see Table 3.7). The ^{1}H spectrum from the boy with Duchenne dystrophy shows a higher CH_2/H_2O ratio when compared with that obtained from the normal, age-matched control (from Younkin et al.,[89] Edgell Communications Inc.; reprinted with permission).

Table 3.7. Relative Phosphorus Metabolite Concentrations in the Fresh, Resting Gastrocnemii of Normal Boys and of Boys with Duchenne Dystrophy[a]

	Duchenne	Normal	Significance
Number of subjects	14	10	
Age (years)	9.7 ± 3.1	10 ± 4	not sig.
PCr[b]	32.4 ± 5.7	42.3 ± 2.6	$p < 0.001$
Pi[b]	11.4 ± 4.1	6.8 ± 1.0	$p < 0.001$
PDE[c]	9.7 ± 3.1	0.4 ± 1.3	$p < 0.001$
βATP[b]	14.2 ± 1.7	15.1 ± 1.3	not sig.
pH$_i$	7.3 ± 0.3	7.1 ± 0.04	$p < 0.01$
PCr/Pi	3.2 ± 1.1	6.4 ± 1.3	$p < 0.001$

[a] From Younkin et al.,[89] Edgell Communications Inc. Reprinted with permission.
[b] Metabolites expressed as percentage of (total mobile phosphate-PDE).
[c] PDE expressed as percentage of total mobile phosphate.

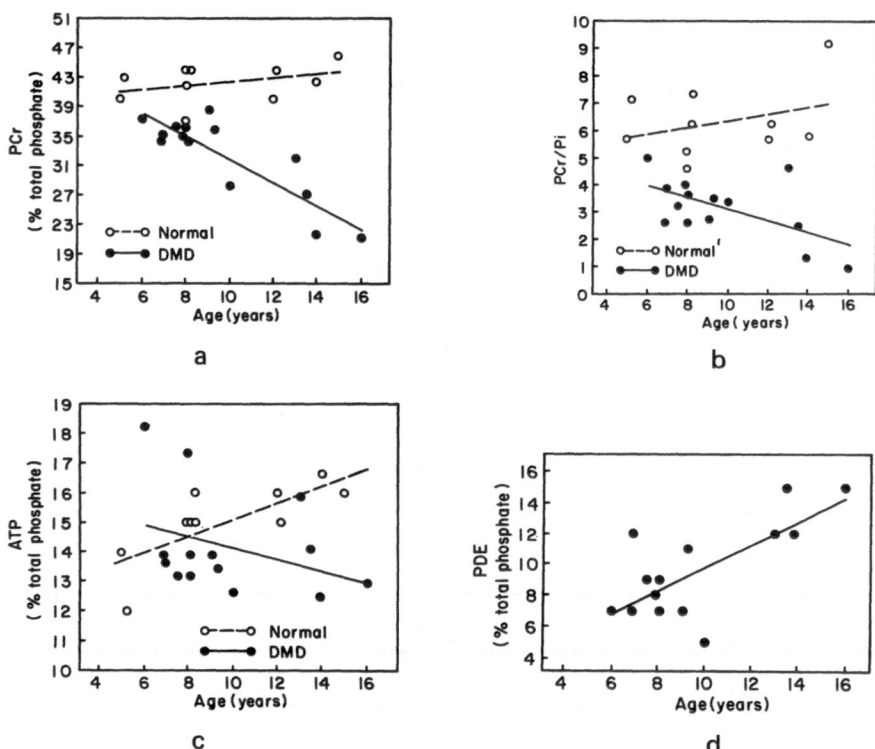

Figure 3.15. Relationships between phosphorus metabolites (as a percentage of total mobile phosphate) and age in normal boys (O, dashed lines) and Duchenne dystrophy patients (●, solid lines). In Duchenne subjects, both PCr (a) and PCr/Pi (b) were found to decrease significantly with age. PDE (d) showed a significant increase with age in these patients. In normal boys, only ATP (c) changed significantly, increasing with age (from Younkin et al.,[89] Edgell Communications Inc.; reprinted with permission).

3.3.1.2b. Peripheral Vascular Disease and Heart Failure. [31]P spectra obtained from the resting, unexercised calf muscles of patients with severe claudication have shown metabolite levels similar to those found in normal controls.[90] However, pH_i was significantly higher in the severe claudicants $(7.06 \pm 0.03;$ mean $\pm$ S.D.) when compared with that obtained from normal subjects (7.01 ± 0.02). A separate study of resting forearm muscle in patients with chronic congestive heart failure showed no significant difference in metabolite levels or pH_i when compared with normal subjects.[91]

3.3.1.2c. Muscle Denervation. Large changes in the resting metabolism of completely denervated first dorsal interosseous (FDI) hand muscles have

been demonstrated.[92] Studies of 13 patients showed significantly lowered PCr and elevated Pi concentrations. Denervated subjects also showed significantly high PME and PDE signals, and pH_i was increased. Further research has corroborated these results.[93] Spectra obtained from the resting forearm flexors of patients with electromyographically verified denervation of these muscles had lower PCr/Pi ratios (4.76 ± 2.50 versus 6.50 ± 1.55; mean $\pm$ S.D.; $p < 0.01$) and higher pH_i (7.09 ± 0.06 versus 7.05 ± 0.04; mean $\pm$ S.D.; $p < 0.01$) when compared with healthy controls. Five patients with severe denervation had extremely low PCr/Pi ratios (1.94 ± 0.70), and these also gave the highest pH_i (7.15 ± 0.04). However, patients with disuse atrophy due to forearm fracture but with intact innervation did not show these changes immediately after cast removal. In the denervated patients, the changes correlated with the degree of denervation, and metabolite levels improved with recovery.

3.3.1.2d. Myonecrosis. Studies of myonecrosis in the quadriceps femoris using spectroscopic localization (two-dimensional phase encoding) have revealed greatly reduced signal amplitude when compared with spectra obtained from nearby healthy muscle tissue.[60]

3.3.1.3. Studies of Muscle During Exercise

One of the most important aspects of the application of MRS to the study of skeletal muscle is that it is very easy to investigate metabolic changes occuring during exercise. This can be performed aerobically or, by application of an inflated sphygmomanometer cuff, anerobically; thereby different pathways for energy production can be tested. ^{31}P spectra with adequate SNR can be collected with a time resolution of the order of 1 min or less. Metabolite changes during both exercise and recovery can be followed. Comprehensive studies of skeletal muscle have shown that great care has to be taken in order to obtain objective results.[94] The surface coil to be used must be of an appropriate diameter and carefully positioned so that signal is obtained only from the particular muscle which performs the work during exercise. If these precautions are not taken, spectra may be contaminated by signals from other muscles which may not be working at the same rate or may not be playing a functioning role in the type of exercise undertaken. This can lead to "split" Pi peaks due to the muscles concerned having different pH_is. Furthermore, it will not be possible to relate changes in metabolite levels directly to the force generated.

3.3.1.3a. Normal Exercising Muscle. The changes in metabolite levels during exercise are characterized by a fall in PCr with a concomitant increase in Pi and a fall in pH_i, as shown in Figures 3.16a and 3.17a.

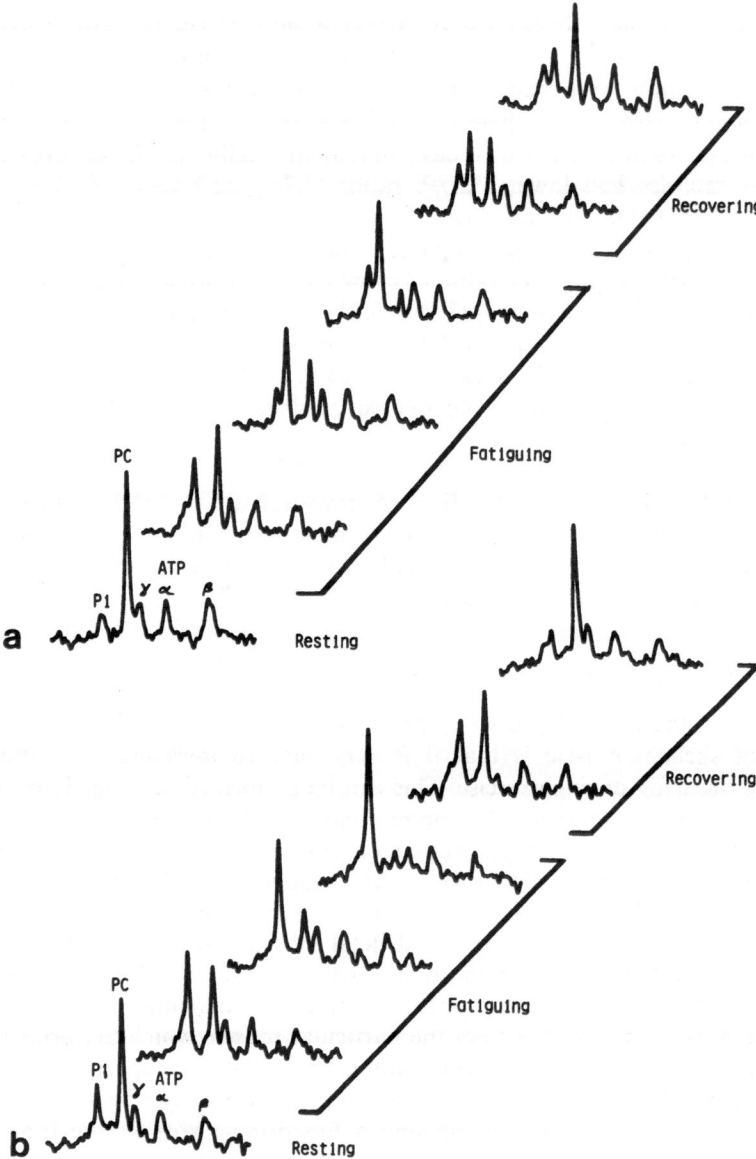

Figure 3.16. ^{31}P spectra collected from the first dorsal interosseous muscle (FDI) of the hand of a normal subject (a) and a McArdle's-syndrome (myophosphorylase-deficient) patient (b) during rest, fatiguing exercise, and recovery. The spectra were acquired at 32.5 MHz using a 2.5-cm surface coil placed over the FDI. Each spectrum is the result of accumulating 96 scans with a 2.256 sec pulse interval and a 90° flip angle at the coil's center. The fatiguing spectra were collected directly following an anerobic maximum voluntary contraction of 15 sec in (a) and 7 sec in (b) (from Cady et al.,[102] The Physiological Society; reprinted with permission).

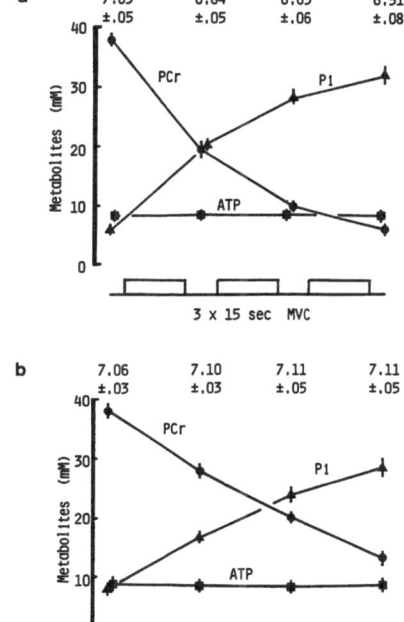

Figure 3.17. Cytosolic metabolite concentrations and intracellular pH determined from the spectra shown in Figure 3.16. (a) [PCr], [Pi], and [ATP] for the normal subject. pH_i is given at the top of the figure. (b) Cytosolic metabolite concentrations and pH_i for the myophosphorylase-deficient subject. Note that, although changes in PCr and Pi similar to those observed in the normal subject were produced in the myophosphorylase-deficient patient by approximately half the MVC × time integral, the latter does not show intracellular acidosis. This is due to the myophosphorylase-deficient patient's inability to generate lactate during exercise (from Cady et al.,[102] The Physiological Society; reprinted with permission).

Detailed studies have shown that severe exercise produces greater hydrolysis of PCr and greater intracellular acidosis when compared with light exercise.[95] Furthermore, resynthesis of PCr takes place more slowly when the exercise has been severe. If the exercise produces enough fatigue, the ATP begins to fall[96] (see Figure 3.18). PCr reaches a lower level in subjects showing ATP depletion ($17 \pm 5\%$ of control concentration compared with $26 \pm 5\%$), and ATP-depleted muscles have lower pH_is than severely exercised muscles with normal ATP levels (6.12 ± 0.04 and 6.37 ± 0.09, respectively). Reduction of ATP is also associated with slower recovery of PCr, Pi, ATP, and pH_i control values. The restoration of the pre-exercise ATP level is particularly slow.

Studies carried out on healthy subjects aged 70 to 80 years have shown no difference in the metabolic response to exercise when compared with controls aged 20 to 45 years.[97] On this basis it has been suggested that changes found in muscles of the elderly are not due to alterations in energy production.

Recently, ^{31}P-saturation transfer studies have been used to measure enzymatic flux through the creatine phosphokinase reaction in the direc-

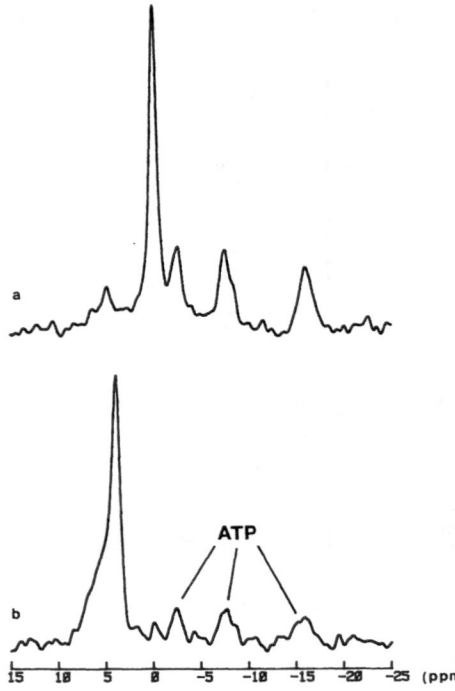

Figure 3.18. ATP depletion in fatigued skeletal muscle. The resonance identifications are the same as for Figure 3.13. (a) A ^{31}P spectrum obtained at 32.5 MHz from the fresh, resting first dorsal interosseous muscle of the hand of a normal subject. The data resulted from 128 scans using a 2.5-cm coil with a 90° pulse at the coil's center and a 2.256-sec pulse interval. Metabolite concentrations as a fraction of the total mobile phosphate (P_{tot}) were as follows: PCr/P_{tot}, 0.52; ATP/P_{tot}, 0.12. The intracellular pH was 7.07. (b) Following fatiguing aerobic exercise (100% maximum voluntary contractions of 1-sec duration at 1-sec intervals), PCr/P_{tot} and ATP/P_{tot} had fallen to 0.05 and 0.06 respectively. The Pi peak showed a clear acid shift, and the intracellular pH was 6.39. The metabolite data indicate reductions in PCr and ATP to about 10% and 50% of their fresh, resting concentrations, respectively. (Data collected in collaboration with D. Newham.)

tion of ATP synthesis in both exercising and fresh, resting forearm muscle.[98] Under conditions of steady-state work, it was reported that the flux from PCr to ATP was $64 \pm 10\%$ (mean $\pm$ S.D.) of the resting value.

3.3.1.3b. Studies of Muscle Fatigue. Because of its non-invasive nature, ^{31}P MRS is a useful research tool for the investigation of the metabolic causes and consequences of muscular fatigue in humans. Initial studies of skeletal-muscle fatigue have been made in pioneering research on anaerobic frog gastrocnemii at 4°C.[99] The relationships between force generation and [PCr], [Cr], [Pi], [ATP], [ADP (free)], and [H$^+$] were investigated. None of the metabolites studied gave a simple relationship between concentration and force generation. ATP is intrinsically involved in muscle contraction. However, a significant decrease in developed force was observed before any reduction in [ATP] occurred. This phenomenon has also been seen in studies on human subjects,[100, 101] although in a further human study[103] ATP and force (measured relative to maximal) had both decreased to $86 \pm 19\%$ and $92 \pm 7\%$ respectively of their control values after only 1 min of exercise. A subsequent re-analysis of the anaerobic frog-muscle data concentrated on the effects of hydrogen ions and Pi.[104] Pi exists in both the monobasic ($H_2PO_4^-$; Pi^-) and dibasic (HPO_4^{2-}; Pi^{2-}) forms. Rapid exchange takes place between the two species, and only one peak is seen in the ^{31}P spectrum. However the relative fraction of each form of Pi depends on [H^+], and this accounts for the pH-dependent chemical shift. As a result of the re-appraisal of the frog-muscle data it was found that the combination of increased [H^+] and [Pi] produced a large Pi^- fraction. [Pi^-] was found to be inversely related to the force of contraction. It was suggested that Pi^- may be a direct inhibitor of the actomyosin ATPase system. Further research on rabbit-psoas skinned fibers[105] has shown that the depression of maximal calcium-activated force by Pi correlates well with [Pi^-]. Studies on normal human subjects[101,103] have given indications that increased [Pi^-] leads to reduced force generation, although [H^+] may be important as well.[101] Recent work has compared the metabolic responses to exercise of normal subjects and a patient with myophosphorylase deficiency (MPD; also known as McArdle's syndrome) who produces no H^+ from glycolysis.[102] (The spectra obtained and metabolite levels measured in this study are shown in Figures 3.17 and 3.18, respectively.) The relationships between force generated by the MPD subject and [Pi^-] and [H^+] were different from the results for normal controls. However, the calculated [ADP] gave a similar relationship between force loss and ADP accumulation for both normal controls and the MPD subject. It was concluded that neither H^+ nor Pi^- could be the sole cause of force loss during fatigue, although it was difficult to suggest how ADP could affect force generation. Further studies investigated the metabolic causes of slow relaxation of fatigued muscle in both normal controls and in the same MPD subject.[106] After a 21-sec ischemic maximal voluntary contraction (MVC), relaxation in the MPD subject slowed to about half the rate in fresh muscle. There was no change in pH_i in the MPD subject during the exercise. This demonstrated that there is a mechanism responsible for the slowing of relaxation that is independent of H^+ accumulation. The normal subjects showed a slow recovery of relaxation compared with the MPD data. The results suggested that the slow recovery is related to acidosis. It was concluded that at least two processes cause slow relaxation in fatigued muscle, and only one of the processes is pH_i dependent.

Three phases of recovery from fatiguing exercise have been suggested from results using MRS and other techniques.[101] The first is indicated by the recovery of the *M*-wave, which indicates impaired muscle-membrane excitation and impulse propagation. This phase lasts about 4 min after the commencement of recovery. The second phase correlates with the metabolic state of the muscle as determined by ^{31}P MRS and is associated with a reduced MVC. The high-energy phosphates, pH_i and MVC, have recovered, usually within 20 min. The third phase is the period of recovery of neuromuscular efficiency, which is defined as the force divided by the rectified, integrated electromyogram. Full recovery of neuromuscular efficiency takes up to 60 min and is independent of changes in high-energy phosphates and pH_i. The reduced efficiency is probably related to impaired excitation–contraction coupling.

3.3.1.3c. Exercise Studies of Abnormal Muscle. One of the first applications of MRS to the study of muscle pathology was the investigation of McArdle's syndrome.[4] This is a rare, inborn metabolic error, characterized by a lack of glycogen phosphorylase activity in skeletal muscle. As a consequence of this enzyme deficiency, glycogen is not broken down during exercise, and lactate production is therefore inhibited. ^{31}P spectra obtained from resting MPD muscle show phosphorylated metabolite levels comparable with normal controls. However, there may be a tendency for the pH_i determined from fresh, resting muscle to be slightly alkaline in MPD subjects. One of the most obvious differences between ^{31}P MRS results obtained from normal subjects and those from MPD patients is that the latter do not exhibit intracellular acidosis in response to exercise. In fact, there may be a tendency to alkalosis during exercise (as indicated in Figure 3.17), even though comparable changes in high-energy phosphate levels have occurred in both normal and MPD muscle.[102] MPD subjects also exhibit excessive loss of PCr when compared with normal exercising muscle. This is indicated in Figure 3.17, in which the MVC × time integral for the MPD patient is half that of the normal subjects. It is also possible that, during ischemic exercise, a much larger PME signal (mainly due to sugar phosphates) is detected in normal muscle than in MPD subjects, as shown in Figure 3.16. Further results concerning muscular fatigue[102] and relaxation[106] in an MPD patient have been mentioned in the previous section.

Another interesting enzyme deficiency which produces a very different metabolic response to exercise when compared with normal subjects is phosphofructokinase deficiency (PFK). This enzyme problem is similar to MPD in that both conditions result in a failure of anerobic glycolysis, lactate accumulation is inhibited, and no acid shift of muscle pH_i occurs. However, the utilization of blood-borne glucose is not possible with PFK. Spectra collected from fresh, resting PFK muscle are similar to those obtained from normal controls.[79,107] However, two important differences are observed during exercise. Because the enzyme deficiency inhibits lactate production, pH_i does not show the acid shift readily observable in normal exercising muscle. Furthermore, the PME peak becomes very noticeable in PFK subjects as exercise progresses and can be much stronger than Pi. This is due to the accumulation of sugar phosphates and reflects the level of the enzyme deficit in the glycolytic pathway. Recovery from maximal leg exercise in a PFK subject has been shown to occur in the presence of a nearly constant Pi signal, with large reductions in PME (sugar phosphates) complimentary to PCr regaining its pre-exercise level.[107]

The postviral exhaustion–fatigue syndrome has also been studied using ^{31}P MRS.[108] During forearm exercise, spectra obtained from a patient showed abnormally early acidosis. The decrease in pH$_i$ was excessive when compared with the associated changes in high-energy phosphates. On the termination of exercise, pH$_i$ recovered normally. It was suggested that the enhanced acidosis may have been caused by a metabolic disorder affecting the regulation of the production of lactic acid. The normal recovery of pH$_i$ made impaired lactate clearance unlikely.

^{31}P MRS has been used to investigate the skeletal-muscle metabolism of patients with chronic congestive heart failure.[109] Studies of fresh, resting forearm muscle have shown Pi/PCr and pH$_i$ values similar to those acquired from normal controls. However, results obtained from heart-failure patients during fatiguing exercise indicated greater-than-normal PCr depletion and excessive acidosis.

Studies of the exercising gastrocnemius in peripheral-vascular-disease patients with severe claudication have shown greater PCr depletion and larger falls in pH$_i$ during exercise when compared with normal controls.[110] Subjects with severe symptoms also had slower recovery of PCr and pH$_i$. Patients with mild or severe disease showed slower recovery of calculated ADP levels. The pain associated with claudication was not related to phosphorus metabolite concentrations or to pH$_i$. Two patients in the study, who underwent surgical correction of their arterial stenoses, exhibited initial metabolic abnormalities, which became normal post-operatively. After treatment, these subjects were able to complete a full 10 min of exercise without claudication symptoms.

3.3.2. ^{1}H Spectroscopy of Human Muscle

The use of ^{1}H MRS for the in vivo investigation of human skeletal-muscle metabolism requires techniques whereby the strong signals from tissue-water protons and fatty acyl-chain protons can be reduced. In addition, due to the narrow resonance bandwidth, the application of some form of spectral editing may be necessary in order to resolve the signals from the metabolites of interest. Initial studies, using simple surface coil localization only, have revealed proton spectra which consist of a strong water resonance and a signal (usually smaller) from subcutaneous and intramuscular fat which varies in amplitude from subject to subject, as shown in Figure 3.14.[79, 82, 87–89] Patients with Duchenne dystrophy, in which muscle is progressively replaced by fatty and connective tissue, have reduced water–fat ratios in their ^{1}H spectra. However, similar spectra can be observed in normal subjects simply by locating the surface coil over a region with a substantial thickness of subcutaneous adipose tissue. In order to obtain less ambiguous results, more sophisticated methods have to be used.

Various localization techniques have been used including SPARS,[68] STEAM,[111] and DEPTH pulses[70] in combination with binomial water suppression.[67] Using both SPARS and STEAM, localized spectra show resolved Cr and choline–carnitine peaks in addition to residual-water and lipid signals.[112] Spectra collected from the gastrocnemius in normal subjects using the combined DEPTH-pulse and water-suppression approach showed resonances mainly from protons in saturated fatty acyl chains.[69] Studies of patients with spina bifida produced spectra in which strong resonances from unsaturated fatty acyl-chain protons were present.

Recent work has focussed on the application of localization and spectral editing for the detection and quantitation of lactate[113] and the use of the C2 and C4 resonances of carnosine for pH_i estimation.[114] In the muscle lactate studies, the Cr + PCr resonance at 3.1 ppm was used as an internal concentration reference for absolute quantitation. One volunteer was studied on seven separate occasions after exhaustive finger-flexion exercise, and the mean, measured lactate concentration was 14 mM. Spectroscopic studies of this nature give the possibility of directly comparing lactate concentrations with intracellular pH_i.

3.3.3. ^{13}C Spectroscopy of Muscle

Two problems with the application of ^{13}C spectroscopy to the investigation of muscle metabolism in vivo are the presence of strong lipid signals due to subcutaneous and intramuscular fat and the multiplet structure of peaks due to proton coupling.

Detailed spectroscopic studies of surgical samples obtained from normal muscle and from a range of pathologies have been made.[115] Proton decoupling was used to collapse multiplets to singlets and to improve the SNR. Signals due to fat remaining in the samples were reduced by isopentane extraction. Figure 3.19 shows spectra obtained from normal muscle and from a patient with cerebral palsy. The most important differences reported between spectra acquired from isopentane-extracted normal- and diseased-muscle samples was the strength of the methylene-carbon resonance at 30.5 ppm. This resonance is small in normal tissue but large in myogenic and neurogenic muscle disorders. Differences were also found between normal and diseased samples in their creatine content and their ability to produce lactic acid.

Recent in vivo studies have employed simple pulse–acquire and DEPTH pulses to detect the signal from the C1 carbon of glycogen, which resonates

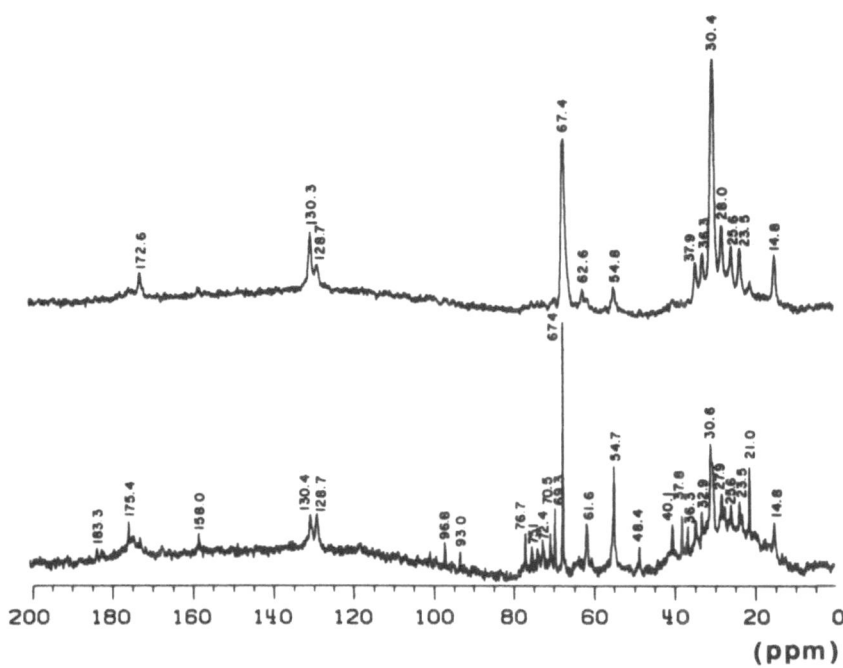

Figure 3.19. ^{13}C spectra of human skeletal muscle-biopsy samples recorded at 118.2 MHz and between 15 and 18°C. The lower trace is from normal muscle. The upper trace is from a patient with cerebral palsy. The spectra were acquired by accumulating 51,624 and 39,506 scans, respectively, with a pulse interval of just over 1 sec and using a 45° pulse. The 67.4 ppm signal in both spectra is from a chemical shift reference (dioxane). Resonance identifications in the abnormal muscle (upper trace; C* indicates the resonating carbon): 172.6 ppm, esterified carboxyl carbon; 130.3 and 128.7 ppm, unsaturated carbons [$(CH_2)_n-C*H=$ and $=C*H-CH_2-CH=$, respectively]; 62.6 ppm, $C1$ and $C3$ of glycerol; 30.4 ppm, $(CH_2)_n$; 28.0 ppm, $(C*H_2-CH=)$; 25.6 ppm, $OCOCH_2C*H_2$; 23.5 ppm, $C*H_2-CH_3$; 14.8 ppm, (CH_3); and 54.8, the methyl carbon of creatine. The normal-muscle spectrum includes, in addition, the resonances of α- and β-glucose carbons from 86.8 to 70.5 ppm and at 61.6 ppm; the lactic acid methyl carbon at 21.0 ppm; alcoholic carbon at 69.3 ppm; and carboxylic carbon at 183.3 ppm (from Barany et al.,[115] Academic Press Inc.; reprinted with permission).

as a coupled doublet at about 101 ppm.[116] The use of the DEPTH pulse greatly reduced the signals from subcutaneous fatty acyl chains and permitted the detection of the glycogen doublet with adequate SNR in about 13 min. Dynamic changes in muscle-glycogen levels were investigated by collecting spectra from a trained runner 1 hr prior to, immediately following, and the morning after a 13-mile run. Directly after the run the glycogen, had fallen to 30% of its prerun level, but it had recovered to 80% of this level by the next morning. Proton-decoupled spectra that demonstrated enhanced SNR were also obtained.

3.4. STUDIES OF HUMAN CARDIAC METABOLISM

The investigation of human cardiac metabolism by in vivo MRS has been delayed due to the fact that localization techniques are required. For ^{31}P MRS they are needed primarily in order to avoid spectral contamination by signal from the chest-wall musculature. For ^{1}H MRS, subcutaneous fat is an additional problem. The task of reducing signals from surface tissues seems to have been achieved satisfactorily. However, unless the localization method produces a well-defined selective volume, contamination of spectra by blood and lung tissue is a possibility. In order to obtain spectra that originate entirely from cardiac muscle, care must be taken to ensure that the selected volume is located so that it is always within the heart tissue during the cardiac cycle. Alternatively, data acquisition must be synchronized to the cardiac cycle. Fortunately, contributions to the cardiac spectrum from PCr and NTP in the blood are negligible. However, the signal from blood 2,3-DPG, a doublet resonating at around 6 ppm (PCr at 0 ppm), is detected in the PME peak. Lung tissue has a low density, and considerable spectrum-line broadening is caused by multiple gas–tissue interfaces.[117] Lung-tissue metabolites may therefore not contribute very much to high-resolution spectra obtained from a well-positioned selective volume. If these points are noted, then, even with nonideal localization methods, cardiac spectra worthy of clinical and physiological interpretation can be collected.

3.4.1. ^{31}P Studies of Human Cardiac Metabolism

The earliest localization method applied to the study of the human heart in vivo was the DRESS[30] technique. In initial investigations on normal volunteers,[118] ^{1}H imaging, using whole-body and surface coils, was employed in order to position the subject so that the part of the heart required was in the field of view. In subsequent studies,[119] DRESS was used to investigate four patients with acute anterior myocardial infarction, and the results were compared with data obtained from seven healthy controls. All the patients were undergoing treatment for their condition. The spectra were collected using a 1.5-tesla whole-body-imaging spectrometer, and a 6.5-cm surface coil was used for both transmission and reception. So that data could be sampled from a similar part of the myocardium throughout the collection, the electrocardiogram (ECG) was used both to give a trigger signal to synchronize FID acquisition and to monitor heart rhythm. The interpulse time [with RF pulses (ECG-synchronized to the end of systole)] was 875 ± 125 msec. In the normal subjects, data were obtained at two to five different times in the cardiac cycle in order to investigate changes in metabolite levels. Optimization of static magnetic field homogeneity was accomplished using ^{1}H DRESS with the localized slice within the heart. This had the twofold effect of optimizing the homogeneity within the tissue of interest and thereby improving the localization because the resultant homogeneity was not as good outside this region. Spectra were localized to 1-cm slices parallel to the chest wall at 0.5-cm intervals using electronic depth control. The total study time was between 45 and 90 min. Spectral analyses, using the fitting of Gaussian or Lorentzian peak profiles to the data acquired from normal controls, gave the following metabolite ratios: PCr/βATP $= 1.6 \pm 0.4$ (mean $\pm$ S.D.); PCr/Pi $= 8.8 \pm 4.8$; Pi/βATP $= 0.16 \pm 0.14$; PCr/PME $= 6.5 \pm 5$; and PCr/PDE $= 3.4 \pm 2.3$. The pH$_i$ in the same subjects was 7.15 ± 0.2. The ratios obtained from normal subjects did not vary significantly at different phases in the cardiac cycle. Significant (approximately threefold) reductions in the PCr/Pi and increases in Pi/ATP were observed in spectra derived transmurally or endocardially when compared with epicardial and normal control spectra.

Variations of rotating-frame imaging are other localization methods which have been applied to ^{31}P MRS studies of the human heart.[120,121] The techniques rely on the existence of a linear RF-field gradient generated by the transmitter pulse. In order to obtain a substantial region in which the RF amplitude varies linearly as a function of depth, a specially designed double coil was used.[122] The larger, outer coil (diameter 15 cm) was used for transmission only. The inner, receiver coil (diameter 6.5 cm) was offset forward from the transmitter coil plane by 2 cm. This ensured that, to a first-order approximation, the sensitive volume of the receiver coil is within the linear region of the RF field. In one variation, spectra localized to a slice parallel to the receiver coil are acquired by applying RF pulses given by

$$\theta = (2n + 1)(\pi/2)(-1)^n \qquad (n = 0, 1, 2,...) \qquad (3.1)$$

where the pulse for $n = 1$ provides maximal excitation in the volume of interest. A combination of FIDs obtained using pulses given by equation (3.1) results in a spectrum originating from a localized slice in front of the coil. In a second variation of the technique, successive increments of the pulse width give data that are phase encoded, and a two-dimensional Fourier transform of the data set results in spectra from slices at various depths into the subject. The mean PCr/ATP ratios obtained from the myocardium of normal subjects were 1.55 ± 0.20 ($\pm$ S.D.; $n = 6$) and 1.53 ± 0.25 ($n = 4$) for the two approaches to the technique, respectively. These values compare very well with those found using DRESS.[119]

A modified version of ISIS using a surface coil has recently been applied to the collection of spectra from the human heart.[123] Positioning of the sensitive volume was guided by the use of ^{1}H imaging. In order to reduce the contamination of spectra by signals from tissue outside the volume of interest, four saturation pulses were inserted in the ISIS sequence immediately after the acquisition of data. The PCr/ATP ratio obtained was approximately 1.5.

3.4.2. ¹H Studies of the Human Heart

Homogeneity of the magnetic field is reported to be a major problem when attempting to obtain highly resolved ¹H spectra from the in situ human heart.[69] Motional effects due to breathing and the cardiac cycle need to be reduced by laying the subject prone on top of the surface coil and synchronizing the homogeneity optimization pulse to the ECG. The ¹H spectra acquired, using a combination of binomial water suppression[67] and DEPTH pulses,[70] from the hearts of six healthy volunteers showed major resonances from PCr + Cr and taurine as well as smaller resonances due to other amino acids and fatty acyl chains.

3.5. STUDIES OF THE HUMAN LIVER

A minimum requirement for the study of the metabolism of the in situ human liver by ³¹P, ¹H, or ¹³C MRS is the use of a localization technique which substantially reduces signals from surface tissues such as the abdominal musculature and subcutaneous fat. In fact, because of its very low PCr content, the liver has often been used as a test tissue for localization methods. If PCr is not detectable in the spectra, its absence has often been taken as an indication that signals from superficial muscle (assumed unfatigued, so that PCr has a high concentration) have been greatly reduced by the method of localization.

3.5.1. ³¹P Studies of the Human Liver

A typical ³¹P spectrum of the human liver, obtained in vivo, is shown in Figure 3.20d. It was obtained using three-dimensional ISIS.[124] Many other studies carried out on the human liver have utilized less-sophisticated methods of localization, but in many instances the spectra acquired have been of similar appearance so that of Figure 3.20d.

One of the earliest ³¹P MRS studies of the human liver reported the investigation of liver tumors (neuroblastomas) discovered at birth in two female infants.[125] Spectra were obtained at 32.5 MHz using a 5-cm-diameter surface coil positioned over the liver. When the lower edge of the liver was less than 5 cm below the right costal margin, the coil also overlay

Figure 3.20. ^{31}P spectra, obtained from human liver, with and without ISIS localization. The spectra were acquired using a 2.1-tesla whole-body magnet. A concentric ^{1}H/^{31}P surface coil was employed to collect the data. The outer ^{1}H loop and the inner ^{31}P loop were 12 cm and 9 cm in diameter, respectively. Pulse lengths were calibrated and adjusted to give optimum signal. (a) The spectrum obtained with surface-coil localization only. PCr from muscle tissue shows a strong resonance at about -2 ppm. (b) The one-dimensional ISIS spectrum. The PCr peak is smaller, due to the reduction of signal from superficial musculature. (c) A two-dimensional ISIS spectrum. The PCr resonance is reduced further. (d) The full three-dimensional ISIS spectrum. The sensitive volume was a $3 \times 3 \times 3$-cm^3 cube centered 4 cm from the coil's center. A pulse interval of 1.2 sec was used, and 512 scans were accumulated (from Jue et al.,[124] National Academy of Sciences of the USA; reprinted with permission).

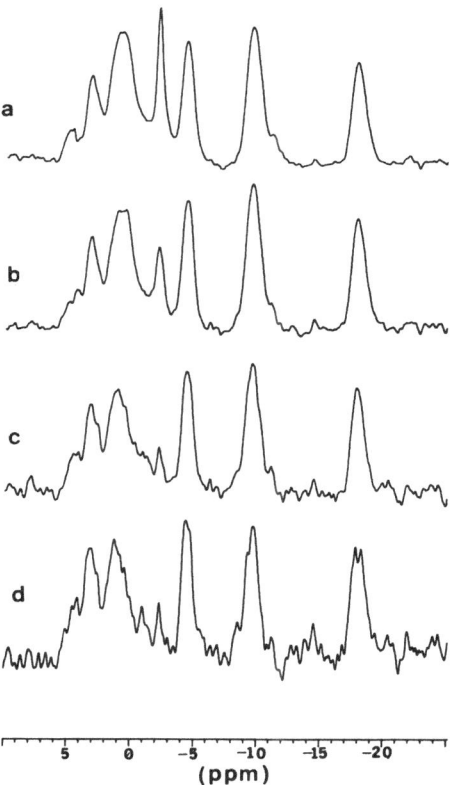

a portion of the rib cage. The pulse length was optimized to give maximum signal from a standard solution, and a pulse interval of 3 sec was used. Because no other method of localization apart from the surface coil was employed, contamination of the spectra by signals from superficial tissues could not be avoided. However, spectra collected from the quadriceps and liver of a normal, three-month-old female control were very different in appearance, although the liver spectrum exhibited a PCr resonance which must have originated from muscle. The metabolite ratios determined from the normal liver were as follows: PME/ATP = 0.3; PCr/ATP = 1.3; Pi/ATP = 0.9; and PDE/ATP = 2.3. The main conclusion obtained from spectra acquired from the infants with neuroblastoma was that the PME/ATP ratio was a good marker of tumor growth or regression. In one of the patients, PME/ATP was 1.3 in the first study but rose to 2.3 after two months. In the following weeks the patient's condition improved in response to therapy, and subsequent spectra showed steady reductions in PME/ATP, which had a value of 0.6 after a further seven months. In addition to the in vivo studies, high-resolution spectra were obtained from neuroblastoma extracts produced from biopsy material. This made possible the identification of the major constituents of the tumor PME resonance (PEt, PCho, and 2,3-DPG).

Spectroscopic studies of adult liver[34, 126–128] have been made using a 1.6-tesla whole-body magnet and employing magnetic field profiling (see Chapter 5) as a localization technique. Data were collected with a 10-cm-diameter surface coil placed on the lateral thoracic wall over the liver and a 180° flip angle at the coil center. PCr was undetectable in spectra obtained at two different pulse intervals (1 sec and 10 sec), and its absence was taken to indicate that magnetic field profiling was giving adequate localization. In studies on seven normal subjects[126] using a pulse interval of 0.5 sec, the PME/ATP, Pi/ATP, and PDE/ATP ratios were 0.53 ± 0.03, 0.97 ± 0.04, and 1.9 ± 0.1, respectively (mean $\pm$ S.E.M.). The pH_i determined from the same seven normal subjects was 7.18 ± 0.08 (mean- $\pm$ S.E.M.). From the difference in chemical shift between the γ and β ATP resonances, it was also possible to estimate that about 86 % of the cytosolic ATP was magnesium bound, and this correspond to a free-magnesium pool of approximately 300 μmole/liter.

Intravenous fructose-tolerance tests were conducted, during which spectra were collected with 5-min time resolution. 5 min after a bolus injection of 250 mg/kg of body weight, the PME resonance (mainly sugar phosphates) had increased by a factor of seven, whereas Pi and ATP had decreased by 80 % and 66 % of their control levels, respectively. Sugar-phosphate metabolism was observed to be complete within 20 min. Hereditary fructose intolerance (HFI) has been studied in five patients with HFI and in 8 heterozygotes for HFI.[128] Spectra obtained from subjects with HFI following ingestion of small amounts of fructose (two of the subjects were given an oral load of 20 mg/kg of body weight) showed increased PME and decreased Pi resonances. Ingestion of 50 gm of fructose by obligate heterozygotes produced greater depression of the Pi level and a larger elevation of the PME level than was observed in normal controls.

Two liver tumors (a hepatoblastoma and an endometrial adenocarcinoma) were studied using the same technique.[127] Both the tumors exhibited raised PME levels, and pH_i in the hepatoblastoma was higher (7.31) than that determined in normal controls. The adenocarcinoma showed a substantial PCr resonance and little PDE. After embolization of this tumor, twofold increases in the Pi/ATP and Pi/PCr ratios and a reduction in pH_i from 7.15 to 6.80 were observed. These alterations were compatible with ischemia caused by the embolization.

Other recently developed localization techniques have given improved rejection of signal outside the sensitive volume and greater positional flexibility than the methods used in the studies described so far. Those that have been applied to the in vivo investigation of human-liver metabolism include fast rotating-gradient spectroscopy (FROGS),[129] improved-depth selective single-surface coil spectroscopy (IDESS),[130] two- and four-dimensional chemical shift imaging,[1,35,131,132] and ISIS.[123,124] (For details of these techniques, see Chapter 5.) Both FROGS and IDESS have produced liver spectra from normal volunteers in which the PCr signal from superficial muscle was adequately suppressed. The spectra obtained have characteristics very similar to those acquired by magnetic field profiling. Two-dimensional chemical-shift imaging also gives good rejection of muscle PCr signal when spectra are selected from a suitable depth.[1,35] In data obtained from three contiguous slices, each pair separated by 20 mm, the surface spectrum originated mainly from skeletal muscle and had a strong PCr resonance.[35] The spectrum collected from the deepest slice showed very little PCr and had a PME/ATP ratio of 0.2. A spectrum acquired at a similar depth from a patient with sclerosing cholangitis gave a PME/ATP ratio of 1.0. It was suggested that the increased PME level reflected structural damage within the liver rather than the accumulation of sugar phosphates. Four-dimensional chemical shift imaging produces a spectroscopic "checkerboard" output for each phase-encoded slice similar to that shown for the brain in Figure 3.9. (For the brain data, slice selection was achieved by use of a selective RF pulse, and three-dimensional chemical-shift imaging was used.) Localized spectra have been acquired with the four-dimensional technique from patients suffering with acute fatty liver of pregnancy and Caroli syndrome[131] and from a patient with carcinoid liver metastases.[132] Data collected from the latter subject showed elevated PME/ATP and decreased PDE/ATP ratios. The regions of spectroscopic abnormality corresponded with the positions of lesions found by CAT and MRI. The full three-dimensional ISIS method of localization also gives good suppression of outer volume signals as shown in Figure 3.20d.[123,124]

3.5.2. ^{13}C Studies of Human Liver Metabolism

^{13}C MRS has recently been exploited as an in vivo investigational tool in human subjects. Spectra obtained with surface coils, but without further localization, show large signals from fatty acids and, unless low level decoupling RF is applied, the resonances appear as multiplets. Certain useful spectral windows exist, however, and one of these (90 to 120 ppm) is convenient for the study of liver glycogen.[133] The glycogen resonance which occurs in this spectral window is located at 100.5 ppm. An SNR of 15 to 1 for this resonance is obtainable in vivo using a 10-cm surface coil and a 180° pulse at the coil center and collecting approximately 4000 scans with a pulse interval of 0.4 sec.

3.6. STUDIES OF HUMAN KIDNEYS

These organs are more difficult to study in vivo than the liver. Their smaller size and greater depth presents problems with obtaining adequate signal and positioning the sensitive volume. So far only ^{31}P spectra have been acquired from humans. The relatively poor SNR obtained is demonstrated by the ISIS spectrum shown in Figure 3.21b.[124]

The first investigations of human kidney metabolism were carried out on excised, ice-stored, perfused organs.[134] Most of the kidneys were studied because they included tumorous tissue, but normal kidneys were also examined. MRS data were acquired in a 1.9-tesla magnet simply by placing a 5-cm-diameter surface coil directly on the perfused organ. The spectra obtained were similar in appearance to that shown in Figure 3.21b, but PCr was completely undetectable and the PME resonance was somewhat stronger. Nine hypernephromas and two Wilm's tumors were studied. Seven of the former and one of the latter showed an additional resonance at about 4.2 ppm, which was interpreted as Pi in a highly acidic environment (pH 6.1 to 6.5, compared to pH 7.0 to 7.2 for normal perfused kidneys). Five tumors were tested by adding chemotherapeutic agents to the

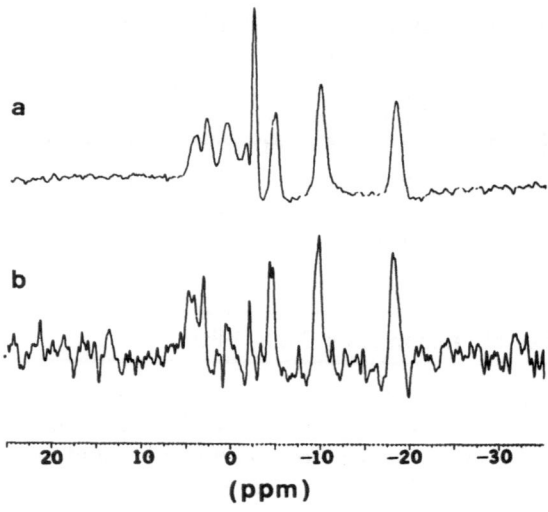

Figure 3.21. [31]P spectra of in situ human kidney obtained using a surface coil without localization (a) and with ISIS localization (b). The spectrum in (b) was collected using a three-dimensional ISIS technique with a $4 \times 4 \times 4$-cm^3 sensitive volume centered 6 cm from the plane of the coil. 512 scans were acquired with a 1.2-sec pulse interval. Other collection conditions were the same as for the liver spectra in Figure 3.20 [from Jue et al.,[124] National Academy of Sciences of the USA; reprinted with permission].

perfusate. In one of these tumors, the 4.2 ppm peak rose rapidly, and this was interpreted as due to drug sensitivity.

Surface coil ISIS has been used recently to obtain in vivo spectra as shown in Figure 3.21.[123,124] SNR and resolution were worse than those exhibited by the spectra collected from excised, perfused kidneys. Adequate work has not yet been carried out to define normal metabolite ranges for these organs in vivo. The investigation of in situ human kidneys by MRS is at its beginnings, although the localization tools exist and future work will provide much more information.

3.7. STUDIES OF HUMAN TESTES

With the development of high-field whole-body magnets capable of performing spectroscopic studies, it is somewhat surprising that there has been little investigation of human testicular metabolism. These exterior organs are very amenable to study using solenoidal or surface coils, and further localization methods may not be necessary in order to obtain clinically useful results. However, preliminary ^{31}P studies have been undertaken recently.

A ^{31}P spectrum of normal human testicle tissue, shown in Figure 3.22, is very similar to that obtained from rat testis (shown in Figure 1.15). Both infertile patients and normal subjects were examined in the study in which the human spectrum was obtained.[135] A 3-cm × 3.5-cm two-turn surface coil was used to acquire the MRS data in a 1.5-tesla whole-body magnet. The coil was placed on the anterior aspect of the testis to be studied. Biopsy material was obtained from all the patients. The PME/βATP ratio was found to correlate well with biopsy results. In the presumed normal subjects, the PME/βATP ratio was 1.84 ± 0.09 ($n = 5$). The spectra acquired from patients whose biopsy showed normal spermatogenesis gave PME/βATP $= 1.80 \pm 0.10$ ($n = 10$), whereas those whose biopsy indicated

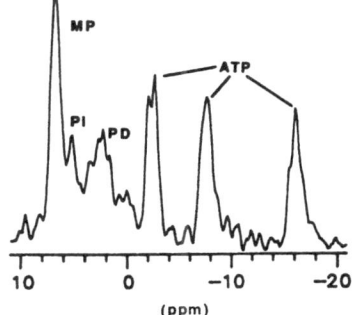

Figure 3.22. A ^{31}P spectrum obtained from normal human testicle. The data were collected in a 1.5-tesla whole-body magnet using a 3 cm × 3.5-cm, two-turn surface coil tunable to both ^{1}H and ^{31}P frequencies. The spectrum was acquired with the coil placed on the anterior aspect of the testis, and 200 scans were accumulated with a 2-sec pulse interval. Resonance identifications: MP, phosphomonoester; PD, phosphodiester [from Chew et al.,[135] Society for Magnetic Resonance in Medicine; reprinted with permission].

atrophy and lack of spermatogenesis had $PME/\beta ATP = 1.43 \pm 0.06$ ($n = 6$). The difference between the mean $PME/\beta ATP$ ratios measured in the normal and decreased spermatogenesis groups was significant. Preliminary results have also been reported from similar studies on nine normal subjects and one patient with a unilateral varicocele who was studied pre- and postoperatively.[136]

3.8. IN VITRO STUDIES OF HUMAN BODY FLUIDS

The analysis of fluids using very-high-resolution spectrometers offers great potential in a clinical setting. Because the samples can be very small, excellent magnetic homogeneity can be obtained, and the large number of resonances detected provide "fingerprints" for various pathologies. In contrast to in vivo studies, for which the ^{31}P nucleus has so far proven most useful, ^{1}H spectroscopy has been the main investigative tool for work on fluids.

3.8.1. ^{1}H Studies of Urine

^{1}H MRS of urine has been applied to the investigation of metabolic diseases including propionic acidemia,[137,138] methylmalonic acidemia,[137,138] and phenylketonuria,[139] although other physiological and pathological states have also been examined.[140] The technique is rapid (about 10 min), no preliminary extraction or derivative preparations are required, and only 0.5 ml samples are needed. The method allows for fast monitoring of metabolic alterations with a time scale short enough to permit efficient therapeutic management. In ^{1}H spectra, collected from normal urine at frequencies between 400 and 500 MHz, numerous metabolites including creatinine, citrate, hippurate, glucose, ketone bodies, and amino acids have been identified,[140] as shown in Figures 3.23a and b. Metabolic abnormalities can produce very different spectra when compared with those obtained from the urine of normal subjects, as shown in Figures 3.24a and b, in which the propionylglycine and methylmalonate resonances at about 1.2 ppm are particularly strong. In normal subjects, hard exercise results in greatly elevated lactate signals, which are barely detectable in pre-exercise urine, and spectra acquired after fasting show resonances from ketone bodies and other metabolites.[140] ^{1}H MRS is also a very useful method whereby the excretion of metabolites resulting from the ingestion of pharmaceuticals can be monitored. Acetaminophen (N-acetyl-4-aminophenol or paracetamol) has been investigated in this way using both one- and two-dimensional spectroscopy, as shown in Figure 3.25.[140–143]

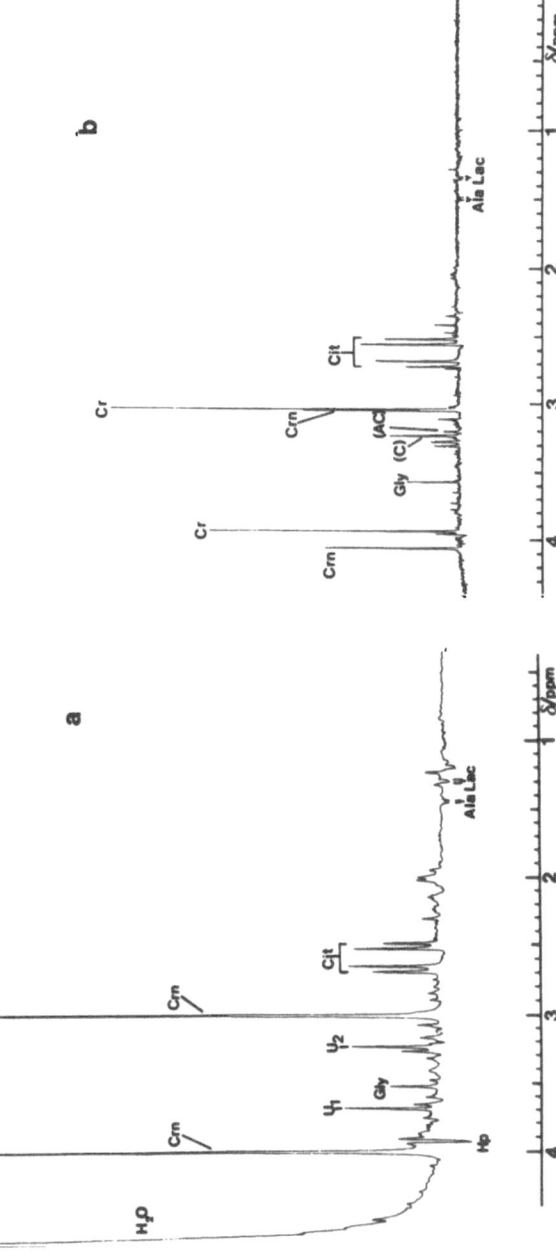

Figure 3.23. ^{1}H spin-echo spectra, acquired at 400 MHz, of urine from a normal adult male (a) and a normal two-year-old boy (b). 112 scans and 136 scans were accumulated for spectrum (a) and spectrum (b), respectively. The spectra were collected at 20°C using a $90° - \tau - 180° - \tau -$ acquire data sequence, with $\tau = 60$ msec and a sequence-repetition time of 1 sec. Resonance identifications: AC, acetyl-carnitine; Ala, alanine; Cit, citrate; Crn, creatinine; Cr, creatine; Gly, glycine; Hp, hippurate; Lac, lactate; U_1 and U_2, unidentified (RA Iles, personal communication). [Reprinted with permission from Iles et al.,[137] *Clinical Chemistry* (1985) Volume 31, Page 1796, Figure 1. Copyright American Association for Clinical Chemistry.]

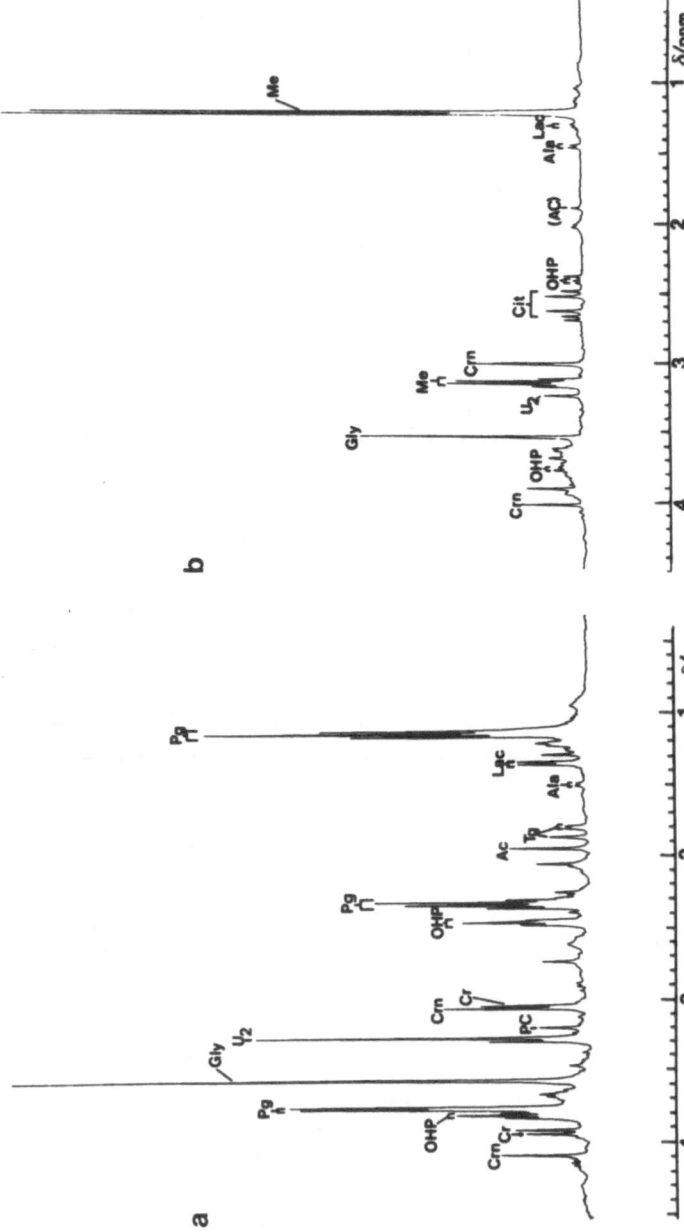

Figure 3.24. ¹H spectra obtained at 400 MHz from the urine of patients with organic acidurias. (a) A spectrum acquired from the urine of a patient with propionic acidemia. The data were collected at 20°C with 112 scans using a 30° pulse. Resonance identifications in addition to those in Figure 3.23: Ac, acetate; OHP, 3-hydroxypropionate; PC, propionylcarnitine; Pg, propionylglycine; and Tg, tiglylglycine. (b) A spectrum from the urine of a patient with methylmalonic aciduria. 280 scans were acquired using the same conditions as for spectrum (a). Additional resonance identifications: Me, methylmalonate; U₂, betaine. [Reprinted with permission from Iles et al.,[137] *Clinical Chemistry* (1985) Volume 31, Page 1797, Figures 3 and 4. Copyright American Association for Clinical Chemistry.]

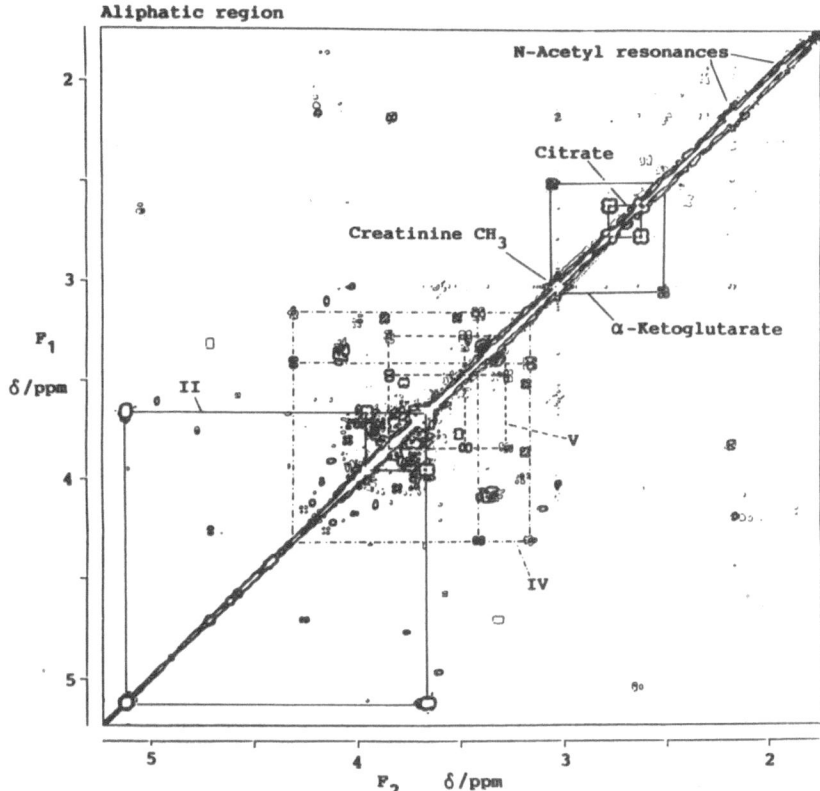

Figure 3.25. The aliphatic region of a two-dimensional ^{1}H shift-correlated (COSY) spectrum of urine acquired at 500 MHz and collected 3 hr after acetaminophen (paracetamol) ingestion. The pulse sequence used was: $90° - \tau_1 - 90° -$ acquire data (acquisition time $= \tau_2$) and τ_1 was incremented in sequential experiments. After two-dimensional Fourier transformation of the data set, a map is obtained as shown in the figure. The two-dimensional spectrum has the advantage that peaks which would overlap in the one-dimensional spectrum are now separated according to their spin couplings. Resonance identifications: II, acetaminophen glucuronide; IV and V, acetaminophen N-acetylcysteinyl and cysteinyl conjugates, respectively. IV and V were not resolved in one-dimensional spectra. [Reprinted with permission from Bales et al.,[142] *Clinical Chemistry* (1985) Volume 31; Page 761, Figure 5. Copyright American Association for Clinical Chemistry.]

3.8.2. ^{1}H Studies of Amniotic Fluid

The investigation of amniotic fluid samples by ^{1}H spectroscopy at 360 MHz has demonstrated the detection of several physiologically relevant compounds, including glutamic acid, leucine, and isoleucine (which have been correlated with CNS disorders) and lactate (involved in fetal acidosis).[144] Acid-extracted samples were used because these produced spectra with less baseline "roll" and more resolution. This effect was due to the elimination of proteins from the samples. Much more information was present in the spectra than could be interpreted with the resolution obtained using the collection techniques applied. Further diagnostic information should be obtainable with spin–echo or two-dimensional spectroscopy.

3.8.3. ¹H Studies of Human Blood Plasma

Single-pulse ¹H spectra from blood plasma exhibit sharp resonances from mobile metabolites with underlying broad features from albumin and immunoglobulins. Strong signals are present in the 0.8 to 1.3 ppm region of the spectrum, and these have been assigned to $-CH_3$ and $-CH_2$ groups in mobile fatty-acid components of chylomicrons and lipoproteins. Spin–echo spectroscopy has enabled specific identifications to be made for individual lipoproteins[145] and has shown that a pool of MRS-invisible lactate exists in blood plasma.[146]

Attention has recently been drawn to the line widths of lipoprotein resonances, and it has been claimed that these correlate with the presence of cancer.[147] Recent work has shown that other factors can influence the widths of these peaks. Freezing and thawing of plasma samples broadens these resonances, while the consumption of food by the donor prior to sample collection sharpens them.[148] Carefully controlled investigation of the lipoprotein peaks in plasma obtained from normal subjects and cancer patients revealed no significant differences in $-CH_3$ and $-CH_2$ linewidths. Other recent research, although still claiming a difference in width of the lipoprotein resonance between cancer patients and normal controls, has shown that hyperlipidemics without malignant disease exhibit similar peak-width changes.[149] A further report has indicated differences in the chemical shifts of the methylene plasma-lipid–proton resonances between healthy human adults, adults with tumors, and neonates.[150] In cancer patients and neonates, the resonances are shifted in the same direction when compared with normal adults. It was suggested that the chemical shifts reflected different lipoprotein compositions in the groups of subjects studied.

3.9. STUDIES OF HUMAN RED BLOOD CELLS

This is another body tissue which is readily examinable by MRS using several nuclei including ^{31}P, ^1H, and ^{23}Na.

3.9.1. Results from ^{31}P Spectroscopy

The important discovery by Moon and Richards[151] of the pH_i sensitivity of the 2,3-DPG and Pi chemical shifts in blood erythrocytes has already been mentioned in Chapter 2. ATP signals in the spectra of human erythrocytes were demonstrated soon after this discovery, and it became possible to monitor changes in erythrocyte metabolism.[152] Studies of the metabolism of fresh blood revealed a gradual decrease of 2,3-DPG and a concomitant increase in Pi. 2,3-DPG–depleted cells exhibited an increase in the signal of this metabolite when incubated with inosine and pyruvate. Other detailed studies have been made of blood metabolism, including investigations of the binding of ATP and 2,3-DPG to hemoglobin,[153] intra-erythrocyte ATP–Mg^{2+} binding,[154] the rate of metabolism of various phosphate compounds in normal and abnormal erythrocytes,[155] and 2,3-DPG degradation in human adult and cord erythrocytes.[156] The investigation of ATP–Mg^{2+} binding revealed that $84 \pm 4\%$ and $78 \pm 4\%$ of the ATP is complexed to Mg^{2+} in the aerobic and anerobic states of the cell, respectively.[154] The intracellular concentrations of free Mg^{2+} in normal red blood cells were estimated to be 0.25 ± 0.07 mM in the aerobic and 0.67 ± 0.15 mM in the anerobic states. In the last study,[156] neonatal

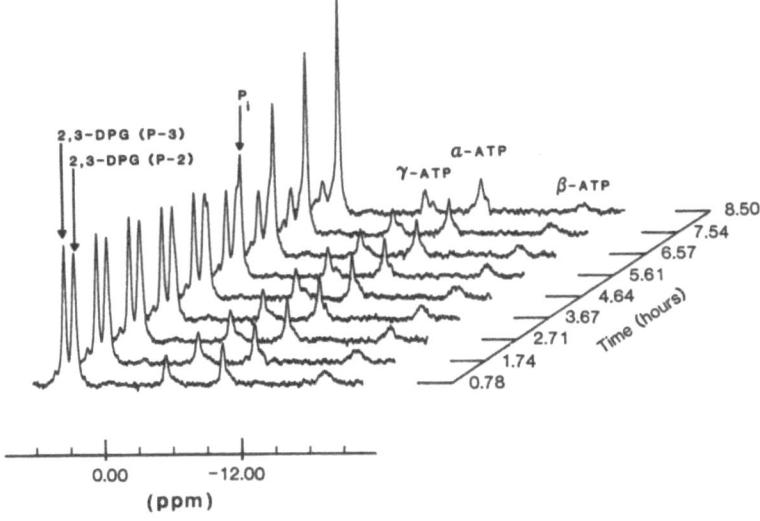

Figure 3.26. Consecutive ^{31}P spectra of cord blood erythrocytes acquired at 145.8 MHz. The data were collected with the sample at 37°C and with 95% O_2/5% CO_2 at one atmosphere. The chemical shift scale is referenced to 85% H_3PO_4 at 0 ppm. Each spectrum is the result of accumulating 2000 scans with a 1.74-sec pulse interval and a 90° pulse. Each collection took nearly 1 hr. The time axis gives the midpoint of each collection (from Zwerling et al.,[156] Academic Press Inc.; reprinted with permission).

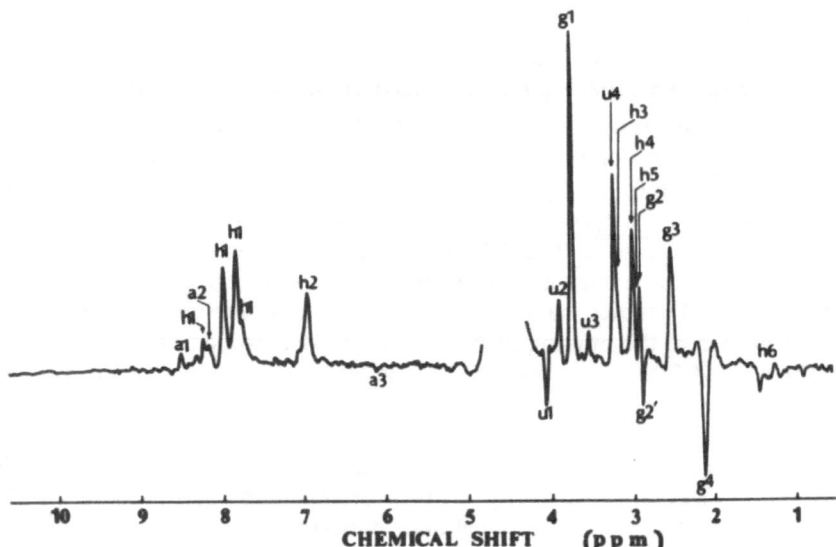

Figure 3.27. A ¹H spin–echo spectrum of glucose-depleted red blood cells at 37°C. The data were collected at 270 MHz with a spin-echo delay of 60 msec, and the sequence was repeated at 1-sec intervals. Resonance identifications: a1 and a2, C8 and C2 protons of purine, respectively; a3, C1′ proton of ribose; g1 to g4, glutathione residues; h1 to h6, hemoglobin residues; and u1 to u4, unassigned (from Brown et al.,[157] Elsevier Science Publishers; reprinted with permission).

cord and adult erythrocytes were incubated at 37°C, and the percent change in 2,3-DPG was monitored. For the first 4.0 ± 0.9 hrs (mean $\pm$ S.D.), cord erythrocytes showed no change in 2,3-DPG concentration, whereas for adult erythrocytes this phase lasted for 6.0 ± 1.0 hrs. Subsequently cord and adult erythrocytes exhibited reductions in 2,3-DPG at rates of $11.6 \pm 0.4\%$/hr and $11.4 \pm 0.5\%$/hr relative to the initial concentration, respectively. There was no significant difference between the cord and adult erythrocytes for both the plateau and decay phases. A concomitant increase in Pi was seen during the 2,3-DPG decay phase, although ATP appeared to remain unchanged as shown in Figure 3.26. The existence of two distinct phases for 2,3-DPG reduction corroborates the results of previous work on adult erythrocytes.[154]

3.9.2. ¹H Spectroscopy of Red Blood Cells

Fresh, washed red cells have also been studied using ¹H spin–echo techniques, in which broad resonances due to relatively immobile molecular species (i.e., those with short T_2s) are greatly reduced when compared with those which are highly mobile (i.e., those with long T_2s).[157] The signal from water was suppressed by a frequency-selective pulse applied at all times apart from the period of data acquisition. Over twenty resonances were detected (see Figure 3.27), and these were mostly identified as due mainly to ATP, lactate, pyruvate, glutathione, and hemoglobin residues, although some peaks were of unassigned origin. The pH_i was estimated to be 7.4 using the chemical shift of the C(2) histidine resonance from hemoglobin.

3.9.3. ^{23}Na Spectroscopy of Red Blood Cells

^{23}Na spectra obtained from fresh cells, under conditions preserving the sodium transport system, showed that the signal from intact cells is 98% of that obtainable after complete hemolysis.[158] MRS quantitation of the intracellular sodium content in intact cells gave a value equal to 92% of that determined by flame photometry of ashed samples. It was concluded that a pool of MRS-invisible sodium does not exist in human erythrocytes. The ^{23}Na spectrum of blood shows only one resonance, unless a chemical shift reagent such as dysprosium(III)tripolyphosphate[159] is used (this is unable to penetrate the cellular membrane). If such a reagent is utilized, then the signals from intracellular and extracellular sodium are resolvable.

REFERENCES

1. Cox, I. J., Bryant, D. J., Ross, B. D., et al.: Spectral resolution in clinical magnetic resonance spectroscopy, *Magn. Reson. Med.* 5:186–190, 1987.
2. Radda, G. K.: The use of NMR spectroscopy for the understanding of disease, *Science* 233:640–645, 1986.
3. Chance, B., Eleff, S., and Leigh, J. S.: Noninvasive, nondestructive approaches to cell bioenergetics, *Proc. Natl. Acad. Sci. USA* 77:7430–7434, 1980.
4. Ross, B. D., Radda, G. K., Gadian, D. G., et al.: Examination of a case of suspected McArdle's syndrome by ^{31}P nuclear magnetic resonance, *N. Engl. J. Med.* 304:1338–1342, 1981.
5. Chance, B., Eleff, S., Leigh, J. S., et al.: Mitochondrial regulation of phosphocreatine/inorganic phosphate ratios in exercising human muscle. A gated ^{31}P NMR study, *Proc. Natl. Acad. Sci. USA* 78:6714–6718, 1981.
6. Gadian, D. G., Radda, G. K., Ross, B. D., et al.: Examination of a myopathy by phosphorus nuclear magnetic resonance, *Lancet* 2:774–775, 1981.
7. Cady, E. B., Costello, A. M. de L., Dawson, M. J., et al.: Non-invasive investigation of cerebral metabolism in newborn infants by phosphorus nuclear magnetic resonance spectroscopy, *Lancet* 1:1059–1062, 1983.
8. Bottomley, P. A., Hart, H. R., Edelstein, W. A., et al.: NMR imaging/spectroscopy system to study both anatomy and metabolism, *Lancet* 2:273–274, 1983.
9. Pape, K. E. and Wigglesworth, J. S.: *Haemmorhage, Ischaemia and the Perinatal Brain*, London, Heinemann (Spastics International Medical Publications), 1979.

10. Volpe, J. J.: *Neurology of the newborn*, Philadelphia, Saunders, 1987.
11. Younkin, D. P., Delivoria–Papadopolous, M., Leonard, J., et al.: Unique aspects of human newborn cerebral metabolism evaluated with 31-P NMR spectroscopy, *Ann. Neurol.* **16**:581–586, 1984.
12. Hope, P. L., Costello, A. M. de L., Cady, E. B., et al.: Cerebral energy metabolism studied with phosphorus NMR spectroscopy in normal and birth-asphyxiated infants, *Lancet* **2**:366–370, 1984.
13. Veech, R. L., Harris, R. L., Veloso, D., et al.: Freeze-blowing: A new technique for the study of brain in vivo, *J. Neurochem.* **20**:183–188, 1973.
14. Duffy, T. E., Kohle, S. J., and Vannucci, R. C.: Carbohydrate and energy metabolism in perinatal rat brain: Relation to survival in anoxia, *J. Neurochem.* **24**:271–276, 1975.
15. Vannucci, R. C. and Vannucci, S. J.: Cerebral carbohydrate metabolism during hypoglycemia and anoxia in newborn rats, *Ann. Neurol.* **4**:73–79, 1978.
16. Chapman, A. G., Westerberg, E., and Siesjo, B. K.: The metabolism of purine and pyrimidine nucleotides in rat cortex during insulin-induced hypoglycemia and recovery, *J. Neurochem.* **36**:179–189, 1981.
17. Glonek, T., Kopp, S. J., Kot, E., et al.: P-31 nuclear magnetic resonance analysis of brain: The perchloric acid extract spectrum, *J. Neurochem.* **39**:1210–1219, 1982.
18. Gyulai, L., Bolinger, L., Leigh, J. S., et al.: Phosphorylethanolamine—the major constituent of the phosphomonoester peak observed by 31P-NMR on developing dog brain, *FEBS Lett.* **178**:137–142, 1984.
19. Brenton, D. P., Garrod, P. J., Krywawych, S., et al.: Phosphoethanolamine is major constituent of phosphomonoester peak detected by ^{31}P NMR in newborn brain, *Lancet* **1**:115, 1985.
20. Pettegrew, J. W., Kopp, S. J., Dadok, J., et al.: Chemical characterization of a prominent phosphomonoester resonance from mammalian brain. ^{31}P and ^{1}H NMR analysis at 4.7 and 14.1 tesla, *J. Mag. Res.* **67**:443–450, 1986.
21. Agrawal, H. C. and Himwich, W. A.: Amino acids, proteins and monoamines of developing brain, in Himwich, W. A. (ed.): *Developmental Neurobiology*, Springfield, Ill., C. C. Thomas, 1970, p. 298.
22. Okumura, N., Otsuki, S., and Kameyama, A.: Studies on free amino acids in human brain, *J. Biochem.* (Tokyo) **47**:315–320, 1960.
23. Chapman, A. G., Westerberg, E., and Siesjo, B. K.: The metabolism of purine and pyrimidine nucleotides in rat cortex during insulin-induced hypoglycemia and recovery, *J. Neurochem.* **36**:179–189, 1981.
24. Gonzalez–Mendez, R., Litt, L., Koretsky, A. P., et al.: Comparison of ^{31}P NMR spectra of in vivo rat brain using convolution difference and saturation with a surface coil. Source of the broad component in the brain spectrum, *J. Mag. Res.* **57**:526–533, 1984.
25. Stubbs, M., Vanstapel, F., Rodrigues, L. M., et al.: Phosphate metabolites in rat skin, *NMR in Biomedicine* **1**:50–55, 1988.
26. Pettegrew, J. W., Minshew, N. J., Diehl, J., et al.: Anatomical considerations for interpreting topical 31P-NMR, *Lancet* **2**:913, 1983.
27. Tofts, P. S., Cady, E. B., Delpy, D. T., et al.: Surface coil NMR spectroscopy of brain, *Lancet* **1**:459, 1984.
28. Evelhoch, J. L., Crowley, M. G., and Ackerman, J. J. H.: Signal-to-noise optimization and observed volume localization with surface coils, *J. Mag. Res.* **56**:110–124, 1984.
29. Bottomley, P. A., Hart, H. R., Edelstein, W. A., et al.: Anatomy and metabolism of the normal human brain studied by magnetic resonance at 1.5 tesla, *Radiology* **150**:441–446, 1984.
30. Bottomley, P. A., Foster, T. B., and Darrow, R. D.: Depth-resolved surface-coil spectroscopy (DRESS) for in vivo ^{1}H, ^{31}P, and ^{13}C NMR, *J. Mag. Res.* **59**:338–342, 1984.

31. Bottomley, P. A., Edelstein, W. A., Hart, H. R., et al.: Spatial localization in ^{31}P and ^{13}C NMR spectroscopy in vivo using surface coils, *Magn. Reson. Med.* 1:410–413, 1984.
32. Brindle, K. M., Smith, M. B., Rajagopalan, B., et al.: Spectral editing in ^{31}P NMR spectra of human brain, *J. Mag. Res.* 61:559–563, 1985.
33. Ng, T. C., Majors, A. W., and Meany, T. F.: In vivo MR spectroscopy of human subjects with a 1.4-T whole body MR imager, *Radiology* 158:517–520, 1986.
34. Oberhaensli, R. D., Galloway, G. J., Hilton–Jones, D., et al.: The study of human organs by phosphorus-31 topical magnetic resonance spectroscopy, *Br. J. Radiol.* 60:367–373, 1987.
35. Bailes, D. R., Bryant, D. J., Bydder, G. M., et al.: Localized phosphorus-31 NMR spectroscopy of normal and pathological human organs in vivo using phase encoded techniques, *J. Mag. Res.* 74:158–170, 1987.
36. Tofts, P. S. and Wray, S.: Changes in brain phosphorus metabolites during the postnatal development of the rat, *J. Physiol.* (London) 359:417–429, 1985.
37. Reynolds, E. O. R., Wyatt, J. S., Azzopardi, D., et al.: New non-invasive methods for assessing brain oxygenation and haemodynamics, *Br. Med. Bull.* 44:1052–1075, 1988.
38. Azzopardi, D., Wyatt, J. S., and Hamilton, P. A.: Phosphorus metabolites and intracellular pH in the brains of normal and small-for-gestational age infants investigated by magnetic resonance spectroscopy, *Pediatr. Res.* 25:440–444, 1989.
39. Minton, S. D., Steichen, J. J., and Tsang, R. C.: Bone mineral content in term and preterm appropriate-for-gestational age infants, *J. Pediatr.* 95:1037–1042, 1979.
40. Dobbing, J. and Sands, J.: Quantitative growth and development of human brain, *Arch. Dis. Child.* 48:757–767, 1973.
41. Lawson, B., Anday, E., Guillet, R., et al.: Brain oxidative phosphorylation following alteration in head position in preterm and term neonates, *Pedatr. Res.* 22:302–305, 1987.
42. Cady, E. B.: Absolute quantitation of phosphorus metabolites in the cerebral cortex of the newborn human infant and in the forearm muscles of young adults using a double tuned surface coil, *J. Mag. Res.* (in press), 1990.
43. Thulborn, K. R. and Ackerman, J. J. H.: Absolute molar concentrations by NMR in inhomogeneous B1. A scheme for analysis of in vivo metabolites, *J. Mag. Res.* 55:357–371, 1983.
44. Cohen, M. M. and Lin, S.: Acid soluble phosphates in the developing rabbit brain, *J. Neurochem.* 9:345–352, 1962.
45. Cerdan, S., Subramanian, V. H., Hilberman, M., et al.: ^{31}P NMR detection of mobile dog brain phospholipids, *Magn. Reson. Med.* 3:432–439, 1986.
46. Delpy, D. T., Cope, M. C., Cady, E. B., et al.: Cerebral monitoring in newborn infants by magnetic resonance and near infrared spectroscopy, *Scandinavian Journal of Clinical Laboratory Investigation* 47(Suppl. 188):9–17, 1987.
47. Hamilton, P. A., Hope, P. L., Cady, E. B., et al.: Impaired energy metabolism in brains of newborn infants with increased cerebral echodensities, *Lancet* 1:1242–1246, 1986.
48. Azzopardi, D., Wyatt, J. S., Cady, E. B., et al.: Prognosis of newborn infants with hypoxic-ischaemic brain injury assessed by phosphorus magnetic resonance spectroscopy, *Pediatr. Res.* 25:445–451, 1989.
49. Griffiths, R.: *The Abilities of Babies.* London, University of London Press, 1954.
50. Bottomley, P. A., Smith, L. S., Leue, W. M., et al.: Slice-interleaved depth-resolved surface-coil spectroscopy (SLIT DRESS) for rapid ^{31}P NMR in vivo, *J. Mag. Res.* 64:347–351, 1985.
51. Bottomley, P. A., Drayer, B. P., and Smith, L. S.: Chronic adult cerebral infarction studied by phosphorus NMR spectroscopy, *Radiology* 160:763–766, 1986.
52. Gordon, R. E., Hanley, P. E., Shaw, D., et al.: Localization of metabolites in animals using ^{31}P topical magnetic resonance, *Nature* 287:736–738, 1980.

53. Gordon, R. E., Hanley, P. E., and Shaw, D.: Topical magnetic resonance, *Progress in Nuclear Magnetic Resonance Spectroscopy* **15**:1–47, 1982.

54. Bottomley, P. A. and Hardy, C. J.: PROGRESS in efficient three-dimensional spatially localized in vivo ^{31}P NMR spectroscopy using multidimensional spatially selective (*p*) pulses, *J. Mag. Res.* **74**:550–556, 1987.

55. Ordidge, R. J., Connelly, A., and Lohman, J. A. B.: Image selected in-vivo spectroscopy (ISIS). A new technique for spatially selective NMR spectroscopy, *J. Mag. Res.* **66**:283–294, 1986.

56. Segebarth, C. M., Baleriaux, D. F., Arnold, D. L., et al.: MR image-guided P-31 MR spectroscopy in the evaluation of brain tumor treatment, *Radiology* **165**:215–219, 1987.

57. Silver, M. S., Joseph, R. I., Chen, C–N., et al.: Selective population inversion in NMR, *Nature* **310**:681–683, 1984.

58. Bottomley, P. A., Charles, H. C., Roemer, P. B., et al.: Human in vivo phosphate metabolite imaging with ^{31}P NMR, *Magn. Reson. Med.* **7**:319–336, 1988.

59. Hardy, C. J., Bottomley, P. A., Roemer, P. B., et al.: Rapid ^{31}P spectroscopy on a 4-T whole-body system, *Magn. Reson. Med.* **8**:104–109, 1988.

60. Lenkinski, R. E., Holland, G. A., Allman, T., et al.: Integrated MR imaging and spectroscopy with chemical shift imaging of P-31 at 1.5 T: Initial experience, *Radiology* **169**:201–206, 1988.

61. Tropp, J. S., Sugiura, S., Derby, K. A., et al.: Characterization of MR spectroscopic imaging of the human head and limb at 2.0 T, *Radiology* **169**:207–212, 1988.

62. Bottomley, P. A. and Hardy, C. J.: Rapid, reliable in vivo assays of human phosphate metabolites by nuclear magnetic resonance, *Clin. Chem.* **35**:392–395, 1989.

63. Roth, K., Hubesch, B., Meyerhoff, D. J., et al.: Noninvasive quantitation of phosphorus metabolites in human tissue by NMR spectroscopy, *J. Mag. Res.* **81**:299–311, 1989.

64. Kwee, I. L. and Nakada, T.: Phospholipid profile of the human brain: ^{31}P NMR spectroscopic study, *Magn. Reson. Med.* **6**:296–299, 1988.

65. Bottomley, P. A., Edelstein, W. A., Foster, T. H., et al.: In vivo solvent-suppressed localized hydrogen nuclear magnetic resonance spectroscopy: A window to metabolism? *Proc. Natl. Acad. Sci. USA* **82**:2148–2152, 1985.

66. Luyten, P. R. and den Hollander, J. A.: Observation of metabolites in the human brain by MR spectroscopy, *Radiology* **161**:795–798, 1986.

67. Hore, P. J.: Solvent suppression in Fourier transform nuclear magnetic resonance, *J. Mag. Res.* **55**:283–300, 1983.

68. Luyten, P. R., Marien, A. D. H., Sijtsma, B., et al.: Solvent-suppressed spatially resolved spectroscopy. An approach to high-resolution NMR on a whole body MR system, *J. Mag. Res.* **67**:148–155, 1986.

69. Barany, M., Langer, B. G., Glick, R. P., et al.: In vivo H-1 spectroscopy in humans at 1.5 T, *Radiology* **167**:839–844, 1988.

70. Bendall, M. R. and Gordon, R. E.: Depth and refocusing pulses designed for multipulse NMR with surface coils, *J. Mag. Res.* **53**:365–385, 1983.

71. Hanstock, C. C., Rothman, D. L., Jue, T., et al.: Volume-selected proton spectroscopy in the human brain, *J. Mag. Res.* **77**:583–588, 1988.

72. Bodenhausen, G., Freeman, R., and Turner, D. L.: Suppression of artifacts in two-dimensional J spectroscopy, *J. Mag. Res.* **27**:511–514, 1977.

73. Morris, G. A. and Freeman, R.: Selective excitation in Fourier transform nuclear magnetic resonance, *J. Mag. Res.* **29**:433–462, 1978.

74. Hanstock, C. C., Rothman, D. L., Prichard, J. W., et al.: Spatially localized ^{1}H NMR spectra of metabolites in the human brain, *Proc. Natl. Acad. Sci. USA* **85**:1821–1825, 1988.

75. Rothman, D. L., Hanstock, C. C., Ogino, T., et al.: Edited ^{1}H human brain spectra of amino acids at 2.1 T, *Proceedings*, seventh Annual Meeting, Society for Magnetic Resonance in Medicine, San Francisco, 1988, p. 254.

76. Hanstock, C., Rothman, D., Shulman, R., et al.: Ethanol observed in human brain by proton magnetic resonance spectroscopy, *Proceedings*, seventh Annual Meeting, Society of Magnetic Resonance in Medicine, San Francisco, 1988, p. 1071.

77. Frahm, J., Bruhn, H., Gyngell, M. L., et al.: Localized high-resolution proton NMR spectroscopy using stimulated echoes: Initial applications to human brain in vivo, *Magn. Reson. Med.* 9:79–93, 1989.

78. Bruhn, H., Frahm, J., Gyngell, M. L., et al.: Cerebral metabolism in man after acute stroke: New observations using localized proton NMR spectroscopy, *Magn. Reson. Med.* 9:126–131, 1989.

79. Edwards, R. H. T., Dawson, M. J., Wilkie, D. R., et al.: Clinical use of nuclear magnetic resonance in the investigation of myopathy, *Lancet* 1:725–731, 1982.

80. Wilkie, D. R., Dawson, M. J., Edwards, R. H. T., et al.: ^{31}P NMR studies of resting muscle in normal human subjects, in Pollack, G. H. and Sugi, H. (eds.): *Contractile Mechanisms in Muscle*, New York, Plenum Press, 1984, p. 333.

81. Radda, G. K., Bore, P. J., and Rajagopalan, B.: Clinical aspects of ^{31}P NMR spectroscopy, *Br. Med. Bull.* 40:155–159, 1984.

82. Edwards, R. H. T., Griffiths, R. D., and Cady, E. B.: Topical magnetic resonance for the study of muscle metabolism in human myopathy, *Clin. Physiol.* 5:93–109, 1985.

83. Kushmerick, M. J., McFarland, E. W., Conley, K. E., et al.: Characterization of fiber types in normal human muscle by ^{31}P-NMR spectroscopy, *Proceedings*, seventh Annual Meeting, Society of Magnetic Resonance in Medicine, San Francisco, 1988, p. 462.

84. Aldridge, R., Cady, E. B., Jones, D. A., et al.: Muscle pain after exercise is linked with an inorganic phosphate increase as shown by ^{31}P NMR, *Biosci. Rep.* 6:663–667, 1986.

85. Thomsen, C., Jensen, K. E., Henriksen, O., et al.: Glucose induced thermogenesis in human skeletal muscle studies by ^{31}P MR-spectroscopy, *Proceedings*, seventh Annual Meeting, Society of Magnetic Resonance in Medicine, San Francisco, 1988, p. 302.

86. Dawson, M. J.: Quantitative analysis of metabolite levels in normal human subjects by ^{31}P topical magnetic resonance, *Biosci. Rep.* 2:727–733, 1982.

87. Newman, R. J., Bore, P. J., Chan, L., et al.: Nuclear magnetic resonance studies of forearm muscle in Duchenne dystrophy, *Br. Med. J.* 284:1072–1074, 1982.

88. Griffiths, R. D., Cady, E. B., Edwards, R. H. T., et al.: Muscle energy metabolism in Duchenne dystrophy studied by ^{31}P-NMR: Controlled trials show no effect of allopurinol or ribose, *Muscle and Nerve* 8:760–767, 1985.

89. Younkin, D. P., Berman, P., Sladky, J., et al.: ^{31}P NMR studies in Duchenne muscular dystrophy: Age-related metabolic changes, *Neurology* 37:165–169, 1987.

90. Hands, L. J., Bore, P. J., Galloway, G., et al.: Muscle metabolism in patients with peripheral vascular disease investigated by ^{31}P nuclear magnetic resonance spectroscopy, *Clin. Sci.* 71:283–290, 1986.

91. Wilson, J. R., Fink, L., Maris, J., et al.: Evaluation of energy metabolism in skeletal muscle of patients with heart failure with gated phosphorus-31 nuclear magnetic resonance, *Circulation* 71:57–62, 1985.

92. Frostick, S. P., Taylor, D. J., Yonge, R. P., et al.: A study of muscle denervation using phosphorus-31 magnetic resonance spectroscopy. *Proceedings*, fifth Annual Meeting, Society of Magnetic Resonance in Medicine, Works in Progress, Montreal, 1986, p. 69.

93. Zochodne, D. W., Thompson, R. T., Driedger, A. A., et al.: Metabolic changes in human muscle denervation: Topical ^{31}P NMR spectroscopy studies, *Magn. Reson. Med.* 7:373–383, 1988.

94. Taylor, D. J., Bore, P. J., Styles, P., et al.: Bioenergetics of intact human muscle. A ^{31}P nuclear magnetic resonance study, *Molecular Biology in Medicine* 1:77–94, 1983.

95. Arnold, D. L., Matthews, P. M., and Radda, G. K.: Metabolic recovery after exercise and the assessment of mitochondrial function in vivo in human skeletal muscle by means of ^{31}P NMR, *Magn. Reson. Med.* 1:307–315, 1984.

96. Taylor, D. J., Styles, P., Matthews, P. M., et al.: Energetics of human muscle: Exercise-induced ATP depletion, *Magn. Reson. Med.* 3:44–54, 1986.

97. Taylor, D. J., Crowe, M., Bore, P. J., et al.: Examination of the energetics of aging skeletal muscle using nuclear magnetic resonance, *Gerontology* 30:2–7, 1984.

98. Rees, D., Smith, M. B., Harley, J., et al.: In vivo functioning of creatine phosphokinase in human forearm muscle, studied by ^{31}P NMR saturation transfer, *Magn. Reson. Med.* 9:39–52, 1989.

99. Dawson, M. J., Gadian, D. G., and Wilkie, D. R.: Muscular fatigue investigated by phosphorus nuclear magnetic resonance, *Nature* 274:861–866, 1978.

100. Miller, R. G., Giannini, D., Milner–Brown, H. S., et al.: Effects of fatiguing exercise on high-energy phosphates, force, and EMG: Evidence for 3 phases of recovery, *Muscle Nerve* 10:810–821, 1987.

101. Miller, R. G., Boska, M. D., Moussavi, R. S., et al.: ^{31}P nuclear magnetic resonance studies of high energy phosphates and pH in human muscle fatigue, *J. Clin. Invest.* 81:1190–1196, 1988.

102. Cady, E. B., Jones, D. A., Lynn, J., et al.: Changes in force and intracellular metabolites during fatigue of human skeletal muscle, *J. Physiol.* (London) 418:311–325, 1989.

103. Wilson, J. R., McCully, K. K., Mancini, D. M., et al.: Relationship of muscular fatigue to pH and diprotonated Pi in humans: A ^{31}P-NMR study, *Journal of Applied Physiology* 64:2333–2339, 1988.

104. Wilkie, D. R.: Muscular fatigue: Effects of hydrogen ions and inorganic phosphate, *Federal Proceedings* 45:2921–2923, 1986.

105. Nosek, T. M., Fender, K. Y., and Godt, R. E.: It is diprotonated inorganic phosphate that depresses force in skinned skeletal muscle fibers, *Science* 236:191–193, 1987.

106. Cady, E. B., Elshove, H., Moll, A., et al.: The metabolic causes of slow relaxation in fatigued human skeletal muscle, *J. Physiol.* (London) 418:327–337, 1989.

107. Chance, B., Eleff, S., Bank, W., et al.: ^{31}P NMR studies of control of mitochondrial function in phosphofructokinase-deficient human skeletal muscle, *Proc. Natl. Acad. Sci. USA* 79:7714–7718, 1982.

108. Arnold, D. L., Bore, P. J., Radda, G. K., et al.: Excessive intracellular acidosis of skeletal muscle on exercise in a patient with a post-viral exhaustion/fatigue syndrome, *Lancet* 1:1367–1369, 1984.

109. Wilson, J. R., Fink, L., Maris, J., et al.: Evaluation of energy metabolism in skeletal muscle of patients with heart failure with gated phosphorus-31 nuclear magnetic resonance, *Circulation* 71:57–62, 1985.

110. Hands, L. J., Bore, P. J., Galloway, G., et al.: Muscle metabolism in patients with peripheral vascular disease investigated by ^{31}P nuclear magnetic resonance spectroscopy, *Clin. Sci.* 71:283–290, 1986.

111. Frahm, J., Merboldt, K., and Hanicke, W.: Localized proton spectroscopy using stimulated echoes, *J. Mag. Res.* 72:502–508, 1987.

112. Narayana, P. A., Jackson, E. F., Hazle, J. D., et al.: In vivo localized proton spectroscopic studies of human gastrocnemius muscle, *Magn. Reson. Med.* 8:151–159, 1988.

113. Pan, J. W., Hamm, J. R., Hetherington, H. P., et al.: Quantitation of lactate in human muscle by ^{1}H NMR, *Proceedings*, seventh Annual Meeting, Society of Magnetic Resonance in Medicine, San Francisco, 1988, p. 106.

114. Pan, J. W., Hamm, J. R., Rothman, D. L., et al.: Intracellular pH of human muscle by ^{1}H NMR, *Proceedings*, seventh Annual Meeting, Society of Magnetic Resonance in Medicine, San Francisco, 1988, p. 251.

115. Barany, M., Doyle, D. D., Graff, G., et al.: Natural abundance ^{13}C NMR spectra of human muscle, normal and diseased, *Magn. Reson. Med.* 1:30–43, 1984.

116. Avison, M. J., Rothman, D. L., Nadel, E., et al.: Detection of human muscle glycogen by natural abundance ^{13}C NMR, *Proc. Natl. Acad. Sci. USA* 85:1634–1636, 1988.

117. Pillai, R. P., Buescher, P. C., Pearse, D. B., et al.: ^{31}P NMR spectroscopy of isolated perfused lungs, *Magn. Reson. Med.* 3:467–472, 1986.

118. Bottomley, P. A.: Noninvasive study of high-energy phosphate metabolism in human heart by depth-resolved ^{31}P NMR spectroscopy, *Science* 229:769–772, 1985.

119. Bottomley, P. A., Herfkens, R. J., Smith, L. S., et al.: Altered phosphate metabolism, in myocardial infarction: P-31 MR spectroscopy, *Radiology* 165:703–707, 1987.

120. Blackledge, M. J., Rajagopalan, B., Oberhaensli, R. D., et al.: Quantitative studies of human cardiac metabolism by ^{31}P rotating-frame NMR, *Proc. Natl. Acad. Sci. USA* 84:4283–4287, 1987.

121. Radda, G. K.: The use of NMR spectroscopy for the understanding of disease, *Science* 233:640–645, 1986.

122. Styles, P., Scott, C. A., and Radda, G. K.: A method for localizing high-resolution NMR spectra from human subjects, *Magn. Reson. Med.* 2:402–409, 1985.

123. Matson, G. B., Twieg, D. B., Karczmar, G. S., et al.: Application of image-guided surface coil P-31 spectroscopy to human liver, heart and kidney, *Radiology* 169:541–547, 1988.

124. Jue, T., Rothman, D. L., Lohman, J. A. B., et al.: Surface coil localization of ^{31}P NMR signals from orthotopic human kidney and liver, *Proc. Natl. Acad. Sci. USA* 85:971–974, 1988.

125. Maris, J. M., Evans, A. E., McLaughlin, A. C., et al.: ^{31}P nuclear magnetic resonance spectroscopic investigation of human neuroblastoma in situ, *N. Engl. J. Med.* 312:1500–1505, 1985.

126. Oberhaensli, R. D., Galloway, G. J., Taylor, D. J., et al.: Assessment of human liver metabolism by phosphorus-31 magnetic resonance spectroscopy, *Br. J. Radiol.* 59:695–699, 1986.

127. Oberhaensli, R. D., Hilton–Jones, D., Bore, P., et al.: Biochemical investigation of human tumours in vivo with phosphorus-31 magnetic resonance spectroscopy, *Lancet* 2:8–11, 1986.

128. Oberhaensli, R. D., Rajagopalan, B., Taylor, D. J., et al.: Study of hereditary fructose intolerance by use of ^{31}P magnetic resonance spectroscopy, *Lancet* 2:931–934, 1987.

129. Sauter, R., Mueller, S., and Weber, H.: Localization in in vivo ^{31}P NMR spectroscopy by combining surface coils and slice-selective saturation, *J. Mag. Res.* 75:167–173, 1987.

130. Segebarth, C., Luyten, P. R., and den Hollander, A.: Improved depth-selective single surface-coil ^{31}P NMR spectroscopy using a combination of B_1 and B_0 selection techniques, *J. Mag. Res.* 75:345–351, 1987.

131. Cox, I. J., Bryant, D. J., Collins, A. G., et al.: Four-dimensional chemical shift MR imaging of phosphorus metabolites of normal and diseased human liver, *J. Comput. Assist. Tomogr.* 12:369–376, 1988.

132. Cox, I. J., Sargentoni, J., Calam, J., et al.: Four-dimensional phosphorus-31 chemical

shift imaging of carcinoid metastases in the liver, *NMR in Biomedicine* 1:56–60, 1988.

133. Jue, T., Lohman, J. A. B., Ordidge, R. J., et al.: Natural abundance ^{13}C NMR spectrum of glycogen in humans, *Magn. Reson. Med.* 5:377–379, 1987.

134. Ross, B., Marshall, V., Smith, M., et al.: Monitoring response to chemotherapy of intact human tumours by ^{31}P nuclear magnetic resonance, *Lancet* 1:641–646, 1984.

135. Chew, W. M., Hricak, H., and McClure, R. D.: In vivo human testicular function assessed by ^{31}P MRS, *Proceedings*, seventh Annual Meeting, Society of Magnetic Resonance in Medicine, San Francisco, 1988, p. 322.

136. Achten, E., Van Cauteren, M., Wisanto, A., et al.: Phosphorus-31 magnetic resonance spectroscopy of the human testicle: Normal values and a case of varicocele, *Proceedings*, seventh Annual Meeting, Society of Magnetic Resonance in Medicine, San Francisco, 1988, p. 58 (works in progress).

137. Iles, R. A., Hind, A. J., and Chalmers, R. A.: Use of proton nuclear magnetic resonance spectroscopy in detection and study of organic acidurias, *Clin. Chem.* 31:1795–1801, 1985.

138. Iles, R. A., Chalmers, R. A., and Hind, A. J.: Methylmalonic aciduria and propionic acidemia studied by proton nuclear magnetic resonance spectroscopy, *Clin. Chim. Acta* 173:173–189, 1988.

139. Yamaguchi, S., Koda, N., Eto, Y., et al.: Rapid screening of metabolic disease by proton NMR urinalysis, *Lancet* 2:284, 1984.

140. Bales, J. R., Higham, D. P., Howe, I., et al.: Use of high-resolution proton nuclear magnetic resonance spectroscopy for rapid multi-component analysis of urine, *Clin. Chem.* 30:426–432, 1984.

141. Bales, J. R., Sadler, P. J., Nicholson, J. K., et al.: Urinary excretion of acetaminophen and its metabolites as studied by proton NMR spectroscopy, *Clin. Chem.* 30:1631–1636, 1984.

142. Bales, J. R., Nicholson, J. K., and Sadler, P. J.: Two-dimensional proton nuclear magnetic resonance "maps" of acetaminophen metabolites in human urine, *Clin. Chem.* 31:757–762, 1985.

143. Bales, J. R., Bell, J. D., Nicholson, J. K., et al.: Metabolic profiling of body fluids by proton NMR: Self poisoning episodes with paracetamol (acetaminophen), *Magn. Reson. Med.* 6:300–306, 1988.

144. Gillies, R. J., Powell, D. A., Nelson, T. R., et al.: High resolution proton NMR spectroscopy of human amniotic fluid, *Proceedings*, fourth Annual Meeting, Society of Magnetic Resonance in Medicine, London, 1985, p. 789.

145. Bell, J. D., Sadler, P. J., Macleod, A. F., et al.: ^{1}H NMR studies of human blood plasma, *FEBS Lett.* 219:239–243, 1987.

146. Bell, J. D., Brown, J. C. C., Kubal, G., et al.: NMR-invisible lactate in blood plasma, *FEBS Lett.* 235:81–86, 1988.

147. Fossel, E. T., Carr, J. M., and McDonagh, J.: Detection of malignant tumors: Water-suppressed proton nuclear magnetic resonance spectroscopy of plasma, *N. Engl. J. Med.* 315:1369–1376, 1986.

148. Bell, J. D., Brown, J. C. C., Norman, R. E., et al.: Factors affecting ^{1}H NMR spectra of blood plasma: Cancer, diet and freezing, *NMR in Biomedicine* 1:90–94, 1988.

149. Holmes, K. T., Mackinnon, W. B., May, G. L., et al.: Hyperlipidemia as a biochemical basis of magnetic resonance plasma test for cancer, *NMR in Biomedicine* 1:44–49, 1989.

150. Eskelinen, S., Hiltunen, Y., Jokisarri, J., et al.: ^{1}H NMR studies on human plasma lipids from newborn infants, healthy adults, and adults with tumors, *Magn. Reson. Med.* 9:35–38, 1989.

151. Moon, R. B. and Richards, J. H.: Determination of intracellular pH by ^{31}P magnetic resonance, *J. Biol. Chem.* **248**:7276–7278, 1973.
152. Henderson, T. O., Costello, A. J. R., and Omachi, A.: Phosphate metabolism in intact human erythrocytes: Determination by phosphorus[31] nuclear magnetic resonance spectroscopy, *Proc. Natl. Acad. Sci. USA* **71**:2487–2490, 1974.
153. Costello, A. J., Marshall, W. E., Omachi, A., et al.: Interactions between hemoglobin and organic phosphates investigated with ^{31}P nuclear magnetic resonance spectroscopy and ultrafiltration, *Biochim. Biophys. Acta* **427**:481–491, 1976.
154. Gupta, R. K., Benovic, J. L., and Rose, Z. B.: The determination of the free-magnesium level in the human red blood cell by ^{31}P NMR, *J. Biol. Chem.* **253**:6172–6176, 1978.
155. Tehrani, A. Y., Lam, Y. F., Lin, A. K., et al.: Phosphorus-31 nuclear magnetic resonance studies of human red blood cells, *Blood Cells* **8**:245–261, 1982.
156. Zwerling, H. K., Diamond, J. N., Levy, G. C., et al.: Phosphorus-31 nuclear magnetic resonance measurements of 2,3-DPG degradation in human adult and cord blood erythrocytes, *Magn. Reson. Med.* **3**:10–14, 1986.
157. Brown, F. F., Campbell, I. D., Kuchel, P. W., et al.: Human erythrocyte metabolism studies by ^{1}H spin echo NMR, *FEBS Lett.* **82**:12–16, 1977.
158. Yeh, J. C., Brindley, F. J., and Becker, E. D.: Nuclear magnetic resonance studies on intra-cellular sodium in human erythrocytes and frog muscle, *Biophys. J.* **13**:56–71, 1973.
159. Gupta, R. K. and Gupta, P.: Direct observation of resolved resonances from intra- and extracellular sodium-23 ions in NMR studies of intact cells and tissues using dysprosium(III)tripolyphosphate as paramagnetic shift reagent, *J. Mag. Res.* **47**:344–350, 1982.

Practical Aspects of Clinical Magnetic Resonance Spectroscopy Systems

4.1. AN OVERVIEW OF THE SYSTEM

The clinical magnetic resonance system, as used for both spectroscopy and imaging, splits up into three fundamental units: the magnet, including shim coils, used to optimize magnetic field homogeneity, and gradient coils); the RF "front end," comprising the coil unit and preamplifier; and the RF "back end," which consists of the RF transmitter(s), the RF receiver, the console, and the associated computer for acquisition, analysis, and storage of data. A block diagram of a possible MRS system is shown in Figure 4.1. The preamplifier is always placed as near to the sample coil as possible for the following reasons. First, the signal must be amplified before significant amounts of noise are added in from the cabling. Second, isolation of the preamplifier from the transmitter is essential. This requires the use of quarter-wavelength coil-to-preamplifier cable sections. For clinical MRS systems, these cable lengths are about 1 m or less.

In order to reduce interference, either the "front end" or, in some installations, the entire system can be placed in a Faraday cage, which consists of a highly conductive, grounded screen. The cage prevents the system from picking up unwanted RF emissions from external sources. If the "front end" only is screened, then all cables, apart from those carrying magnetic resonance signals, must pass through a bank of filters in order

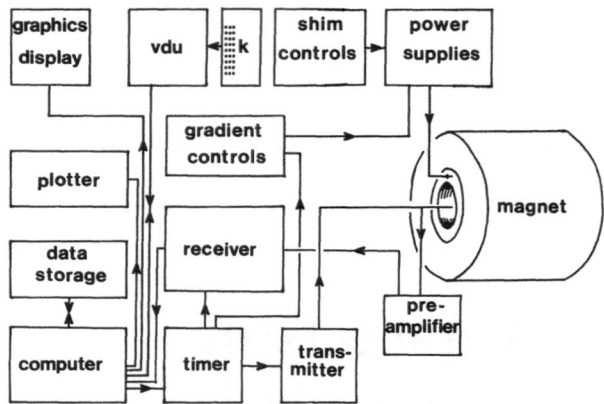

Figure 4.1. A schematic diagram showing the basic building blocks of a clinical magnetic resonance system: k, operator keyboard; vdu, visual display unit. The power supplies include amplifiers driven by pulses of variable magnitude and duration. These produce pulsed magnetic field gradients used in imaging and localized spectroscopy. The computer section includes a digitizer, which converts the receiver's output voltage to numerical form.

that interference not be transmitted into the Faraday cage. Furthermore, in order to minimize interference pickup and the unwanted radiation of signals into the immediate environment, the connecting cabling itself must be screened coaxial cable. This consists of an inner conductor, carrying the signal, which is parallel with and centrally positioned in a grounded cylindrical outer conducting sheath that acts as a shield to outgoing and incoming radiation. Some high-quality cables have two outer shields in order to increase the screening efficiency. The space between the outer sheath and the signal conductor is occupied by an insulator. Such cabling possesses inductance and capacitance per unit length and hence also has a characteristic impedance when oscillating signals such as RF are concerned. The cable's impedance depends on the conductor's dimensions and the dielectric constant of the insulation material between the signal conductor and the outer sheath. If the length of a section of screened cabling is nearly a wavelength, then the RF power transmitted along it can be thought of as a wave motion. For the complete transmission of signal at equipment-cabling interfaces, the input to the equipment must have an impedance that matches that of the cabling; otherwise, some of the power is reflected back along the cable. Many spectrometers are designed specifically to use cabling with a characteristic impedance of 50 Ω, and only cable and connectors with this value should be used.

The overall layout of an installation has to be such that both humans and equipment can live in harmony with each other in the abnormal

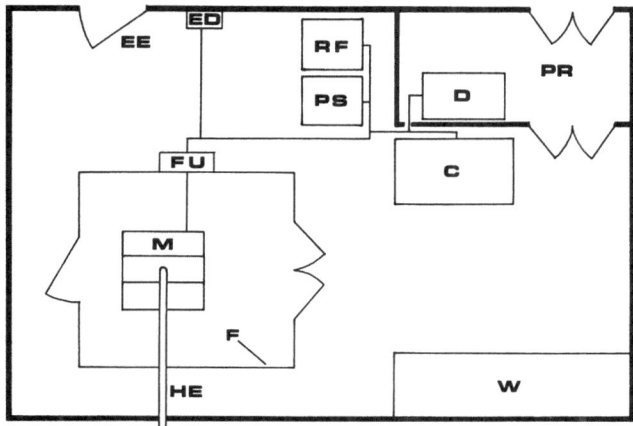

Figure 4.2. A possible floor plan for a clinical magnetic resonance installation. C is the operator console; D is a data storage area; ED is an emergency discharge unit; EE is an emergency exit; F is a Faraday screen for interference reduction; FU is a filter unit to prevent RF interference from entering the screened area via cabling; HE is a tube to exhaust helium boil-off both under normal operation and in the event of a quench; M is the magnet; PR is a patient-reception area and staff entrance, which may incorporate safety devices to screen for personal magnetic objects; PS contains the shim and gradient power supply; RF is the radio-frequency electronics; W is the user workspace.

environment of a strong magnetic field, which may extend for some considerable distance from the magnet. The two items of equipment most sensitive to excess magnetic fields are video display units and magnetic data-storage devices such as discs and tapes. The former need to be positioned some distance away or magnetically screened in order to avoid image distortion. Recorded data is, of course, extremely valuable, and great care must be taken that it is stored and used well away from the magnet, preferably in another room. Paths which can bring equipment containing significant magnetic components (e.g., transformers, gasbottles) close to the magnet should be avoided as completely as possible. (A substantial "front end" Faraday cage can play the dual role of acting as a barrier to magnetic items and as a screen against unwanted RF.) Facilities such as emergency magnet-discharge systems and exits should be accessible from as many directions as possible in order that no delay be encountered when they need to be used. A possible floor plan is shown in Figure 4.2.

4.2. THE MAGNET

The collection of high-quality spectra depends crucially on the quality of the magnet. Inhomogeneity of the field can cause many problems, including broadened and distorted spectral lines. Field-strength instability can introduce drift of the resonant frequency. For spectroscopy, as opposed to imaging, it is usually necessary to use field strengths of more than 1.5 tesla in order to obtain useful resolution and SNR. Current official safety recommendations usually limit the maximum field strength for clinical use to 2.5 tesla in the United Kingdom[1] and 2.0 tesla in the United States.[2] Because of the relatively short relaxation times for many nuclei encountered in body tissues, leading to broader peaks, it is not necessary to have such stringent field-homogeneity requirements as are needed for high-resolution analytical spectrometry. For in vivo applications, a homogeneity of 1 part in 10^7 is quite adequate to produce quantifiable spectra. Due to the dimensions of human organs and lesions, it is essential for this homogeneity to be maintained over a region up to 10 cm or more in diameter. It has been possible to design magnets which meet these criteria of stability, strength, and homogeneity throughout a large enough volume. So far three main types have been used in clinical MR imaging and spectroscopy: permanent, resistive, and liquid-helium superconductor.[3, 4] At the moment, it is uncertain what impact the recently discovered ceramic higher-temperature superconductors will have.

Careful consideration needs to be applied to the siting of magnet systems, including consideration of floor loading, distance to nearby ferromagnetic structural items and, particularly in the case of horizontal-bore superconducting magnets, the stray magnetic field. The strength of the field should be reduced substantially (currently to less than 5×10^{-4} tesla) before any areas with unrestricted public access are reached. If space is restricted, then magnetic shields can be utilized. These consist of substantial sheets erected around or above the magnet as ceiling plates or barn- or box-type structures, as dictated by the shielding requirements. Some magnets are now constructed with self-shielding built in.

4.2.1. Permanent Magnets

The permanent magnets used in magnetic resonance are like greatly enlarged, stronger, and carefully designed versions of the traditional horseshoe magnets used in simple school physics experiments. They are not really useful in spectroscopy because of the difficulty encountered in producing a homogeneous and strong field over a large enough volume. However, permanent-magnet systems for imaging (which has less-stringent requirements) are available. These magnets are claimed to be cost-effective for imaging purposes when compared with other techniques for generating suitable fields.

The main advantages of permanent magnets are

- Running costs are lower, because no electric power or coolants are required.

- The field axis can be vertical. This greatly reduces fringe fields, thereby minimizing the required floor space and reducing risks from extraneous magnetizable objects (tools, etc.).
- It is possible to design systems with all-round access and visibility, allowing for greater patient comfort.

In addition to the problem of low homogeneous field strength, permanent magnets have the following disadvantages:

- The poles of the magnet are very massive and may require substantial floor strengthening.
- Although there is a reduced fringe field, if a life-threatening accident does occur it may be difficult to reduce the magnetic field rapidly.

4.2.2. Resistive Electromagnets

A resistive electromagnet fundamentally consists of a coil of a suitably selected conductor. However, because of the high power consumption needed to generate a strong field, an efficient cooling system is also required. Conventional windings of insulated copper wire cannot be used due to the poor thermal dissipation across the insulation. A method commonly used for the construction of resistive magnets is shown in Figure 4.3a, in which several sets of coils are combined to produce a high degree

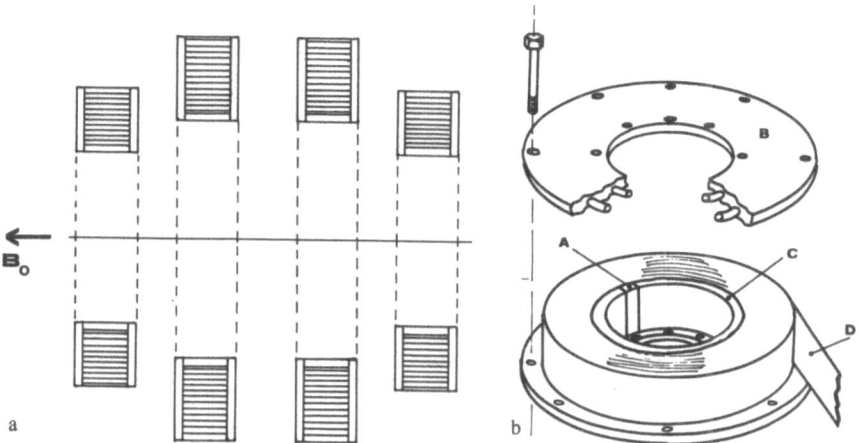

Figure 4.3. (a) A cross-sectional diagram of a commonly used combination of four resistive coils for the production of an adequately homogenous magnetic field. (b) The construction of a single resistive coil (as used in the construction shown in (a), with particular attention to the dissipation of thermal energy by use of water cooling. A is the electrical terminal; B, the water-cooled cheek; C, the coil former; D, the anodized aluminum ribbon [(b) is from Hanley,[3] Churchill Livingstone (Longman Group Ltd); reprinted with permission].

of homogeneity. In order to solve the problem of power dissipation, each individual coil consists of a disc of wound anodized-aluminum ribbon sandwiched between two water-cooled plates, as indicated in Figure 4.3b. The aluminum ribbon carries the current that generates the magnetic field. Electrical insulation between turns is provided by the anodization. This method of construction allows for good thermal contact between the aluminum and the water-cooled plates and for good thermal conduction across the coil.

Due to the high power requirements for magnetization and cooling, these magnets are limited, for reasons of economy, to the lower range of field strengths (about 0.25 tesla), although they have found application in MRI. They can be designed with either a vertical or a horizontal field axis. Such magnets have a distinct advantage when compared with permanent systems in that they can be switched off when not in use and in emergencies.

4.2.3. The Liquid-Helium Superconducting Magnet

Because a liquid-helium superconducting magnet is capable of generating a strong field that is homogeneous over a large volume, such magnets have dominated the application of MRS to living systems. They rely on the ability of certain materials to lose their electrical resistance when cooled close to absolute zero ($-273°C$). The phenomenon of superconductivity was first discovered in 1911 by Kammerlingh Onnes in Leiden, when he found that mercury lost its resistivity at about $-269°C$. The first known superconductors were all pure metals, and it was possible to achieve magnetic fields of only about 0.1 tesla before the superconductivity was "quenched." In the 1950s, a second generation of superconducting materials was discovered, consisting of alloys and intermetallic compounds. These materials do not "quench" until fields of the order of 10 to 16 tesla are reached.

Most superconducting components for magnets are made of niobium–titanium alloy ($Nb_{44\% \text{ to } 50\%}Ti$), although niobium–zirconium ($NbZr_{25\% \text{ to } 50\%}$) and the intermetallic compounds of niobium and tin (Nb_3Sn) and vanadium and gallium (V_3Ga) have been used. Nearly all the magnets that have been employed for biomedical magnetic resonance have used a composite wire manufactured by drawing niobium–titanium alloy strands about 10 μm in diameter embedded in a copper matrix. This material is stable and strong enough to withstand the great electromagnetic stresses that occur in large high-field magnets. A diagram of the interior construction of a superconducting magnet is shown in Figure 4.4.

The critical temperature, at which superconductivity disappears, is

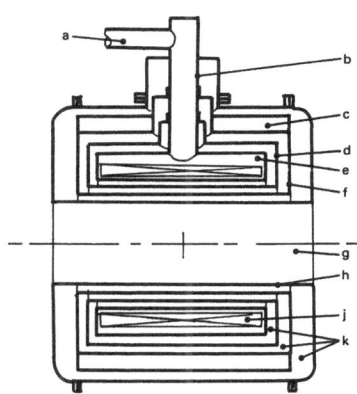

Figure 4.4. A cross-sectional diagram of the interior construction of a helium-cooled superconducting magnet system. Constituent parts of the cryostat are as follows: a, helium gas vent; b, neck tube; c, liquid nitrogen; d, gas-cooled shield; e, liquid helium; f, liquid-nitrogen–cooled shield; g, room-temperature bore; h, outer vacuum case; j, magnet coils; k, vacuum [from Hanley,[3] Churchill Livingstone (Longman Group Ltd); reprinted with permission].

only a few degrees above absolute zero, and liquid helium is required to maintain the superconducting material at a low enough temperature. Liquid helium is relatively expensive and, in order to have an economical superconducting system, it is essential to minimize the rate at which this scarce commodity evaporates. For this reason, superconducting magnets are built as cryostats (i.e., similar to Dewar flasks). Conservation of liquid helium is achieved by minimizing the conductive, convective, and radiative inflow of thermal energy. With this aim in view, the liquid-helium tank, which contains the superconducting coil, is situated in a very high vacuum and connected to the outside world only by support materials of low heat conduction such as glass–fiber composites. Once the magnet is energized, the system is also electrically isolated from the exterior. This is possible due to the persistent nature of the flowing current in a superconductor and is fortunate, because otherwise the connecting leads would be a major source of conductive thermal input. Most of the heat conducted down the neck tube (which is required for the venting of boiled-off gas and for refilling) is expended in warming up the emerging cold helium gas. Convection is possible in the neck tube. However, if it is correctly designed, thermal transfer in this manner is negligible and involves only a little mixing between convection cells of helium gas which has already boiled off. Very little thermal energy is transferred to the liquid. One of the largest sources of heat input comes from thermal radiation, and this can be minimized in various ways. The heat transfer between two surfaces is proportional to their emissivity and is also related to the difference in temperature. Interior metal surfaces are highly polished and kept as clean as possible in order to produce high reflectivity and hence low emissivity, thus reducing radiative transfer. Multiple layers of reflective metallic foil are used as an additional "superinsulation" screen around the cryostat's interior. The biggest reduction of radiative thermal input is achieved by surrounding (but not touching) the helium can with another cylindrical tank containing liquid nitrogen. This acts as a cooled radiation shield; in addition, the liquid helium is within a local environment which is much colder than room temperature. The use of liquid nitrogen has a considerable effect on reducing boil-off and, although an additional expense to be taken into account when considering running costs, more than pays for itself in reducing helium consumption.

4.2.3.1. Routine Maintenance of a Superconducting Magnet

Most routine maintenance can be carried out quite easily by local technical staff. The two main tasks are monitoring the cryogen levels and refilling with liquid helium or liquid nitrogen whenever needed.

4.2.3.1a. Monitoring Cryogen Levels. It is vital that the magnet have effective content metering for liquid helium and liquid nitrogen, so that refill requirements can be anticipated and problems can be detected at the earliest possible moment. The risk of the helium level getting low and a quench taking place, which may cause damage to the superconducting solenoid, should be avoided at all costs. Most systems are fitted with resistive sensors, which measure the helium and nitrogen levels according to the change in resistance which occurs as more of the probe is exposed to the warmer gas above the liquid. In order to limit the boil-off which results from the thermal input due to the small current required to measure the cryogen level, the helium sensor utilizes brief pulses at intervals on the order of a minute. The chances of problems occurring are greatly reduced if monitoring is carried out on a daily basis and an accurate record of cryogen levels is kept. It is also useful to plot the day-to-day helium levels so as to allow rapid detection of increased boil-off rates. The installation of a helium gas-flow meter in the vent system is extremely useful and allows very sensitive boil-off monitoring.

4.2.3.1b. Cryogen Refills. Liquid nitrogen usually needs to be replenished on a weekly basis. It is important to ensure that this is done, as otherwise the effectiveness of the liquid-nitrogen–cooled radiative shield will be reduced and the helium boil-off rate will increase. Liquid nitrogen fills are easy to perform and consist merely of attaching the storage Dewar to the appropriate neck tube on the magnet via a strong hose. Many Dewars are self-pressurizing but, as an alternative, the transfer can be driven by compressed air. Great care must be taken to ensure that there are no spills of liquid nitrogen onto the magnet's surface, as such spills can cause the freezing of various seals which may lead to leaks and "softening" of the cryostat's vacuum.

Refilling with liquid helium requires more care and is usually done at intervals of one to two months. Filling with liquid helium should not be left to the last minute but should be carried out while there is still a substantial quantity left in the magnet. There is a risk that, if the level gets too low, the magnet may quench during the fill. Transfer from the storage vessel to the magnet is through a special thermally insulated tube which consists of two concentric, flexible pipes with a high vacuum between them. The transfer tube is carefully introduced into the neck of the helium Dewar until the tip almost, but not quite, touches the bottom of the container. If the tube enters too quickly, excessive boil-off can result. The Dewar is pressurized by warm helium boil-off gas introduced into the storage vessel by some means (e.g., from a balloon attachment which functions as a bellows). Air or other gases must not be allowed to enter the Dewar or the cryostat; otherwise, they may freeze solid and cause blockages of either the liquid or the gas vents. On pressurizing the storage vessel, helium gas will begin to emerge from the other end of the transfer tube. This will shortly begin to get cold, and water vapor will condense from the air, producing a cloud around the nozzle. Eventually a narrow, bluish cone of extremely cold helium gas will appear; this indicates that helium liquid is beginning to transfer. At this stage the transfer tube is inserted into the refill neck tube of the magnet and remains there until the fill is complete. The transfer tube should never be inserted into the magnet before the cone of cold helium gas has been seen; otherwise, warm gas will enter the can containing the superconducting solenoid and a quench may result. Under no circumstances should anyone's unprotected skin be allowed to come into contact with anything that has recently been immersed in liquid helium or liquid nitrogen, as a very painful cryogenic burn may result.

Some routine tasks involving storage Dewars may be necessary from time to time. Vents blocked by frozen air or frozen condensed moisture may need to be cleared. A suitable rod should be available for this purpose (it should be T-shaped so as to be incapable of dropping into the storage Dewar). The helium remaining in the Dewar after the fill may also require checking. It is possible to purchase resistive level meters, but these are expensive. However, a very cheap and effective alternative exists. This consists of a length of narrow stainless-steel tube at one end of which is fitted a small cone with its flared orifice covered by a latex diaphragm. The other end of the tube is slowly introduced into the storage vessel until it touches the bottom of the container. A reference position is marked on the tube. The tube is now slowly withdrawn with the tip of the index finger touching the latex diaphragm. A slow oscillation of the diaphragm is felt due to helium boiling off. When the tip of the tube reaches the surface of the helium, this oscillation becomes noticeably faster, and this position is also marked on the tube. The difference between these two positions gives the depth of liquid helium in the vessel.

4.2.3.2. Problems with Superconducting Magnets

One of the main problems encountered when running a superconducting system is excess helium boil-off. This may be caused by allowing the liquid-nitrogen level to get too low, or barometric pressure effects may show up if a sensitive helium-flow meter is used. However, there are two more serious causes to consider.

The interior construction of a superconducting magnet is not completely stable. Every time a liquid nitrogen or liquid helium fill is carried out, various internal components expand and contract due to temperature changes, and the relative positions of parts can alter. This can lead to the possibility of a "touch" between a surface at liquid helium temperature and one at a much higher temperature. This touch will produce a thermal bridge across which conduction can take place, causing elevated helium boil-off. The only solution if the situation becomes intolerable (i.e., the magnet cannot be reliably maintained without risk of a quench), is to de-energize the magnet and then allow the liquid helium to boil off. After the cryostat has warmed up, it is then dismantled in order to establish the exact nature of the problem. However, there is no guarantee that the "touch" will be found because it may only exist when the magnet is in its cold state. Due to the expense incurred because of the wastage of liquid helium, this attempt at solving the problem should be approached as a last resort since the probability of finding the cause may not be high.

Another possible reason for high helium boil-off may be "softening" of the vacuum. This could be due to helium leaking into the vacuum space via seals (O-rings) or through cracks in the metal fabric (particularly in welded joints). Due to its small atomic size, helium is the most likely contaminator' of the vacuum space. Because cracks may open up only during cryogen fills or at particular cryogen levels, the boil-off could increase discretely from one fill to another. Any helium entering the vacuum space will initially "cryo-pump" onto the cold-helium tank, preserving the high vacuum. Eventually, though, the vacuum loses its high thermal insulation properties and the helium boil-off increases. The presence of helium in the vacuum space can be tested for by attaching a leak tester to the cryostat. (This item of equipment is basically a mass spectrometer designed to detect helium in small quantities.) The exterior seals can also be tested by blowing helium gas over them and using the leak tester to detect any increase in the vacuum space's helium content. If helium is found at a sufficient concentration in the vacuum space, then the possibility of a "touch" being the cause of the elevated boil-off is reduced. Yet dismantling the system in order to find the leak can be expensive, and there is no guarantee that the cause will be removed. It may be decided that the optimum solution is to periodically pump out the vacuum space, as this can dramatically reduce the rate of helium loss for a period of several months.

Another sort of problem can be brought about by allowing the influx of air into either the helium-fill neck tube or the magnet-energizing neck tube. In the very cold environment within the cryostat, air and moisture can freeze into a solid plug and block up electrical sockets at the bottom of the energizing neck tube. This can cause difficulties with the insertion of energizing leads and hence with running down the magnetic field. An obstruction in the energizing neck tube can be removed by using a jet of warm helium gas which will vaporize the plug. Other gases should not be used, as they will freeze and make the blockage worse. It is preferable that these problems should be dealt with by experienced technicians and not by on-site personnel.

4.2.4. Shim, Gradient, and Profiling Coils

In addition to the main, permanent resistive or superconducting magnet, it is essential to have other built-in resistive coils, which generate smaller, operator-controlled magnetic fields. Some of these are used for shimming (i.e., making the magnetic field as uniform as possible within the sample so that all of the nuclei of a particular type resonate at the same frequency and hence produce well-resolved spectral peaks). For some applications it may also be necessary to have gradient coils. These produce magnetic fields that vary linearly across the sample and are of use in imaging and in localizing spectra to a particular volume of tissue. In addition, some earlier systems may have profiling coils,[5, 6] which produce a homogeneous field only in a localized volume the diameter of which is under operator control. Well-resolved, narrow spectral peaks are obtained from this homogeneous region. Outside this volume, the field changes rapidly with position, and the nuclei resonate over a broad range of frequencies. The spectrum from this outer region of nonuniform field consists of a broad, unresolved hump, which can be removed by computer processing to leave only signal which has originated from the central region of the homogeneous volume. This and other spectral-localization methods will be dealt with in more detail in Chapter 5.

4.3. THE SPECTROMETER

The spectrometer consists of the RF and digital sections of the system. This comprises the RF probe, the preamplifier(s), the receiver and transmitter(s), and a suitable computer to perform the functions of equipment control and acquisition, analysis, and storage of data. Many modern spectrometers require a substantial number of equipment racks for this hardware. A significant thermal output can be generated, and provision for air-conditioning may be necessary.

4.3.1. The RF Probe

The RF probe, apart from a layer of suitable electrical insulation such as poly-tetrafluroethane (commonly known as PTFE or Teflon), will be quite close to the subject. It consists of at least one RF coil, which is connected to variable capacitors allowing for fine tuning of the resonant frequency and matching the coil's impedance to that of the receiver. (Often, one RF coil is used to perform the dual functions of pulse transmission and signal reception. Under some circumstances it may be beneficial to use a much larger, separate transmitter coil in order to obtain a more uniform RF field in the sensitive volume of a smaller receiver coil.) In use, the probe will be located within the bore of the magnet and has therefore to be designed so that there is easy operator access to the tuning and matching controls. It must, of course, be made of completely nonmagnetic materials, and apart from the sample coil the circuitry will usually be housed in an RF-screened box. Suitable screening can easily be made from high-purity copper foil soldered together and fitted in a ready built plastic instrument housing. The screen should be grounded to the magnet by a fairly substantial electrical connection (e.g., copper braid). To reduce signal losses, conductor lengths should be minimized and components should be soldered directly to one another using the grounded screening as a substantial "ground plane." Stray capacitances should be avoided, and care should be taken to prevent "arcing" in circuitry handling transmitter pulses.

4.3.2. The RF Coils

The tuned RF coils used in magnetic resonance studies consist basically of a combination of an inductor and a capacitor. An inductor is a conducting structure which induces a magnetic field when current flows through it. For the purposes of magnetic resonance, the current varies at RF, and so will the induced magnetic field. If the inductor is placed in an already existing RF field, then an alternating current of the same frequency will be induced in it producing a voltage at the coil terminals. This voltage can be detected by the spectrometer's receiver. For many applications the inductor consists of one or more circular loops of highly conductive material, such as annealed copper, arranged so that the coil can be placed flat on the skin of the subject. These are called surface coils. As a general rule, the direction of the magnetic field produced by the RF transmitter's pulse must be at right angles to the main static field in order to obtain maximum signal. For the surface coils just mentioned, this implies that the plane of the coil should be parallel to the bore of the superconducting magnet.

Several other methods of construction have been developed, and these find use in particular applications. "Solenoid" coils (see Figure 4.5a) have been used extensively for test-tube studies and for some in vivo work. For use in superconducting magnets, they have the disadvantage that the coil's axis has to be perpendicular to the magnetic axis; this makes the handling

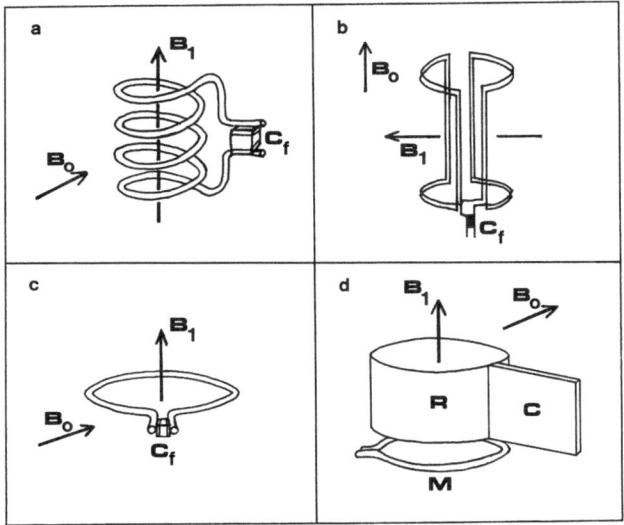

Figure 4.5. A selection of coils used for various applications in magnetic resonance spectroscopy. B_0 is the main magnetic field produced by the superconducting, resistive, or permanent magnet; B_1 is the RF magnetic field generated by the transmitter pulse; C_f is a fixed tuning capacitor, usually of the ceramic chip variety. (a) A solenoid coil, primarily for test-tube work and, due to the colinearity of B_0 with the bore axis in superconducting systems, more useful in permanent or resistive magnets. A nearly homogenous B_1 field is produced inside the coil, and hence the nuclear magnetic moment is flipped uniformly throughout the enclosed sample. (b) A two-turn saddle coil, also mainly used with test-tube studies but designed for easy application in superconducting systems. This coil also produces a nearly homogeneous B_1 field. (c) A simple, single-turn, planar surface coil, ideal for obtaining high signal from tissues at shallow depths. B_1 is high close to the coil but decreases rapidly with distance, as also does the flip angle. (d) A loop-gap resonator, impedance matched using a secondary coil (M). This feature provides electrical isolation from the spectrometer. The resonator (R) is a conducting loop with a tuning capacitance consisting of a narrow gap made of two plates (C) separated by a layer of insulating material. Signal from the resonator is picked up by the additional movable matching coil (M) connected to the preamplifier.

of living subjects difficult. Sample "spinning" (a technique commonly used in "test-tube" studies to improve the spectral resolution) can also be a problem if the coil's axis is at right angles to the magnet's bore. To avoid these difficulties, the "saddle" coil has been introduced (see Figure 4.5b). This has an axis of symmetry which allows it and the contained sample tube to be positioned parallel to the magnetic field, while the arrangement of conducting loops generates an RF field at right angles. These two coil designs contain the sample completely and hence both have potentially high filling factors (the fraction of volume enclosed by the coil that is filled with sample). They also generate uniform RF fields in the enclosed volume, and therefore the nuclear spins all experience the same flip. In this respect, both solenoid and saddle coils are ideal for application with spin–echo techniques and, as described in Chapter 1, have been adapted many times in order to perform perfusion studies on various tissues. Coils of these types and their uses have been well described in the literature.[7, 8]

The surface coil[9] is a design which has been applied very often in clinical MRS studies. This is a flattened solenoid coil with one or more turns of conductor, which are essentially coplanar, as shown in Figure 4.5c. The surface coil has the main advantage that it can be placed on the skin adjacent to a particular muscle or organ of interest. Strong signals will be picked up from nuclei contained in a hemisphere of radius approximately equal to that of the coil. Surface coils thus have some ability to localize the origin of the signal.

The loop–gap resonator,[10] a variant of the surface coil, is electrically isolated from the spectrometer. The coil-tuning capacitance is provided by a specially designed gap in the coil's conductor. Signal is picked up from the resonator by another coil situated close by (see Figure 4.5d). The whole construction is fundamentally that of an air-cored RF transformer.

The main characteristics of a particular coil are the frequency at which it resonates and the efficiency (quality) factor, which is commonly called Q. All the above coils have to be tuned to the resonant frequency of the nucleus to be studied. They also have to be matched to the impedance of the preamplifier input so that maximum power is transferred from the coil to the receiver. It is also important that the coil be designed to possess a high Q. The higher the Q, the larger the signal detected and the more efficient the coil is at producing a given flip angle (i.e., the pulses required are shorter or of smaller amplitude). Tuning is usually performed via a combination of a fixed and a variable capacitor, while impedance matching is often carried out by use of a variable capacitor only. (A capacitor is, of course, a device for storing charge. It essentially consists of a series of parallel, or concentric and cylindrical, conducting plates separated by a thin insulating layer. This layer allows a large voltage to develop between the plates and, hence, a large charge to accumulate.) If a suitable capacitor is connected in parallel with an inductor (e.g., a RF coil; see Figure 4.6), then a resonance is created at a certain frequency. The coil then has the ability to pick up signals of this frequency relatively efficiently. The resonant frequency has to satisfy the following condition:

$$4\pi^2\nu^2 LC = 1 \tag{4.1}$$

where L and C are the coil's inductance and the required tuning capacitance, respectively, and ν is the resonant frequency. Matching the coil to the receiver's impedance is commonly carried out by means of an additional variable capacitance fitted in series with the coil's inductance, as shown in Figure 4.6. Loop–gap resonators can be matched by adjusting the position of the matching coil (see Figure 4.5d).

The construction of high-Q coils depends on several factors. The Q of a resonant circuit[11] is given by

$$Q = 2\pi\nu L/R, \tag{4.2}$$

where R is the RF resistance of the inductive element (in this case, the coil's conductor itself). The Q is also equivalent to the resonant frequency of the coil divided by the bandwidth capable of being transmitted or received by the coil; the bandwidth is measured between the frequencies above and below the resonant frequency where the signal's amplitude has dropped to

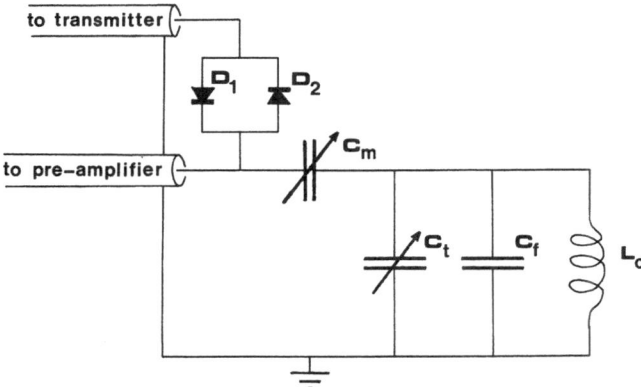

Figure 4.6. A simple tuning and matching circuit for a coil suitable for biomedical magnetic resonance studies. L_c is the coil's inductance; C_f is a fixed tuning capacitor mounted as close to the coil as possible; C_t is a small variable capacitor for fine tuning; C_m is a variable matching capacitor; D_1 and D_2 are crossed diodes, which isolate the preamplifier from residual transmitter voltages while data is acquired but switch on during the transmitter pulses.

70% of the peak value.[7] Obviously, maximizing L while minimizing R will be beneficial in producing a higher Q. However, increased L is often accompanied by raised coil resistance due to the extra length of conductor involved. One way to reduce R, without significantly changing L, is to utilize the most highly conducting materials available, such as pure copper or silver. Minimizing the number of solder joints and the lengths of conductor between components such as variable capacitors is also important.

The definition of Q given by equation (4.2) might lead one to believe that a coil will be more efficient when tuned for a higher frequency. However, the factors affecting the SNR are more complicated when analyzed in detail. The SNR produced by a single 90° pulse is often formulated as described in equation (2.22)[12, 13]

$$\text{SNR} = K \frac{\eta Q^{1/2} \omega_0^{1/2} V_c^{1/2} (\mu_0/4\pi)^{1/2}}{(F 4 k_B T_c \, \Delta v)^{1/2}} M_0 \qquad (4.3)$$

where K is a factor (close to unity) depending on the coil's geometry, η is the filling factor (the fraction of the coil volume filled by sample), Q is the coil's quality factor, ω_0 is the frequency ($2\pi v_0$), V_c is the coil's volume, μ_0 is the permeability of free space, F is the spectrometer's "noise figure," k_B is Boltzmann's constant, T_c is the coil's temperature, Δv is the receiver bandwidth, and M_0 is the nuclear magnetization (see Chapter 2). By inspection of equation (4.3), it is readily seen that the critical factors involving the coil are η, Q, and V_c, and that

$$\text{SNR} \propto \eta (Q V_c)^{1/2} \qquad (4.4)$$

It is important to note that Q and V_c are related because the resistance of a coil depends on the length of conductor used in its construction and therefore on the coil's dimensions. Hence, increasing V_c and introducing a larger value of R in equation (4.2) will probably, depending on the design, also alter Q. Therefore the effects on SNR of changes in V_c are not immediately obvious and require more rigorous analysis which may necessitate detailed consideration of the coil's structure. (For further details see Section 2.4.1.)

The surface-coil and loop–gap resonator (Figures 4.5c and d) have been used in many clinical applications. These will be dealt with in greater detail in the following sections. Versions of solenoid and saddle coils (Figures 4.5a and b) have occasionally been used in situations requiring a homogeneous RF field over a large coil volume. In many studies, the necessary coil volume is so large that the coil design has to be of particularly low inductance in order that the structure may resonate at the required frequency. One such design, the slotted-tube resonator (STR),[14, 15] has been used widely for both imaging and localized spectroscopy. For the latter role, this type of coil is capable of being used either by itself in the transmit–receive mode or, in conjunction with a receive-only surface coil, it can be used to transmit only. This permits the delivery of a uniform spatial distribution of flip angle to the sample. The STR will also be described in more detail in Section 4.3.2.4.

4.3.2.1. Simple Surface Coils

The simplest surface-coil design is that shown in Figure 4.4c; however, it is often convenient for tuning purposes to design the coil so that it has more than one turn. This increases the inductance and hence minimizes the required tuning capacitance, in accordance with equation (4.1). Obviously, one of the most important factors in a surface-coil design is the diameter, and this will have to be decided on with respect to the particular application intended. The inductance, and hence from equation (4.1) the required tuning capacitance for a particular frequency, depends strongly on the coil's radius. The coil's inductance L is given by the following formula:[16]

$$L = an^2 \{ \mu[\ln(8a/b) - 2] + (\mu'/4) \} \qquad (4.5)$$

where μ is the permeability of the medium surrounding the coil (for free space, μ is called μ_0 and is equal to $4\pi \times 10^{-7} \, \mathrm{H \, m^{-1}}$, where H is the henry, a unit of inductance), μ' is the permeability of the material from which the coil is made, a is the coil's radius, n is the number of turns, and b is the radius of the coil's wire. In practice, the term involving μ' can often be ignored because the coil's radius is usually much greater than that of the wire.

Having decided on the coil's radius and knowing the frequency at which the coil has to resonate, it is now possible to estimate the required tuning capacitance. This should be such that stray capacitances and inductances do not become important. An appropriate range of tuning-capacitance values for clinical MRS is 25 to 500 pF. In order to obtain a suitable tuning-capacitance value, it may be necessary to raise the number of turns and so increase the inductance given by equation (4.5). The tuning capacitors should be highly efficient in order to obtain a high-Q coil. They are usually of the microwave ceramic "chip" variety. These are mounted as close to the coil's loop as possible (as shown in Figure 4.5c) so as to restrict resonance to the turns of the coil. In addition to this fixed capacitance, it is necessary to incorporate a high-quality variable capacitor mounted in parallel as close to the coil as is practicable. This usually has a range from 0 to 20 pF and covers this range in four or more turns in order to allow for fine tuning. Because the variable tuning capacitance is in parallel with the fixed capacitance it adds on to give the total tuning capacitance. It may be necessary to try several different fixed tuning capacitances before the coil resonates at the required frequency. A useful first approximation can be calculated using equations (4.1) and (4.5). In order to match the coil's impedance to that of the spectrometer and hence obtain maximum power transfer, a second variable capacitor is often used in series with the coil (see Figure 4.6). The setting of the matching capacitance depends critically on the loading of the coil brought about by the conductive nature of the tissue being studied. As ionic content increases, so does the loading and also the required value for the matching capacitance. The result is the introduction of losses in the coil circuit, which reduce the Q and make the probe less efficient. Because of changes in loading, both the tuning and the matching variable capacitors will need to be finely adjusted from patient to patient in order to optimize the signal. This is usually carried out by use of a variable RF signal generator and an RF bridge, which indicates when the coil is tuned and matched correctly. Equipment of this sort is also required during construction of the coil in order to help select the best value for the fixed tuning capacitance. The resonant frequency of a coil for a particular tuning capacitor can be found using the RF bridge. The resonance can then be adjusted by altering the capacitance according to equation (4.1). This is best carried out with the coil loaded with a sample of 150 mM NaCl to simulate in vivo ionic conditions. To preserve tuning flexibility, resonance should be achieved with the variable tuning capacitance set as close to its midrange value as possible.

Since their first use,[9] a lot of effort has been directed towards characterizing the spatial sensitivity and improving the response of surface coils. If one considers a single-turn, flat circular surface coil of radius a in the zy plane of a laboratory coordinate frame (x, y, z), as shown in Figure 4.7, then the radial and axial components B_r and B_x of the magnetic field at any point $P(x, y, z)$ generated by the current i are given by:[9, 16, 17]

$$B_r = \frac{\mu_0 i}{2\pi} \frac{x}{\rho[(a+\rho)^2 + x^2]^{1/2}} \left[-K(k^2) + \frac{a^2 + \rho^2 + x^2}{(a-\rho)^2 + x^2} E(k^2) \right] \quad (4.6)$$

$$B_x = \frac{\mu_0 i}{2\pi} \frac{1}{[(a+\rho)^2 + x^2]^{1/2}} \left[K(k^2) + \frac{a^2 - \rho^2 - x^2}{(a-\rho)^2 + x^2} E(k^2) \right] \quad (4.7)$$

In the above equations, $\rho = (y^2 + z^2)^{1/2}$ (i.e., the off-axis distance of P), and

$$k^2 = \frac{4a(y^2 + z^2)^{1/2}}{[a + (y^2 + z^2)^{1/2}]^2 + x^2}$$

$E(k^2)$ and $K(k^2)$ are complete elliptical integrals of the first and second

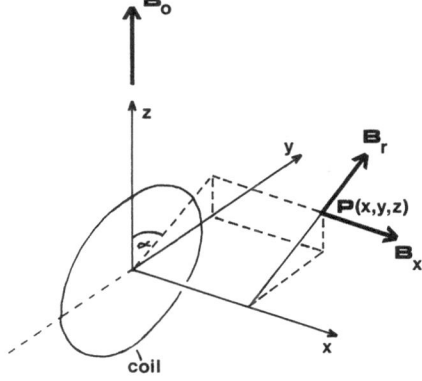

Figure 4.7. The radial and axial components B_r and B_x at the point $P(x, y, z)$ of the magnetic field generated by unit current flowing in a single-turn surface coil using a laboratory coordinate system (x, y, z). B_0 is the main magnetic field and α is the angle between B_r and B_0.

kind.[18] It should be noted that, if only points on the coil axis are considered, then $y = 0$, $z = 0$, and hence ρ and $k^2 = 0$. The elliptical integrals then become $E(0)$ and $K(0)$, which are both equal to $\pi/2$; hence

$$B_x = \frac{\mu_0 i}{2} \frac{a^2}{(a^2 + x^2)^{3/2}} \tag{4.8}$$

This equation is very useful for obtaining a rough estimate of the coil response, albeit only along the axis of symmetry of the coil (the x-axis). Figure 4.8a shows how B_x changes along the x-axis. The important

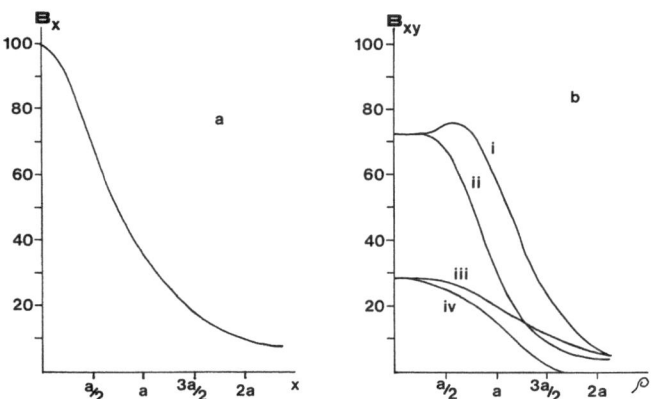

Figure 4.8. (a) The dependence of B_x on position along the coil's axis, measured in units of the coil's radius a, normalized so that $B_x = 100$ at the coil's center. (b) The dependence of B_{xy} on radial position ρ (in units of the coil's radius a). B_{xy} is normalized to 100 at the coil's center ($x = 0$, $\rho = 0$). Curve (i), $x = 0.5a$, $\theta = 90°$; (ii), $x = 0.5a$, $\theta = 0°$; (iii), $x = 1.1a$, $\theta = 90°$; (iv), $x = 1.1a$, $\theta = 0°$. (Data taken from Ackerman et al.[9])

magnetic field generated by the coil is B_{xy}, which is the vector sum of B_x and the y component of B_r (see Figure 4.7). It is B_{xy} which is responsible for the flip of the nuclear magnetic moment (initially along the z-axis) and hence for the detectable signal. Figure 4.8b shows the variation of B_{xy} with radial coordinate. Examination of Figures 4.8a and b reveals that B_{xy} is approximately localized to a hemisphere of radius equal to that of the coil. This indicates the degree of volume selectivity obtainable with a surface coil. Equation (4.8) can also be applied to obtain the theoretical pulse lengths required for particular flip angles at given positions along the x-axis within the sample volume. By considering the transformation of the transmitter-pulse voltage to that on the coil brought about by the match capacitance[19] and the relationship among flip angle, pulse length, and B_x given by[20]

$$\theta = \gamma B_x t/2 \tag{4.9}$$

(where θ is the flip angle, γ is the gyromagnetic ratio, and t is the pulse length), it is possible to show that for a coil of n turns

$$t = \frac{4\theta Z C_m a}{C V_t \gamma \mu_0 n} \frac{(a^2 + x^2)^{3/2}}{a^3}, \tag{4.10}$$

where Z is the cable impedance, C_m is the required match capacitance, C is the total capacitance $(C = C_m + C_t + C_f$; see Figure 4.6), and V_t is the peak-to-peak transmitter voltage. The important deductions from equation (4.10) are

- The pulse length (or amplitude) is proportional to the flip angle required.
- The pulse length necessary to produce a particular flip angle is proportional to C_m and hence is related to the sample loading. In order to maintain a given flip angle at, for instance, the coil center, the pulse length must be increased or reduced as dictated by C_m.

Because all of the constants in equation (4.10) are known or measurable, it is possible to calculate the theoretical pulse length required to produce a particular flip angle at the coil's center $(x = 0)$, provided that C_m has been measured with the sample in position. A further series of simple measurements can be made to check this pulse length. For instance, if one wished to determine the pulse length required to give a 90° flip at the center of a ^{31}P surface coil, the following technique could be used. A sample of concentrated phosphoric acid with dimensions small relative to the coil is placed at the coil center. If possible this sample should be surrounded by a bath containing 150 mM NaCl in order to create coil-loading conditions similar to those for living tissue. An initial FID is collected, using a pulse length shorter than that predicted by theory to give a 90° flip angle, and then Fourier-transformed to give a spectrum. This is phase-corrected to give a flat baseline, and the spectrum is used as a starting point for subsequent collections. A series of spectra are acquired, using increasingly longer pulses, until a point is reached at which the amplitude of the spectrum is minimized and longer pulses produce peak inversion. (During these collections the original phase correction is maintained.) The pulse length required to give the minimum signal generates a 180° flip in the sample (i.e., it completely inverts the magnetic vector, and hence there is no component in the xy plane capable of inducing a signal in the coil). From equation (4.9) it can be seen that the duration of the 90° pulse will be half of the measured length of the 180° pulse.

In order to simplify the formulation, equation (4.10) has been developed to apply only on the coil's axis; however, surface coils generate three-dimensional fields and hence also have three-dimensional sensitivity responses. Using equations (4.6) and (4.7), it is possible to derive the spatial variation of B_{xy} and therefrom to calculate the distributions of either the flip angle[17] or, more interestingly, the signal amplitude[20] to be expected under different pulsing conditions. The spatial distributions of flip angle to be expected using pulse lengths which give coil-center flip angles of 90° and

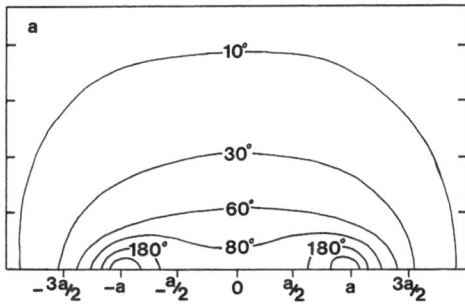

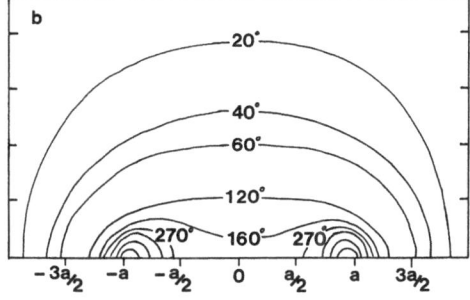

Figure 4.9. Contour plots of the flip angle θ within the xy plane of a surface coil of radius a. The axes indicate distance from the coil's center in units of the coil's radius. (a) For a flip angle of 90° at the coil's center. (b) For a flip angle of 180° at the coil's center. Note how the contours become closely spaced near the coil's conductor (at $\pm a$), in the vicinity of which extremely large flip angles can be produced amounting to many complete rotations of the magnetic vector.

$180°$ are shown in Figure 4.9. It is immediately apparent that the generated flip angle decreases as one gets further away from the coil but also increases rapidly as the coil's conductor is approached. In the close vicinity of the coil's conductor, very large flip angles can be produced. For the case of a $180°$ flip angle at the coil's center, the region in which a $90°$ flip is produced is further away from the coil and, in clinical use, would be deeper within the tissue. Figure 4.10 shows the relation between the signal-amplitude distribution and the interval between pulses of such duration as to give a flip angle of $90°$ at the coil's center. It can be seen that, as the ratio of pulse interval T_R to spin–lattice relaxation constant T_1 becomes larger (i.e., there is more nuclear relaxation between pulses), the signal tends to be acquired from regions more adjacent to the coil. If T_R becomes short when compared with T_1, then signal from close to the coil is reduced, due to saturation, and a more even spatial response is obtained.

So far we have dealt with surface coils tuned for application at one particular frequency, hence obtaining optimum SNR when used for studies involving only one type of nucleus, such as ^{31}P. For many reasons, it is useful to have coils which perform well for more than one nucleus. When performing in vivo studies it is important that one should be able to detect

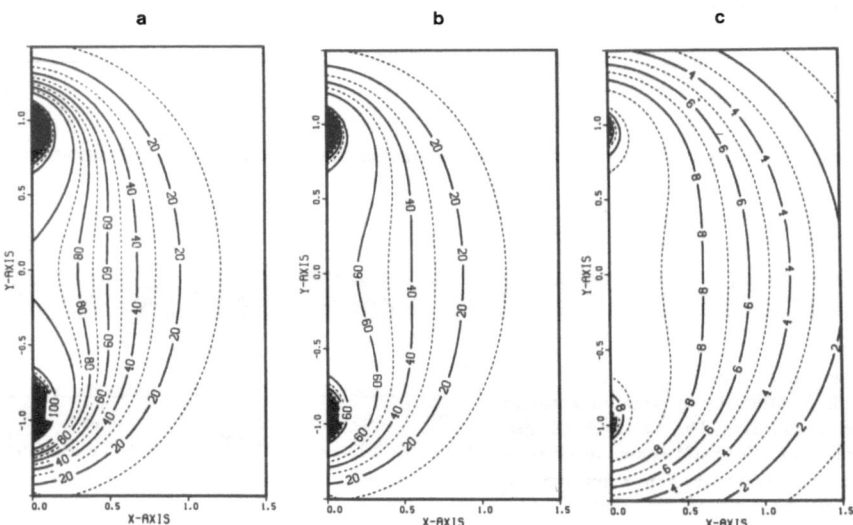

Figure 4.10. Calculated signal-amplitude maps in the xy plane, at $z = 0$, for one pulse under steady-state conditions for different pulse-repetition times T_R. (a) Contour plot for slow pulse repetition ($T_R/T_1 = 5$). (b) Contour plot for $T_R/T_1 = 1$. (c) Contour plot for rapid pulsing ($T_R/T_1 = 0.05$). All signal amplitudes are relative to a total signal amplitude of 100 for one $90°$ pulse at the coil's center (from Evelhoch et al.,[20] Academic Press, Inc.; reprinted with permission).

a reasonably large signal from water protons. This is so that one can improve the homogeneity of the magnetic field using the FID from protons in tissue water which are located in the same volume as the metabolites of interest. Because the concentration of water protons in most body tissues is high and the proton is relatively sensitive, one can detect a reasonably strong signal by simply connecting the coil directly to a ^{1}H-tuned preamplifier and transmitter.[21] Due to the low efficiency of a ^{31}P-tuned coil at the ^{1}H frequency, a very long pulse may be required in order to obtain a significant flip at the coil's center. This pulse length perturbs nuclei over only a narrow range of frequencies. However, to obtain a large enough signal for purposes of shimming, it may be necessary only to flip the nuclear magnetic moment through a small angle, and hence a short pulse can be used. The performance can be further improved simply by adding additional variable tune and match capacitors and a selected length of coaxial cable,[22] as shown in Figure 4.11. This adaptation produces ^{1}H tuning and match which, although suitable for observing tissue-water and strong fat resonances, is not efficient enough for other applications.

4.3.2.2. Improvements in Surface-Coil Design

Many attempts have been made to improve the performance and usefulness of the basic surface coil, and some of these will now be dealt with. The RF technology involved is rapidly advancing, and it is to be anticipated that the future will reveal many useful developments.

According to equation (4.2) it might be expected that the performance of a coil should improve as its resistance drops. Because of this, advantages may be gained from the use of the most highly conductive materials

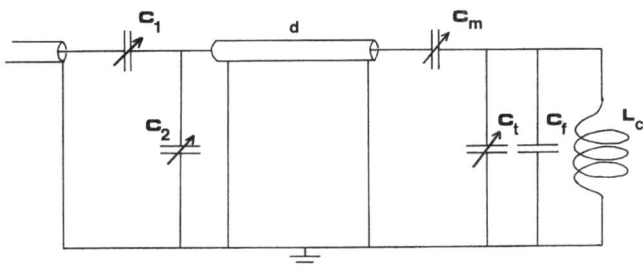

Figure 4.11. A coil primarily tuned for a particular nucleus (^{31}P, for instance) with an add-on circuit which makes it efficient enough at the ^{1}H frequency to be of use for shimming and other applications that do not require optimum sensitivity. L_c is the coil's inductance; C_f is a fixed tuning capacitor mounted close to the coil; C_t is a small variable capacitor for fine tuning; C_m is a small variable matching capacitor; d is a selected length of coaxial cable (approximately a quarter wavelength at the ^{1}H frequency); C_1 and C_2 are add-on variable capacitors which permit matching and tuning of the ^{31}P coil for improved ^{1}H performance.

available for coil construction. Although silver has a higher conductivity than copper at room temperature, the difference is very small and its use may not merit consideration. Cooling of the coil and the preamplifier with liquid helium has been shown to improve the sensitivity of in vitro studies by up to a factor of four.[23] However, the thermal insulation required between the coil and the sample reduces the available coil volume and also makes clinical application difficult. As mentioned in Chapter 2, the root-mean-square RF noise voltage V_n for a coil at temperature T_c is given by

$$V_n = (4k_B T_c R \, \Delta v)^{1/2} \qquad (4.11)$$

where k_B is the Boltzmann's constant, Δv is the receiver's bandwidth, and R is the resistance of the coil. There are two reasons for the improvement in sensitivity obtained when the coil is cooled. Not only is V_n smaller, due to the temperature dependence indicated in equation (4.11), but R is also considerably reduced at low temperature for certain conducting materials, such as copper.

The loading effects introduced when a conductive sample such as human tissue is placed close to various coil designs have been investigated using surface coils constructed from wire and foil[24] (see Figure 4.12). Copper foil is a perfectly good material for coil manufacture providing that it is thicker than the RF penetration depth ("skin depth") described in Chapter 2. If a coil is made from foil, the material can be employed in two basic configurations: flat in the plane of the coil and with the foil perpendicular to the coil plane, as shown in Figures 4.12b and c, respectively. Measurements of the Q values for copper coils of similar dimensions made

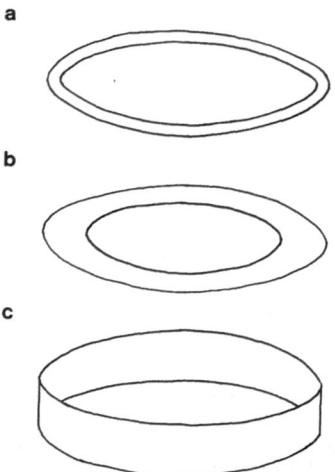

a

b

c

Figure 4.12. Surface coils constructed from (a) Wire with circular cross-section; (b) Foil arranged to be flat in the plane of the coil; (c) Foil similar to a thin slice from a cylinder.

according to these foil designs and from conventional wire indicated that the loading effect and hence the reduction in Q is related to the area of coil material close to the sample. The coil shown in Figure 4.12c had a loaded Q which was about double that for a wire coil of equivalent diameter, whereas the flat foil coil (Figure 4.12b) gave a large reduction in Q.

4.3.2.2a. Balanced Surface Coils. Another method for minimizing the effects of coil loading is to use a balanced matching circuit.[25] This can incorporate two matching capacitors on either side of the coil, as shown in Figure 4.13a. The balanced circuit is symmetrical with respect to ground, and hence the voltages developed on the coil are smaller. This reduces the currents that flow through the sample and hence should also decrease losses. For certain in vivo applications the circuit shown in Figure 4.13a can be modified using a length of transmission line (a twisted pair of insulated wires), as shown in Figure 4.13b. The fixed capacitances close to the coil provide for partial tuning and matching, while the variable capacitances at the other end of the transmission line are required for fine tuning and balanced matching. The SNR obtainable with the balanced configuration was more than double that achieved with a similar, conventionally tuned and matched coil.

An alternative approach to the production of a balanced system is to use inductive coupling[26, 27] as shown in Figure 4.14a. In this method, the tuned coil is isolated from the matching circuit and can be considered to be the primary winding of an air-cored RF transformer. The other winding

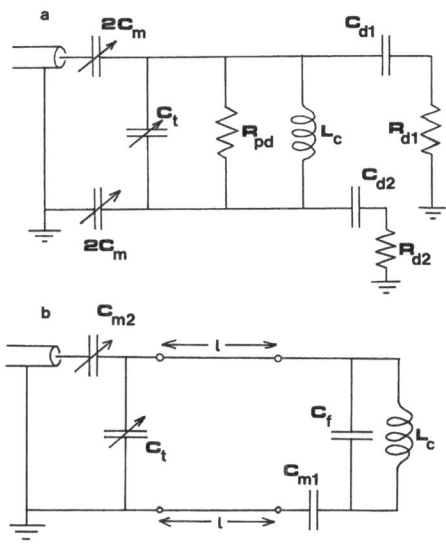

Figure 4.13. Balanced tuning and matching circuits. (a) The theoretical circuit showing the coil-to-ground losses C_{d1}, C_{d2}, R_{d1}, and R_{d2}. L_c is the coil's inductance; C_t is the tuning capacitance; the pair of $2C_m$ capacitances provide the balanced matching; and R_{pd} includes the coil resistance and coil-to-coil dielectric loss within the sample. (b) A practical, balanced tuning and matching circuit for in vivo applications. C_f is the fixed tuning capacitance mounted close to the coil; C_{m1} provides partial matching; l is a length of transmission line (a pair of twisted wires); C_t is for fine tuning; and C_{m2} is for fine matching adjustment.

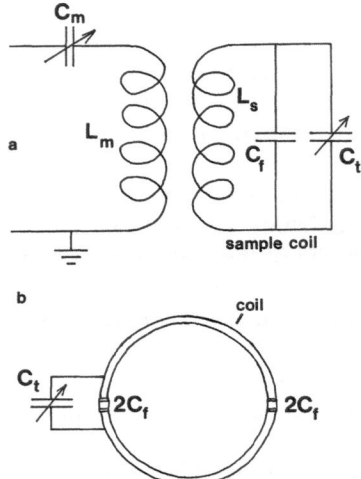

Figure 4.14. Designs for balanced inductive coils. (a) The basic circuit for a balanced, inductively matched coil. L_s is the sample (primary) coil, tuned by capacitances C_f and C_t. L_m is the matching (secondary) coil with capacitance C_m for small matching adjustments. (b) A series-tuned, inductively balanced, sample (primary) coil. This example consists of one loop of wire with two symmetrically located tuning capacitors $2C_f$ and a variable capacitor C_t for fine tuning. The use of tuning capacitors in series lowers the voltage on the coil and theoretically reduces losses. A matching (secondary) coil is also required.

consists of a secondary coil situated close by, and a signal is induced in each by the other due to their mutual inductance. (RF magnetic flux generated by a current in one coil induces a current in the other one.) Because of the electrical isolation, the tuned probe is balanced with respect to ground and hence also has reduced sensitivity loss when loaded. Tuning is accomplished by fixed and variable capacitors in parallel with the primary inductance, while matching is achieved by altering the position of the secondary with respect to the primary coil. Some matching adjustment is also possible by use of a variable matching capacitance in series with the secondary. Additional improvements in performance are claimed[26] if an extra fixed tuning capacitance is incorporated in series, as shown in Figure 4.14b. This reduces dielectric losses by further decreasing the coil's voltage. A novel application of this technique is to allow the surgical implantation of an inductively coupled coil which, of course, requires no cable to connect it to the matching circuit or preamplifier and hence can be hygienically inserted close to the organ of interest.[28] Fine tuning and matching of the probe can be carried out remotely by either tuning the secondary coil with a capacitance or adjusting its position.

4.3.2.2b. Quadrature Coils. A $\sqrt{2}$ improvement in sensitivity has been demonstrated by the use of a pair of orthogonally mounted surface coils connected in quadrature (i.e., with a 90° phase shift between them).[29] The phase shift can be created by use of diode quadrature combiners or by a hybrid coaxial cable ring.[30] For practical reasons most quadrature designs tend to be for larger-volume coils and hence are of greater application in localized spectroscopy or imaging.

4.3.2.2c. Multinuclear Coils. Obviously it would be of great benefit if coils were available that were capable of being tuned and matched to obtain optimum signal at more than one frequency. Many designs have been produced to achieve this goal. One approach is to tune the coil initially for the higher frequency to be used. Then, to obtain resonance at a lower frequency, an additional tuning capacitance can be connected in parallel through a nonmagnetic μ-switch. This approach is quite efficient for large coils. Other methods involve the use of lengths of coaxial cable[31, 32] or additional capacitive and inductive components.[7, 33–37] Many of these approaches result in reduced coil sensitivity at one of the tuned frequencies. Recently attempts have been made to improve the efficiency of these coils. Two methods[35, 36] incorporate the balanced matching technique mentioned previously,[25–27] while another design[37] keeps the number of additional components to a minimum by splitting the multiturn coil inductance into two parts and incorporating a capacitance in parallel with one of these sections. This allows for the selection of a part or all of the coil's inductance depending on the frequency required.

4.3.2.3. Loop–Gap Resonators

Loop-gap resonators are similar in concept to inductively matched surface coils but are constructed so as to incorporate an integral tuning capacitance. The low-field magnetic resonance versions of these probes have evolved from designs specifically intended for use at the higher frequencies (up to 10 GHz) encountered mainly in electron-spin resonance (ESR).[38, 39] At these higher frequencies, the tuning capacitance has to be small and may consist simply of one or more specially designed gaps in the probe, which is often just a short length of cylindrical conductor. However, loop-gap resonators have been designed for use at the lower frequencies encountered in clinical magnetic resonance.[10] In order to obtain a resonance, it is necessary to comply with equation (4.1), and this requires substantially more capacitance than can be provided simply by a gap. To increase the tuning capacitance, the gap can be extended by including relatively large, parallel plates with a dielectric spacer, as shown in Figure 4.5d. Matching and connection to the spectrometer is accomplished in a similar way to that used for inductively matched, balanced surface coils, i.e., by using a secondary movable coil. The improved sensitivity of loop-gap resonators is probably due to the inductive matching which balances them with respect to ground. This reduces efficiency degradation brought about by the conductive tissue sample loading the coil. The unloaded efficiency may also be higher, not only because of reduced coil resistance brought about by the absence of a capacitive matching circuit, but possibly also because of the improved quality of the integral tuning capacitor. Many

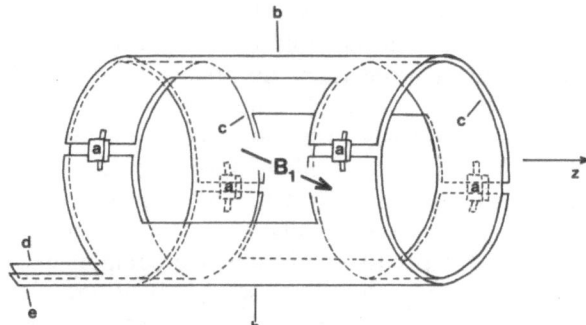

Figure 4.15. A diagram of a slotted-tube resonator (STR). Both the magnetic component B_1 of the pulsed RF field in the rotating frame and the direction of the static magnetic field B_0; (parallel to the z axis) are indicated. Components marked a are ceramic-chip tuning capacitors. These all have the same capacitance. Those marked b are high-conductance H-shaped plates on a cylindrical PTFE former. The tuning capacitors are connected between the H plates at the ends of the four wings to create a resonating structure. Those marked c are the guard rings, which are high-conductance continuous cylindrical sections. These are at ground potential and situated inside the H plates. Their function is to shield the sample from the electric fields generated between adjacent wings of the two H plates. The guard rings improve the efficiency of the STR. Marked d is the ground connection to one of the guard rings. Because the STR is constructed symmetrically and also has a symmetrical distribution of capacitance; the other guard ring is also at ground. At e is the signal feed to the STR.

different and complex designs exist for these resonant structures, including probes for imaging applications[40] and those incorporating quadrature pairs.[41]

4.3.2.4. Slotted-Tube Resonators

The slotted-tube resonator (STR) was originally developed to efficiently provide ^{1}H decoupling RF to samples in high-field magnets.[14] The fundamental property of the STR is that the conducting component has a low inductance, and hence large coils can be constructed. At the RF frequencies encountered in clinical MRS, this means that STRs can be designed to contain samples of sizes comparable to that of the human head.[15] The main uses of these "volume" coils are for imaging and for localized spectroscopy techniques requiring a uniform flip angle over the entire sample. STRs are typically constructed from copper sheets on a Teflon tube former, as shown in Figure 4.15.

4.3.3. The Preamplifier

The signal from the sample coil is on the order of a few μV. It is very important that this signal be amplified as soon as possible, before significant noise or distortion is introduced. The preamplifier is designed to fulfill this function and is mounted close to the coil, as shown in Figure 4.1, thereby minimizing cable lengths. This also allows the use of various techniques to isolate the preamplifier from the transmitter. Ideally the preamplifier should have a low noise figure, which is expressed in decibels (dB). A preamplifier with a noise figure greater than 3 dB will degrade the SNR by more than $\sqrt{2}$. To be acceptable, the preamplifier should have a noise figure less than 1 dB, and many suitable designs have been published with noise figures ranging down to 0.3 dB.[7, 42] The gain of the preamplifier should be such that the signal from the sample coil is increased by a factor of about 1000. Many spectrometers are equipped with preamplifiers tuned for studies on particular nuclear species (i.e., usable only at certain frequencies). Modern instruments are usually broad-band, and preamplifiers capable of adequate performance over a wide range of frequencies are available.

One important requirement is that the preamplifier input be protected from the high-amplitude transmitter pulse, which may otherwise cause damage. One method for doing this[7] employs a specific length of coaxial cable and two pairs of crossed diodes. The important feature of crossed diodes, as shown in Figure 4.16a, is that they do not conduct very well when only small voltages ($\ll 1$ V) are applied. Hence the magnetic resonance signal picked up by the sample coil is incapable of switching them on. However, when a large-amplitude signal, such as the transmitter pulse, is applied, they have high conductivity. This feature of crossed

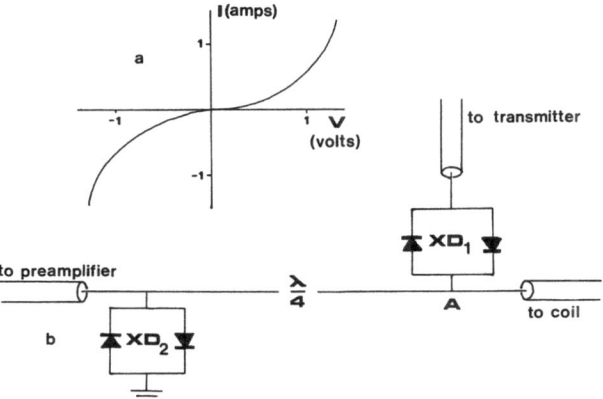

Figure 4.16. The characteristics of crossed diodes and their application for preamplifier protection. (a) The relationship between applied voltage V and current I for crossed diodes. The diodes have a high impedance if the voltage is between 0.7 and -0.7 V. (b) One method for protecting the preamplifier input from the pulse transmitter, using two pairs of crossed diodes (XD_1 and XD_2) and a quarter wavelength of coaxial cable. See text for applications.

diodes can be taken advantage of with the use of a quarter wavelength of coaxial cable, as shown in Figure 4.16b. When the transmitter is off, the crossed diodes XD_1 and XD_2 are not conductive, and the probe and preamplifier are isolated from the transmitter while remaining directly connected to each other. Coaxial cables have the property that, when short-circuited at one end, they reflect a signal back with a 180° phase shift. When the transmitter is switched on, the diodes XD_1 can conduct, and thus connect the transmitter to the probe; at the same time the diodes XD_2 short circuit the preamplifier input to ground, thereby protecting it. The pulse from the transmitter undergoes a 90° shift of phase in reaching the preamplifier input via the quarter-wavelength cable and is reflected with a 180° phase shift at the preamplifier input short-circuited by the diode. This means that the total phase shift at A, including the effect of the return along the quarter-wavelength cable, is 360° and hence, so far as the transmitter and probe are concerned, the quarter-wavelength cable and the preamplifier have no noticeable effect. At the frequencies currently used in clinical magnetic resonance (approximately 20 to 100 MHz), quarter wavelengths of coaxial cable range from about 60 cm for ^{1}H to 2.4 m for ^{13}C, and therefore this protection technique is quite easily applicable.

4.3.4. The Receiver

The function of the receiver is not only to further amplify the signal from the preamplifier but also to change the frequency from RF to the more easily manageable audio frequency (AF). This is often done via a middle stage using an intermediate frequency (IF). Ideally we wish to end up with an AF signal containing frequencies which are the differences between the RF resonance frequencies and a suitable RF reference frequency.

In order to achieve additional amplification of the signal, both the IF and AF sections of the receiver can incorporate amplifiers. The reductions in frequency from RF to IF and then from IF to AF are each accomplished by mixing the signal with suitable reference frequencies obtained from a unit known as the local oscillator (a highly stable, controllable RF generator). If the mixing is performed using a device whose output is the square of the summed inputs (known as a square-law detector), then the output will contain a component with a frequency that is the difference between the input frequencies. This is easily shown by considering the input V_i to consist of a signal of amplitude A with frequency ω_0 and a reference signal of amplitude B and frequency ω_r. ω_r is selected so that it is close to ω_0.

$$V_i = A \sin(\omega_0 t) + B \sin(\omega_r t) \tag{4.12}$$

On squaring this signal (analogous to the operation of the square-law RF detector) and simplifying the algebra, one obtains the following:

$$V_i^2 = \frac{A^2 + B^2}{2} - \frac{A^2}{2}\cos(2\omega_0 t) - \frac{B^2}{2}\cos(2\omega_r t)$$
$$+ AB\cos(\omega_0 - \omega_r)t - AB\cos(\omega_0 + \omega_r)t \qquad (4.13)$$

Because ω_0 is close to ω_r, terms containing $2\omega_0$, $2\omega_r$, and $\omega_0 + \omega_r$ are at very high frequency and can be removed from the detector output simply by filtering. The first term is a constant voltage which again can be removed easily. The remaining fourth term oscillates at the required difference frequency $\omega_0 - \omega_r$. Usually, mixing results in an IF on the order of 10 MHz and an AF in the 10 kHz region. One practical disadvantage of the above scheme is that the term $AB\cos(\omega_0 - \omega_r)t$ is unchanged whether $\omega_0 - \omega_r$ is positive or negative. Hence simple square-law detection by itself cannot tell whether the RF signal has a frequency greater than or less than the reference frequency. In order to avoid this problem, modern pulse-Fourier-transform spectrometers use a quadrature detection technique which employs a pair of devices called phase-sensitive detectors[7] (PSD). The PSD multiplies the signal alternately by ± 1, and the frequency of alternation is that of the reference frequency ω_r. The output from such a device is sensitive to the phase difference between the magnetic resonance and reference signals; it also contains the required components with the reduced frequency $\omega_0 - \omega_r$. A typical PSD for RF use may be constructed from a loop of four suitable diodes.[7] The complete frequency information is preserved by feeding the signal into a pair of PSDs, one of which receives a reference frequency which has been shifted in phase by 90°. The two outputs from the PSDs contain all the information required to determine whether a resonance has a higher or lower frequency than the reference frequency ω_r.

4.3.5. The Analog-to-Digital Converter

The initial RF signal has now been processed in the receiver so that it is entirely AF. The next task is to measure the amplitude as a function of time and input this information into a computer. This is performed by a device known as an analog-to-digital converter (ADC). The ADC samples the input waveform at regular intervals and converts the measured voltages to a sequence of binary numbers, which can then be fed into the computer's memory and stored. A fundamental theorem of digitization (due to Nyquist) states that the sampling rate must be greater than or equal to twice the highest frequency present in the signal. This is illustrated in Figure 4.17, where it is quite clear that the sampling rate used to adequately measure the low-frequency waveform in Figure 4.17a will give the computer ambiguous information about the high-frequency signal in Figure 4.17b. The reason for the radio-receiver part of the spectrometer should now be self-evident; it is obviously far easier to digitize an AF signal and store it in the computer memory than it is to do so with the much more rapidly varying RF voltage. If the signal consisted of RF at 100 MHz, then the ADC would have to sample at 200 MHz (i.e., 2×10^8 samples per second) for the duration of the FID. For a 10-kHz signal a much slower rate will suffice and far less computer memory is required to digitize the FID without loss of important information.

If the range of frequencies covering the required spectrum is known, then an observing band or spectral width (SW) can be defined by the operator. The range of frequencies covered by SW is determined by the time interval T between the ADC signal sampling points. According to the Nyquist theorem,

$$T = \frac{1}{2SW} \tag{4.14}$$

Signals which lie at frequencies beyond the range defined by SW can be reflected back into the observing band, but will be at the wrong frequency and will suffer also from phase distortion when compared to the rest of the spectrum. Such an occurrence is called "aliasing." Noise will also be aliased back into the observing band unless precautions are taken. Usually, AF

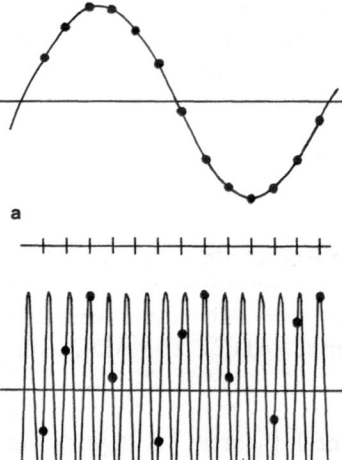

Figure 4.17. The Nyquist theorem can be demonstrated by considering the sampling of the following waveforms: (a) A low-frequency waveform is well sampled and could be closely approximated simply by interpolating between the points. (b) The high-frequency signal is not sampled well enough, and if this information were fed to a computer it would be interpreted as oscillating at a much lower frequency.

filters are incorporated in the system immediately before digitization, and these substantially reduce the amplitude of signals and noise outside the observing band. Such filters are often of the Butterworth type and have some degree of sensitivity variability across the observing band. The cutoffs of these filters are not abrupt "step" functions. Sensitivity decreases over a significant RF band. Hence, in order to restrict the signal cutoff to frequencies outside the observing band, it is usual to use filters with bandwidths slightly larger than the SW.

4.3.6. The Computer System

A computer system suitable for MRS has to be versatile in order to perform the many functions required. The computer should be controllable from the keyboard using a convenient operating system which can be learned without the operator necessarily having to be familiar with the programming language. The computer has to be able to control external devices such as a timer in order to sequence the events required during data collection (e.g., transmitter pulses, data-acquisition periods, interpulse delays, and external triggers to other instrumentation essential to the study). Simultaneously, the computer has to be capable of accepting the digitized FID data from the ADC and accumulating successive readouts in memory by an addition process, thereby improving SNR. Ideally the operator should be able to control, from the computer keyboard, all the experimental conditions, such as observing bandwidth, pulse length, and the like. In this respect the computer will need to be able to "talk" to most of the spectrometer instrumentation via a wired-in link known as a bus. Items of hardware such as the timer are connected to the bus and receive, act on, and acknowledge instructions from the computer. The system will also need to be able to store the accumulated FID data on various media such as magnetic tape, magnetic cartridge discs, or laser discs. As much information as possible concerning the data collection, such as the pulse sequence used, time of beginning of acquisition, etc., should also be stored, in addition to the FID, for reference purposes. For reasons of data security it should be possible to back up data on other tapes or discs, which can be stored elsewhere. The computer can also perform a function in the control of the quality of data acquisition by being able to monitor the root-mean-square noise during FID collection so that comparison may be made with a standard noise source (e.g., a 50-Ω load) and RF interference can be evaluated. Additionally, many spectrometers have facilities that can either assist the operator when shimming the sample by giving a measure of the duration of the ^{1}H FID or even perform the entire shimming procedure automatically (or starting from settings stored in memory).

In addition to its role in controlling data acquisition, the computer must also be capable of very flexible manipulation and analysis of data. Facilities should be available for the performance of such functions as the addition, subtraction, or scaling of FID data. As well as Fourier transforming the FID data to obtain a spectrum, the system should also be able to apply all the commonly used computer techniques[43, 44] for baseline flattening, resolution enhancement, noise reduction by exponential line broadening, and curve fitting to obtain spectral peak areas and positions. Many of these methods will be described in Chapter 6.

A system which is limited to the manufacturer's present programming will not be as useful as one which has a capacity for operator-designed collection and analysis schemes. In this respect the operating system should ideally have the facility for the user to be able to produce amended or new routines in case problems of data handling are encountered or innovative, improved techniques are developed. Direct operator control of phase and baseline corrections for spectra is very important because the input of trial parameters from keyboard alone can be a long and tedious process. Software should also be available for automatic phase and baseline correction. Suitable devices are available, which can be connected to the bus and allow the operator to adjust these parameters simply by turning a knob.

In order to facilitate both data acquisition and analysis, the computer must have adequate display and output facilities, which should include a quality graphics monitor (for FIDs, spectra, and images if required), a plotter, and a printer for hard copies of results. With respect to the workload that a system may be expected to carry, it may also be important to consider the ease with which data can be transferred from the spectrometer to another computer in order that analysis can be carried out elsewhere, thereby creating more data-collection time.

4.4. SAFETY REQUIREMENTS

One of the strong points of MRS above several other diagnostic and investigative techniques is its noninvasive nature. In order that MRS evolve into an accepted clinical tool, it has to be demonstrated in practice that the method is safe and that potential risks can be minimized to negligible and acceptable levels. For many reasons, great care should be taken when considering the installation of an MRS system, especially in a hospital environment.[45] This section will deal with the main safety aspects of clinical MRS and the relevant guidelines published by the National Radiological Protection Board (NRPB) in the United Kingdom[1] and the Department of Health and Human Service (DHHS) in the United States.[2] The reader is referred to the publications from these organizations if further detail than that given here is necessary. From time to time, these guidelines may be updated as safety considerations are reassessed. The NRPB is currently (1989-90) implementing a consultation process with the final aim of producing revised guidelines. In particular, spectroscopic advantages to be obtained by using higher fields may eventually lead to the use of stronger magnets if criteria of safety and patient comfort are met with.

4.4.1. The Static Magnetic Field

There are two main potential hazards from the static magnetic field, which can be categorized as physiological effects on the subject or effects due to the introduction of various conducting or magnetizable objects into the vicinity of the magnet. The NRPB[1] has recommended that, during the course of magnetic resonance studies on human subjects, the whole body or a large portion thereof should not be exposed to a static magnetic field exceeding 2.5 tesla, whereas the DHHS guideline[2] sets the maximum static magnetic field strength at 2 tesla. In addition, the NRPB recommends that staff operating MRS equipment should not be exposed for extended periods to static fields more than 0.02 tesla for the whole body or more than 0.2 tesla for arms or hands. These limits can be increased to 0.2 tesla and 2 tesla respectively if the duration of exposure is less than 15 min. There should be an interval of approximately one hour between successive exposures to fields of this strength.

4.4.1.1. Physiological Effects

A strong static magnetic field can have several effects on living tissue. Because the various extra- and intracellular components may have different magnetic susceptibilities, cells could be distorted due to induced stresses.[46] It is also theoretically possible that applied magnetic fields could change molecular kinetic energies or alter the structure of enzymes so that reaction rates would be different.[47] Static magnetic field effects have been searched for using an animal model.[48] Six sets of mice (156 in total) were raised from birth in a 1.9-tesla field. Exposure times ranged from 360 hr over one month to 624 hr over three months. After each period of exposure, the mice were studied. Both whole-body and organ weights were lower in three or six sets exposed to the field when compared with controls. However, no significant difference was found when comparison was made with controls housed in the magnet suite. It was concluded that decreased growth was caused by other environmental factors. Further investigations showed no significant alterations in gross and microscopic morphology, hematocrit and white-blood-cell counts, or plasma-creatine-phosphokinase, lactic dehydrogenase, cholesterol, triglyceride, or protein concentrations.

It has also been reported that static magnetic fields are capable of inducing a rise in the body temperature of mice.[49] A detectable increase was observed for a field of 0.4 tesla, and the maximum effect was reported at a field of 2 tesla. However, subsequent investigations on human subjects in a 1.5-tesla magnet have shown so significant temperature changes.[50] Theoretically, blood flow could be compromized in a strong, static magnetic field. Serious effects have not been observed at the field strengths within current guidelines.[51]

4.4.1.2. The Introduction of Foreign Objects

The high magnetic field strengths required for spectroscopy (>1.5 tesla) and the large floor area in which metal objects experience a strong field (due to the horizontal axis of the magnet) imply that there is a strong risk from items made of magnetic material. It is very easy for tools or the like to be wrenched out of the users' hands and projected into the bore with considerable force. Often the object that appears to have no magnetizable material content is the most dangerous, because the user is unaware of the hazard. A problem is also raised by eddy currents generated in nonmagnetic conductors (such as prostheses) when the patient is moved in the magnetic field. Forces are produced which tend to oppose the movement, and in certain instances this could be highly dangerous. Heart pacemakers may be affected at some distance from the magnet. Obviously these risks have to be reduced as much as possible by the following precautions:

1. Screening staff and patients who enter the danger zone for objects of magnetizable material, cardiac pacemakers, and metallic prostheses.
2. Eliminating, as far as possible, the use of magnetizable materials in the vicinity of the magnet, especially if they are movable. Essential items should be securely anchored.
3. Making the area around the magnet an area in which magnetic' materials are not used. A safety barrier can be erected around the installation and entry to the danger zone can be through a metal detector. Only nonmagnetic tools should be allowed within the danger area.
4. Providing protective housing for the subjects (especially neonates).
5. Routinely checking relevant details of the medical histories of patients and staff.
6. Providing for quenching the magnetic field in an emergency in easily accessible ways which all personnel should know. Suitable provision for expelling the helium blown off during a quench (controlled or spontaneous) should be available. This can be provided by metal piping from the cryostat to an exterior wall. During a

quench, some exterior parts of the magnet get very cold, and air may liquify and drip off. These parts should be located so that they are normally away from patients and staff. It should be possible to evacuate patients and staff rapidly and safely, whilst avoiding panic.

7. Ensuring that the stray magnetic field is not strong enough to affect cardiac pacemakers in adjacent areas with public access. A plot of the stray magnetic field should be studied so as to ensure that regions of high strength are always in barrier-protected zones.

8. Certain patients may be at risk of cardiac arrest. Staff should be trained to cope with cardiac-arrest emergencies, so that the medical team can work in a safe environment. Under no circumstances should staff ever forget that a strong magnetic field is present.

4.4.2. Rapid Magnetic Field Changes

These can consist of changes in flux density due to applied gradients, to controlled changes in static field strength, or to a quench. The DHHS limits the rate of change of magnetic flux density to less than 3 tesla/sec under all circumstances.[2] However, the NRPB has different safety guidelines which take into account gradient pulses and other causes of field change.[1]

4.4.2.1. Magnetic Gradient Pulses

The NRPB recommends that, for periods of fluctuation less than 10 msec, the following relationship should hold:

$$(dB/dt)^2 \, t < 4 \text{ tesla}^2/\text{sec} \qquad (4.15)$$

where dB/dt is the root-mean-square value of the rate of change of magnetic flux density in any part of the body in tesla/sec and t is the duration of the field change.

4.4.2.2. Controlled Alterations of Field Strength and Quenches

The NRPB guideline[1] states that field changes of more than 10 msec duration should measure less than 20 tesla/sec for staff and patients alike. Gradient pulses are designed to have rise times on the order of 1 msec or less. Hence the above condition is more likely to be tested during a quench. A study of the physiological effects of a quench under controlled conditions has been made.[52] A 1.6-tesla magnet was quenched with an anesthetized pig lying in the bore. Blood pressure, ECG, EEG, and temperature were monitored. The rate of decay of the magnetic field was measured using coils positioned at four points in the bore and a Hall effect device 4 m from the magnet center and off axis. The decline in field strength from 1.6 to 0.3 tesla was at an almost linear rate of 1300 gauss/sec. No obvious physiological effects on the pig were noticed.

4.4.3. Radio-Frequency Fields

The RF power delivered and the consequent heating of the tissues has to be considered in detail. Significant rises in temperature of any part of the body should be avoided. The NRPB states that acceptable exposures should not give rise to an elevation of skin or rectal temperature of more than 1 °C. In addition, the temperature rise of any particular 1-g mass of tissue should also be less than 1 °C. Very firm guidelines exist dealing with the safety limits for power delivered to a specific volume of tissue (4 W/kg[1] or 2 W/kg[2]) and to the whole body (0.4 W/kg[1, 2]). If strict adherence to these maximum power levels is maintained, then the temperature-rise limits should not be exceeded. Particularly careful attention must be paid to the total power deposition from multipulse sequences, to the localizing effects of surface coils, and decoupling[53] if used. Detailed reports of RF-power deposition in MRI, including theoretical estimates and measurement methods, have been published, and these may be of particular importance if multipulse localization sequences are to be used.[54, 55]

Even if the operator has made calculations of RF power deposition for the pulse-sequence and pulse-power levels to be used, additional safety devices, including a fail-safe component, should be incorporated in the RF transmitter line when human subjects are to be studied. These devices should be independent of the spectrometer. They should not depend on the correct functioning of any other equipment. The operator should be aware that software and hardware may not always function properly. Most manufacturers of MRS equipment intended for clinical use already have such safety features built in, and others will usually be prepared to develop such devices. If problems are encountered, monitoring equipment can be constructed without too much difficulty.[56] Additionally, a very simple fail-safe device can be constructed quite easily by inserting a fuse into the transmitter line.[57] Normally this would present quite a high impedance to the transmitter pulse, but the incorporation of a suitable capacitance in series with the fuse can tune this out for a particular frequency. Alternately, a fusing device covering a broad band of frequencies can be constructed by omitting the series capacitance and inserting capacitances to ground at both ends of the fuse instead. Simple calculations and experiments are all that is necessary to determine the necessary ampere rating of the fuse and the required capacitance. This sort of device will blow out quite quickly if the RF power limit is exceeded and thereby protect the patient. It should be noted that transmitter damage may occur if there is no mechanism whereby automatic shut-down can be initiated. For this reason, only a tuned fuse should be considered as fail-safe.

4.4.4. Other Hazards

In some situations particular hazards may arise, and staff should always be alert to covert dangers. For instance some patients require an oxygen-enriched atmosphere, and this could be dangerous if arcing occurs in the tune–match circuitry attached to the surface or volume coil during the transmit pulse. By paying particular attention to insulation testing and by flushing the circuitry with an inert gas, this hazard can be removed.

Some patients (especially neonates) require comprehensive physiological monitoring, including electrocardiogram and apnea and temperature measurements, which can introduce interference problems from RF pickup if electrical cabling is introduced into the bore of the magnet. This can usually be avoided if either suitable RF filters or optical-fiber cabling is used. Of course, monitoring equipment should be kept well away from the magnet. Occasionally drips have to be maintained and blood needs to be sampled, and care has to be taken that no problem arises due to magnetic hypodermic needles or other components.

One report has appeared in the literature concerning hearing loss during MRI.[58] Gradient switching during the imaging process can produce quite a loud click within the bore of the magnet. If rapid imaging or multislice imaging is performed, then these clicks repeat very quickly. Similar gradient switching noise will be encountered with many spectroscopic localization techniques. Close attention should therefore be paid to the severity of gradient noise in MRS as well as MRI. In the published research on the effect of gradient switching, twenty-four patients were tested audiometrically before and after imaging. Fourteen subjects were imaged without ear protection, and of these six suffered temporary mild hearing loss ($\leqslant 15$ dB at at least one frequency). Of ten patients imaged with ear protection, only one subject experienced hearing loss. It was concluded that ear plugs are useful in preventing hearing loss during MRI examination.

4.5. PATIENT HANDLING

The comfort of the patient is vitally important if good spectra are to be obtained. Movement of the subject can detune the coil, reduce the coil-filling factor, degrade magnetic field homogeneity, and render localization techniques ineffective. Suitable padded supports and straps should be used to immobilize the tissue and ensure that little discomfort is experienced.

Particular attention has to be paid to maintenance of circulation and to avoiding pressure on nerves. In studies of delicate tissues (such as the neonatal head) great care must be taken in designing restraints which do not apply much pressure. Neonates present additional difficulty because, unless sedated, they become agitated easily. Sedation is often clinically unacceptable, and hence newborn infants are best studied postfeed, when they are likely to sleep.

There can be problems associated with the materials from which supports are made, because they may be located very close to the coil. Some foam materials contain phosphorus, and can thus give rise to spurious spectral features. The problem is worse for ^1H spectroscopy because of the scarcity of suitable materials. It is essential to test the materials rigorously in advance of construction. Teflon is a very useful material because all the hydrogen atoms have been replaced by fluorine and it also has very good dielectric and machining properties.

Nonmagnetic couches and wheelchairs are essential to facilitate patient handling. These can be designed so as to allow the study to be set up outside the magnet before the patient and MRS probe are inserted into the bore. Examination of neonates can be accomplished using a specially constructed nonmagnetic incubator equipped with standard monitoring facilities.[59] This can be brought from the neonatal-intensive-care unit, wheeled up to the magnet, and fixed in position using locating pins. The infant is then slid into the bore inside a fully enclosing cylindrical cartridge for maximum protection.

Of prime importance is the optimization of data-acquisition technique (flip angle, pulse interval, coil design, etc.) so that the maximum information is acquired in the minimum amount of patient time. Obviously a major factor in the comfort of the patient is the amount of time spent immobilized in the magnet.

REFERENCES

1. Saunders, R. D. and Smith, H.: Safety aspects of NMR clinical imaging, *Br. Med. Bull.* **40**:148–154, 1984.
2. *Guidelines for evaluating electromagnetic exposure risk for trials of clinical NMR systems*: Department of Health and Human Services, Food and Drug Administration, Rockville, Maryland. 1982.
3. Hanley, P. E.: Magnets for medical applications of NMR, *Br. Med. Bull.* **40**:125–131, 1984.
4. Timms, W. E.: An introduction to magnet design and operation, in Lerski, R. A. (ed.): *Physical Principles and Clinical Applications of Nuclear Magnetic Resonance*, London, Institute of Physical Sciences in Medicine, 1985, p. 43.
5. Gordon, R. E., Hanley, P. E., Shaw, D., et al.: Localization of metabolites in animals using ^{31}P topical magnetic resonance, *Nature* **287**:736–738, 1980.

6. Gordon, R. E., Hanley, P. E., and Shaw, D.: Topical magnetic resonance, *Progress in Nuclear Magnetic Resonance Spectroscopy* 15:1–47, 1982.
7. Hoult, D. I.: The NMR receiver: A description and analysis of design, *Progress in Nuclear Magnetic Resonance Spectroscopy* 12:41–77, 1978.
8. Hoult, D. I. and Richards, R. E.: The signal to noise ratio of the nuclear magnetic resonance experiment, *Journal of Magnetic Resonance* 24:71–85, 1976.
9. Ackerman, J. J. H., Grove, T. H., Wong, G. G., et al.: Mapping of metabolites in whole animals by ^{31}P NMR using surface coils, *Nature* 283:167–170, 1980.
10. Grist, T. M. and Hyde, J. S.: Resonators for in vivo ^{31}P NMR at 1.5 T, *Journal of Magnetic Resonance* 61:571–578, 1985.
11. Thompson, H. A.: *AC and Transient Circuit Analysis*, New York, McGraw–Hill, 1955.
12. Abragam, A.: *The Principles of Nuclear Magnetism*, Oxford, Clarendon Press, 1961.
13. Hill, H. W. D. and Richards, R. E.: Limits of measurement in magnetic resonance, Journal of Scientific Instruments, *Journal of Physics E*, Series 2, 1:977–983, 1968.
14. Alderman, D. W. and Grant, D. M.: An efficient decoupler coil design which reduces heating in conductive samples in superconducting spectrometers, *Journal of Magnetic Resonance* 36:447–451, 1979.
15. Leroy–Willig, A., Darrasse, L., Taquin, J., et al.: The slotted cylinder: An efficient probe for NMR imaging, *Magn. Reson. Med.* 2:20–28, 1985.
16. Smythe, W. R.: *Static and Dynamic Electricity*, 3rd ed., New York, McGraw–Hill, 1968.
17. Haase, A., Hanicke, W., and Frahm, J.: The influence of experimental parameters in surface coil NMR, *Journal of Magnetic Resonance* 56:401–412, 1984.
18. Abramowitz, M. and Stegun, J. A.: *Handbook of Mathematical Functions*, Washington D.C., U.S. Department of Commerce, 1964.
19. Cady, E. B., Delpy, D. T., and Tofts, P. S.: Clinical ^{31}P NMR Spectroscopy, in Lerskii, R. A. (ed.): *Physical Principles and Clinical Applications of Nuclear Magnetic Resonance*, London, Institute of Physical Sciences in Medicine, 1985, p. 97.
20. Evelhoch, J. L., Crowley, M. G., Ackerman, J. J. H.: Signal to noise optimization and observed volume localization with circular surface coils, *Journal of Magnetic Resonance* 56:110–124, 1984.
21. Ackerman, J. J. H., Gadian, D. G., Radda, G. K., et al.: Observation of ^{1}H NMR signals with receiver coils tuned for other nuclides, *Journal of Magnetic Resonance* 42:498–500, 1981.
22. Gordon, R. E. and Timms, W. E.: An improved tune and match circuit for B_0 shimming in intact biological samples, *Journal of Magnetic Resonance* 46:322–324, 1982.
23. Styles, P., Soffe, N. F., Scott, C. A., et al.: A high-resolution NMR probe in which the coil and pre-amplifier are cooled with liquid helium, *Journal of Magnetic Resonance* 60:397–404, 1984.
24. Balaban, R. S., Koretsky, A. P., and Katz, L. A.: Loading characteristics of surface coils constructed from wire and foil, *Journal of Magnetic Resonance* 68:556–560, 1986.
25. Murphy–Boesch, J. and Koretsky, A. P.: An in-vivo NMR probe circuit for improved sensitivity, *Journal of Magnetic Resonance* 54:526–532, 1983.
26. Decorps, M., Blondet, P., Reutenauer, H., et al.: An inductively coupled, series tuned NMR probe, *Journal of Magnetic Resonance* 65:100–109, 1985.
27. Froncisz, W., Jesmanowicz, A., and Hyde, J. S.: Inductive (flux linkage) coupling to local coils in magnetic resonance imaging and spectroscopy, *Journal of Magnetic Resonance* 66:135–143, 1986.
28. Schnall, M. D., Barlow, C., Subramanian, V. H., et al.: Wireless implanted magnetic resonance probes for in-vivo NMR, *Journal of Magnetic Resonance* 68:161–167, 1986.
29. Chen, C. N., Hoult, D. I., and Sank, V. J.: Quadrature detection coils – A further $\sqrt{2}$ improvement in sensitivity, *Journal of Magnetic Resonance* 54:324–327, 1983.

30. Sank, V. J., Chen, C. N., and Hoult, D. I.: A quadrature coil for the adult human head, *Journal of Magnetic Resonance* **69**:236–242, 1986.
31. Cross, V. R., Hester, R. K., and Waugh, J. S.: Single coil probe with transmission-line tuning for nuclear magnetic double resonance, *Review of Scientific Instruments* **47**:1486–1488, 1976.
32. Doty, F. D., Inners, R. R., and Ellis, P. D.: A multinuclear double tuned probe for applications with solids or liquids utilizing lumped tuning elements, *Journal of Magnetic Resonance* **43**:399–416, 1981.
33. Schnall, M. D., Subramanian, V. H., Leigh, J. S., et al.: A technique for simultaneous ^{1}H and ^{31}P NMR at 2.2 T in-vivo, *Journal of Magnetic Resonance* **63**:401–405, 1985.
34. Schnall, M. D., Subramanian, V. H., Leigh, J. S., et al.: A new double tuned probe for concurrent ^{1}H and ^{31}P NMR, *Journal of Magnetic Resonance* **65**:122–129, 1985.
35. Chang, L. H., Chew, W. M., Weinstein, P. R., et al.: A balanced-matched double-tuned probe for in-vivo ^{1}H and ^{31}P NMR, *Journal of Magnetic Resonance* **72**:168–172, 1987.
36. Fitzsimmons, J. R., Brooker, H. R., and Beck, B.: A transformer-coupled double resonant probe for NMR imaging and spectroscopy, *Magn. Reson. Med.* **5**:471–477, 1987.
37. Rajan, S. S., Wehrle, J. P., and Glickson, J. D.: A novel double-tuned circuit for in-vivo NMR, *Journal of Magnetic Resonance* **74**:147–154, 1987.
38. Hardy, W. N. and Whitehead, L. A.: Split-ring resonator for use in magnetic resonance from 200–2000 MHz, *Review of Scientific Instruments* **52**:213–216, 1981.
39. Froncisz, W. and Hyde, J. S.: The loop gap resonator: A new microwave lumped circuit ESR sample structure, *Journal of Magnetic Resonance* **47**:515–521, 1982.
40. Kneeland, J. B., Jesmanowicz, A., Froncisz, W., et al.: High-resolution MR imaging using loop-gap resonators, *Radiology* **158**:247–250, 1986.
41. Hyde, J. S., Jesmanowicz, A., Grist, T. M., et al.: Quadrature detection surface coil, *Magn. Reson. Med.* **4**:179–184, 1987.
42. Hoult, D. I. and Richards, R. E.: VHF FET preamplifier with 0.3 dB noise figure, *Electronics Letters* **11**:596–597, 1975.
43. Lindon, J. C. and Ferrige, A. G.: Digitization and data processing in Fourier transform Nuclear Magnetic Resonance, *Progress in NMR Spectroscopy* **14**:27–66, 1980.
44. Hilberman, M., Subramanian, V. H., Haselgrove, J., et al.: In vivo time resolved brain phosphorus nuclear magnetic resonance, *J. Cereb. Blood Flow Metab.* **4**:334–342, 1984.
45. Bore, P. J. and Timms, W. E.: The installation of high-field NMR equipment in a hospital environment, *Magn. Reson. Med.* **1**:387–395, 1984.
46. Persson, B. R. R. ad Stahlberg, F.: Safety aspects of magnetic resonance examinations, *International Journal of Technology Assessment in Health Care* **1**:647–665, 1985.
47. Atkins, P. W.: Magnetic field effects, *Chemistry in Britain* **12**:214–218, 1976.
48. Osbakken, M., Griffith, J., and Taczanowsky, P.: A gross morphologic, histologic, hematologic, and blood chemistry study of adult and neonatal mice chronically exposed to high magnetic fields, *Magn. Reson. Med.* **3**:502–517, 1986.
49. Sperber, D., Oldenbourg, R., and Dransfeld, K.: Magnetic field induced temperature change in mice, *Naturwissenschaften* **71**:100–101, 1984.
50. Shellock, F. G., Schaeffer, D. J., and Gordon, C. J.: Effect of a 1.5 tesla static magnetic field on body temperature of man, *Magn. Reson. Med.* **3**:644–647, 1986.
51. Jehenson, P., Duboc, D., Lavergne, T., et al.: Change in human cardiac rhythm induced by a 2-T static magnetic field, *Radiology* **166**:227–230, 1988.
52. Bore, P. J., Galloway, G. J., Styles, P., et al.: Are quenches dangerous? *Magn. Reson. Med.* **3**:112–117, 1986.
53. Led, J. S. and Peterson, S. B.: Heating effects in Carbon-13 NMR spectroscopy on

aqueous solutions caused by proton noise decoupling at high frequencies, *Journal of Magnetic Resonance* **32**:1–17, 1978.

54. Bottomley, P. A., Redington, R. W., Edelstein, W. A., et al.: Estimating radiofrequency power deposition in body NMR imaging, *Magn. Reson. Med.* **2**:336–349, 1985.

55. Chen, C.–N., Sank, V. J., Cohen, S. M., et al.: The field dependence of NMR imaging. I. Laboratory assessment of signal-to-noise ratio and power deposition, *Magn. Reson. Med.* **3**:722–729, 1986.

56. Funk, A., Rippy, R., and London, R. E.: A pulse reflectometer for routine monitoring of transmitted and reflected power in physiological NMR studies, *Magn. Reson. Med.* **4**:175–178, 1987.

57. Hedges, L. K.: A fuse for magnetic resonance imaging probes, *Magn. Reson. Med.* **9**:278–281, 1989.

58. Brummett, R. E., Talbot, J. M., and Charuhas, P.: Potential hearing loss resulting from MR imaging, *Radiology* **169**:539–540, 1988.

59. Chu, A., Delpy, D. T., and Thalayasingam, S.: A transport and life support system for newborn infants during NMR spectroscopy, in Rolfe, P. (ed.): *Neonatal Physiological Measurements*, London, Butterworth, 1986, p. 409.

Data Acquisition in Clinical Magnetic Resonance Spectroscopy

5.1. FUNDAMENTAL CONSIDERATIONS

The methods used for collecting MR spectra are fundamental to the eventual interpretation of the results. The spectroscopist needs to know what effects the conditions of examination may have on peak areas and phases, chemical shifts, and the efficiency of signal localization. These effects may be important for the full understanding of the information present in the spectra.

One of the prime aims of clinical MR data acquisition is to obtain the required information in the minimum time and with the least discomfort to the patient. Important factors which affect the time required to obtain adequate results include metabolite concentration, RF noise and interference, the spin–lattice relaxation time T_1, the flip angle θ, and the delay between successive pulses T_R. Although the concentrations and T_1s of in situ tissue metabolites are not normally under the operator's control, it is possible to have a direct influence on the other factors just mentioned.

5.1.1. Collecting Data Using Repeated Single Pulses

The simplest data-acquisition protocol to describe consists of a brief RF transmitter pulse followed first by FID digitization and then by an interpulse delay. However, the fundamental criteria for minimizing patient time are basically the same if more-complex pulse sequences are used.

5.1.1.1. Noise and Interference

Obviously, the operator should attempt to reduce these to the lowest possible level before attempting to collect data. If the spectrometer components are matched to each other with an impedance of 50 Ω, then the RMS noise measured with a 50-Ω load on the preamplifier input should be the minimum possible, and this can be used as a quality standard for data collection. The ultimate aim should be to get the RMS noise with the subject in position as close as possible to that measured on the 50-Ω load. The RF noise can be reduced a great deal by locating the whole system, the magnet, or even just the subject[1,2] in a Faraday cage (a grounded RF interference shield constructed from highly conductive metal sheet or mesh—usually copper). Because of the necessity for medical staff to be able to see and perhaps communicate with the patient at all times, it may not be possible to have a cage built entirely from continuous metallic sheet. The enclosure will have to have doors, and it will need to be fitted with mesh viewing windows. The efficiency of such an interference shield will depend on the signal frequency. In particular, because essential apertures and meshes will be less efficient at screening interference as the wavelength gets shorter, meshes for the construction of such screens should be selected with reference to their RF characteristics and the resonant frequencies of the nuclei to be studied. Extremely good electrical contact between the sections of the screen is essential, as also is an efficient grounding point.

Additional worthwhile improvements can be obtained by grounding the subjects at suitable points on their anatomy using one or more conveniently shaped metallic plates or flexible metallized straps adequately coated with conducting gel.

Under most circumstances it should be possible to get the RMS noise' down to the 50-Ω level. Without alterations in the receiver, this is the best that can be done. In order to collect patient data with a better SNR than would be obtained from one FID by transmitting a single pulse with the RMS noise at the 50-Ω level, it is necessary to acquire many individual FIDs and store the summed data in the spectrometer's computer. If N FIDs are collected, then the summed signal is proportional to N, whereas the RMS noise, because it is a random signal, increases only as $N^{1/2}$. Hence

$$\text{SNR} \propto N^{1/2} \tag{5.1}$$

This means that the accumulation of four FIDs will give an SNR twice that from a single FID.

5.1.1.2. Flip Angle and Pulse-Repetition Time

As has been seen in Chapter 2, the signal strength depends on two acquisition parameters, flip angle and pulse-repetition time. As a rule of thumb, the larger the flip angle and the shorter T_R is, the less relaxed are the nuclei and the smaller the signal is. Figure 2.12 shows the relationship between these variables and signal strength for a coil with a homogeneous B_1 field (i.e., uniform θ). Fortunately, the functions change quite slowly, and one does not have to be too precise about the settings for θ and T_R. In most tissues, T_1s for ^{31}P nuclei are on the order of a few seconds, and maximum signal per unit time for a 90° flip angle is obtained with $T_R \approx T_1$. If faster pulsing is used, then some of the signal strength can be recovered by using a smaller flip angle (e.g., 45° or 60°). Most important ^{31}P metabolites in liver have substantially shorter T_1s, permitting the use of much faster pulsing.[3] For surface coils the situation is very much more complicated, because of the spatial variation of flip angle. For a 90° pulse at the coil's center, nuclei close to the coil are less relaxed than those further away. The maps of signal strength given in Figure 4.10 show that, as T_R decreases, the signal picked up from a volume element gets smaller but the sensitivity becomes more uniform over the volume interrogated by the coil. Alternately, a large T_R concentrates the sensitivity closer to the coil. In practice, the signal obtained also depends on the spatial distribution of the metabolites of interest, and it will usually be necessary to adjust θ and T_R a few times for a particular application until the signal strength is optimized. Quite often, for ^{31}P studies using surface coils, it will be found that a 90° pulse at the coil's center is adequate. Such a pulse will give a flip angle of about 60° over a fairly large volume within the tissue, and this may be optimum for a T_R of a couple of seconds. These conditions are a useful starting point. If the tissue of interest is at a greater depth or one is using localization procedures, a larger θ at the coil's center may be necessary in order to obtain a flip angle close to 90° in the required volume. If one wishes to obtain a spectrum from completely relaxed nuclei, then T_R should be at least $5T_1$, which is 10 to 20 sec for ^{31}P tissue metabolites.

At the beginning of a data collection, the nuclei are fully relaxed, and the saturation gradually builds up over the first few pulses until an equilibrium state is reached. This means that the contributions to the FID from the earliest pulses are larger than those from later ones. It is often useful, particularly if one is accumulating data from only a small number of pulses, to allow a few pulses at the beginning of a collection to be transmitted without the data being stored until a steady state has been reached. Analysis of the signal strength under repetitive pulsing reveals that the number of dummy pulses before commencing data accumulation should be at least five.[4]

5.1.1.3. Phase Cycling

If data is collected with the phase of the transmitter pulse and receiver held constant, then the digitized FID may include unwanted interference from the end of the transmitter pulse in its first few points, and there may also be a constant FID offset. One obviously does not want to include in the FID information from anything except the metabolites. The FID offset would soon use up the available data-storage bits and could possibly cause data overflow. These effects and other distortions are avoided if the transmitter-pulse and receiver phases are cycled by increasing them by 90° for each successive pulse. This incrementing of phase is commonly known as CYCLOPS (cyclically ordered phase sequence).[5]

5.1.2. Reduction of Spectrum-Baseline Artifacts

Unless precautions are taken, the filters employed in the spectrometer can introduce spectrum-baseline distortions due to their transient responses. The distortion appears as a baseline curvature, which can be quite severe. The effect can be greatly reduced by correct setting of the switch-on times for the receiver and digitizer after the RF transmitter pulse.[6] This allows data collection to begin when the signals from the filters are all in phase. For commonly used four-pole Butterworth filters, this state can be achieved by digitizing the first data point in the signal π/ω_c sec after the center of the RF transmitter pulse, where ω_c is the filter's cutoff frequency.

5.2. SIGNAL LOCALIZATION

A prime requirement of a clinical spectroscopy system is that the operator should have the ability to define the shape, size, and location of the tissue volume from which the spectrum is acquired. Ideally, the selected volume will have a sharply defined boundary, and signal from outside this volume should make a negligible contribution to the final spectrum. One of the disadvantages of many localization techniques is that the required volume is often small with respect to the sensitive volume of the coil, causing a reduction in SNR. Hence data may have to be collected for a longer time in order to obtain spectra of acceptable quality. For several years, localized studies on human subjects depended solely on the physical properties of surface coils[7-11] or volume coils. For ^{31}P studies, the knowledge that skin, fat, and bone do not contain significant quantities of mobile phosphorylated metabolites[10] has also been of assistance, particularly for the investigation of skeletal muscle. A volume coil (e.g., a solenoidal or saddle coil) is approximately uniformly sensitive for tissue located within the enclosed volume, whereas the sensitivity decreases as the sample is moved further away. For a surface coil, the volume from which signal is detected is contained within (mainly) a hemisphere of radius approximately the same as that of the coil.[11] For both surface coils and volume coils, the sensitive volume does not have a sharply defined boundary, and hence, unless further techniques are employed, localization of the signal is not precise. Use has been made also of anatomical information (e.g., the poor development of cranial musculature in studies of neonatal brain[12, 13]); in studies of the liver, rapid pulsing and the consequent saturation of nuclei close to the coil can reduce signal from superficial muscle while preserving that from the rapidly relaxing hepatic ^{31}P metabolites.[3]

Although these opportunities for localization have proved useful, additional methods are required for clinical spectroscopy to develop into a flexible tool capable of gathering information from whichever part of the body is desired. In recent years a great deal of effort worldwide has been attracted to the provision of clinically practical solutions to the problem of localization. Some of these have found useful application. The number of approaches to the problem of localization grows daily with the publication of either new techniques or variations on and improvements to already existing ones. The following sections can only illustrate the sort of methods that are available, and greater detail is given only to those techniques that are relatively easy to apply and show potential for clinical application.

5.2.1. Surface-Coil-Localization Methods

In addition to the primitive localization provided by the sensitive volume of the surface coil, attempts have been made to use the volume-selective attributes of spin–echo pulse sequences and to take advantage of the region in front of the coil where the RF field varies linearly with distance from the plane of the coil.

5.2.1.1. DEPTH Pulses

The DEPTH pulse utilizes a combination of the properties of surface coils and spin–echo pulse sequences.[14] The fundamental spin–echo sequence can be written as

$$\theta - \tau - 2\theta - \tau - \text{Acquire Echo} - \text{Delay } (T_R) \qquad (5.2)$$

where θ is the flip angle and τ is a time interval during which relaxation (dephasing) of the transverse magnetization occurs (see Chapter 2 for a description of spin–echo formation). Using a solenoidal or saddle coil, θ

can be set close to 90° and the flip angle will be uniform throughout the sample. However, if a surface coil is used, advantage can be taken of the spatial variation of θ within the coil's sensitive volume, as illustrated in Figure 4.9. By changing the duration or amplitude of the θ pulse, it can be arranged that the region of interest experiences a flip close to 90°. After the 2θ pulse, adjacent regions which receive a flip of $(90 + \delta)°$ will have their magnetization inclined at an angle of 3δ to the xy plane. (The z axis of the xyz coordinate system is, as usual, parallel to both the unperturbed magnetization and the static magnetic field B_0.) This magnetization will contribute less to the total signal, and hence the spin echo tends to originate from the region in which θ is close to 90°. This is especially true if the 2θ pulse is phase-cycled through 360° at 90° intervals on successive scans,[15] as in the following sequence:

$$\theta - \tau - 2\theta[\pm x, \pm y] - \tau - \text{Acquire Echo}[\pm] - \text{Delay } (T_R) \quad (5.3)$$

where $[\pm x, \pm y]$ implies that, instead of the 2θ pulse flipping the magnetization solely about one axis, the rotation is about the $+x$, $+y$, $-x$, and $-y$ axes in sequence and $\pm$ indicates that the acquired data are alternately added to and subtracted from the accumulating FID. The phase cycling sequence described by (5.3) is completed in four scans. The above pulse series can be written more shortly as $\theta; 2\theta[\pm x, \pm y]$. Additional depth discrimination can be obtained by including a further phase-cycled 2θ pulse to give the following sequence:

$$\theta; 2\theta[\pm x, \pm y]; 2\theta[\pm x, \pm y] \quad (5.4)$$

Table 5.1 gives the 16 phase combinations required to obtain depth localization using the sequence in (5.4). The theoretical dependences of signal magnitude on θ are shown in Figure 5.1 for the DEPTH pulse given by equation (5.3), and this is compared with that obtained from a single θ pulse. Excellent comparison is obtained between experimental measurements and theory.[14] The flip angle produced by the θ pulse decreases with distance from an axis perpendicular to and through the center of the surface coil. However, because flip angles get larger close to the coil (as shown in Figure 4.9), the region experiencing a 90° flip from the θ pulse approaches the windings at the coil's perimeter. (This becomes particularly important for regions which are off axis by >0.7 times the coil's radius.) The sensitive volume of the DEPTH pulse is therefore shaped like a concavo-convex lens, with its concavity facing the coil. It should be remembered that regions experiencing flips of approximately 270°, 450°, etc. can also contribute to the signal obtained using some DEPTH sequences, raising the possibility of additional sensitive-volume lobes. In general such

Table 5.1. Relative Phases of the Pulse Components of the DEPTH
Pulse θ; $2\theta(\pm x, \pm y)$; $2\theta[\pm x, \pm y]^a$

		Pulse components				
		Add[b]			Subtract[b]	
	θ	$2\theta_1{}^c$	$2\theta_2{}^d$	θ	$2\theta_1{}^c$	$2\theta_2{}^d$
	x	x	x	x	y	x
	x	$-x$	x	x	$-y$	x
	x	x	$-x$	x	y	$-x$
Phases	x	$-x$	$-x$	x	$-y$	$-x$
	x	y	y	x	x	y
	x	$-y$	y	x	$-x$	y
	x	y	$-y$	x	x	$-y$
	x	$-y$	$-y$	x	$-x$	$-y$

[a] From Bendall et al.[14]
[b] Data acquisition.
[c] First 2θ pulse.
[d] Second 2θ pulse.

regions are very close to the coil, and it can often be arranged that they are outside the sample. Further sequences have been developed that, in addition to possessing depth-localization properties, can be used to measure T_1s by a variation of the inversion–recovery method (see Section 5.7.1.2.). These give reliable, localized T_1 measurements[14] with a sensitive volume which experimental determination has shown to agree very well with that predicted from theory.[16] More recent work[17] has shown that a combination of a large and a small surface coil, with only the small coil used for signal reception but both used for pulse transmission, can produce a single, discrete sensitive volume.

An alternative approach to surface-coil depth localization is the harmonically analyzed sensitivity profile (HASP) method,[18] in which, as data accumulation proceeds, the sensitive volume is built up by the use of

Figure 5.1. A plot of signal magnitude as a function of the flip angle θ following a single RF pulse sequence a (θ – Acquire Data) and the depth-pulse sequence b ($\theta - \tau - 2\theta[\pm x, \pm y] - \tau$ – Acquire Data). The plot b (dashed) indicates the enhanced depth-localization properties of the phase-cycled spin–echo sequence (from Bendall and Gordon,[12] Academic Press Inc.; reprinted with permission).

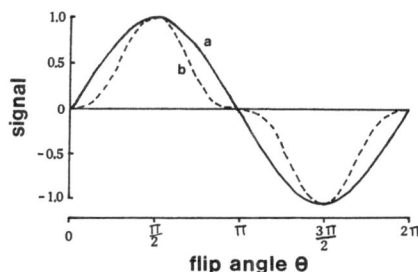

varying pulse lengths and by weighting the individual FIDs as they are added together. No phase cycling is required, and the data-collection time is reduced.

5.2.1.2. Rotating-Frame Imaging

This technique relies on the combination of a homogeneous static B_0 field with the part of the inhomogeneous RF field of a surface coil which decreases approximately linearly with depth.[19-22] (This approach to localization was originally suggested as a novel way of obtaining images.[23]) The method has been applied by locating a receiving surface coil coaxially and parallel to, but slightly offset from, the larger surface coil of the transmitter. This positions the sensitive volume of the smaller coil within the part of the transmitter-coil volume where the RF field decreases almost linearly with depth.[21, 22] (The region in which the RF field has a linear dependence extends from about 0.3 to about 0.8 times the transmitter coil's radius). If one considers an elemental volume of sample within this region and applies a transmitter pulse, the spins will be flipped by an angle dependent on the depth of the elemental volume. The magnetization at other locations will be flipped through different angles depending on the depth. In a rotating frame of reference x', y, z', the magnetization rotates about the magnetic component B_1 of the RF field at a frequency given by

$$v(x) = \gamma B_1(x) \tag{5.5}$$

where γ is the gyromagnetic ratio of the nucleus of interest. The flip angle θ is proportional to the duration t of the transmitter pulse and is given in radians by

$$\theta(x) = 2\pi t v(x) \tag{5.6}$$

The amplitude of the signal obtained from the volume element is proportional to $\sin \theta$ and hence varies sinusoidally with pulse length at a rate that depends on the depth. By acquiring a series of spectra from the whole sample using incremented pulse lengths and then obtaining the FT of the spectra with respect to pulse length, a two-dimensional plot is obtained giving signal strength as a function of both depth and chemical shift. Thus the spectra obtained using the rotating-frame imaging technique are localized to tissue slabs parallel to the plane of the coil but at different depths. The depth resolution and hence the slab thickness depend on the number of pulse increments used and on the linearity of the RF field as a function of depth. The lateral resolution is dictated by the sensitive volume of the receiver's surface coil.

An important requirement for rotating-frame imaging is the provision of a second coil, larger than the receiver coil, in order to generate the linearly depth-dependent RF field during pulse transmission. One problem with this dual coil system is caused by interaction between the coils (which are both tuned to the same frequency) while the transmitter pulse is on. During the transmitter pulse, a large current, induced in the receiver coil, disturbs the homogeneity of the RF field. This can be avoided by including a properly selected length of coaxial cable, terminated by crossed diodes, in parallel with the coil's tuning capacitor.[21] During the transmitter pulse, the diodes conduct, and the cable behaves as an additional inductance in parallel with the tuning capacitance. The receiver coil is therefore detuned during pulse transmission. Between transmitter pulses, the diodes do not conduct and the length of coaxial cable behaves only as a small capacitance, part of the total tuning capacitance of the coil. (Interaction between the coils is not a problem during reception if the transmitter coil is at least three times the size of the receiver coil.) This technique has been recently applied successfully to the investigation of the metabolism of the in situ human heart[24] (see Chapter 3).

5.2.2. Magnetic Field Gradient Techniques

Magnetic field gradients, either static or pulsed, are employed in many of the localization methods which have been suggested. Some methods require selective RF pulses (i.e., pulses capable of exciting only a narrow frequency band), which, in a linear magnetic field gradient, rotate the sample magnetization within a narrow slab of tissue only. (The resonant frequencies of nuclei outside this slab are not present in the bandwidth of the RF pulse.) Nonselective pulses are commonly called "hard" pulses and are brief (usually about 100 μsec in duration). They have a rectangular amplitude envelope (see Figure 5.2a). It is a well-known property of oscillating signals that a continuous sinusoidal waveform contains only one frequency. However, if the duration of the waveform is gradually reduced, the bandwidth of the frequencies contained will increase. Because of its brevity, a hard pulse has a very wide frequency spectrum and is therefore capable of flipping nuclear spins over a wide range of chemical shifts. "Soft" pulses are of much longer duration (up to several msec) and are capable of perturbing spin systems over a much narrower range of frequencies. Soft pulses (Figure 5.2b) have the same rectangular shape as hard pulses, and their shape introduces the disadvantage that the sharply rising and falling edges cause closely spaced side lobes in their frequency spectra. If such a pulse were applied within a magnetic field gradient, spins would be flipped in regions other than the tissue of interest slab. To avoid this problem, selective pulses containing only a specific frequency band are produced by

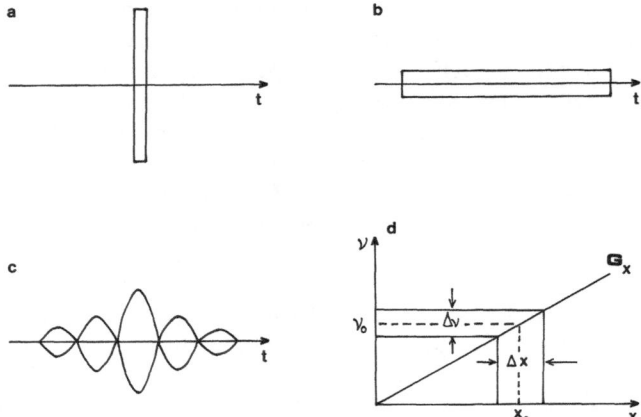

Figure 5.2. The various main types of RF-transmitter pulse envelopes used in MR spectroscopy. It is usually possible to vary amplitude, phase, and pulse duration in order to obtain the required spectral characteristics. (a) A hard pulse, large in amplitude but of short duration (up to a few 100 μs). (b) A soft pulse, which has a low amplitude but is of long duration and therefore has a narrower bandwidth than a hard pulse. (c) A selective pulse, which is of long duration and has an envelope whose shape is designed so that side lobes in its spectrum are minimized and excitation is therefore uniform with respect to frequency over a certain bandwidth. (d) Slice definition using a selective pulse and a magnetic field gradient along the x axis. The selective pulse, has a center frequency offset from resonance by v_0 and a bandwidth Δv. In a gradient of G_x gauss/cm this excites the nuclei in a slice of thickness Δx centered at distance x_0 from the center of the gradient field. v_0, x_0, Δv, and Δx are related by the equations $x_0 = 2\pi v_0/\gamma G_x$ and $\Delta x = 2\pi \Delta v/\gamma G_x$.

multiplying a rectangular soft pulse by an envelope which gradually increases from zero and then slowly decreases (Figure 5.2c). Many designs of varying complexity exist for selective pulses and, for full flexibility, the pulse-programming software should allow comprehensive control of amplitude, phase, frequency, duration, and sequence. Figure 5.2d shows how such a selective RF pulse, due to its narrow frequency bandwidth, will flip spins only in a slab of tissue within which, due to an applied gradient, the magnetic field strength is limited to a certain range of values.

One problem with many pulsed-gradient methods is caused by eddy currents generated in adjacent conductors (e.g., the cryostat). These make it difficult to obtain short rise and fall times for the gradient pulse, and hence a significant residual magnetic field may exist during data acquisition. This can be partially compensated for by pre-emphasis (i.e., shaping the gradient pulses to compensate for the eddy-current effect) or by using gradient coils which are either offset away from the cryostat bore tube or shielded (i.e., designed so that there is negligible exterior magnetic field).

5.2.2.1. Magnetic Field Profiling

This method, also known as topical magnetic resonance (TMR), was one of the first of the more sophisticated attempts to control the position, shape, and size of the localized volume of tissue. In addition to the main static magnetic field B_0, a further static, profiled field is produced by ancillary, resistive coils.[25, 26] The profiled field is designed so that only a specified region in the center of the magnet experiences a homogeneous field, as shown in Figure 5.3. Outside this volume, the field is made to decay rapidly with distance from the magnet's center. The effect of this, when used in conjunction with a volume or surface coil, is that well-resolved spectra are obtained only from the homogeneous region (a in Figure 5.3). In the exterior volume, nuclei experience a large range of field strengths, and hence they resonate over a wide band of frequencies, thereby producing a broad, unresolved hump in the spectrum (b in Figure 5.3). Computer techniques, such as convolution difference[27] or profile correction,[26] can be used to remove the unwanted broad spectral humps, leaving a high-resolution spectrum which contains signals from nuclei in the region of interest only. The broad hump can be separated from the highly resolved components of the spectrum because the FID signal from outside the homogeneous volume decays rapidly. Multiplying the FID by a suitable function can reduce the broad component while preserving the more slowly

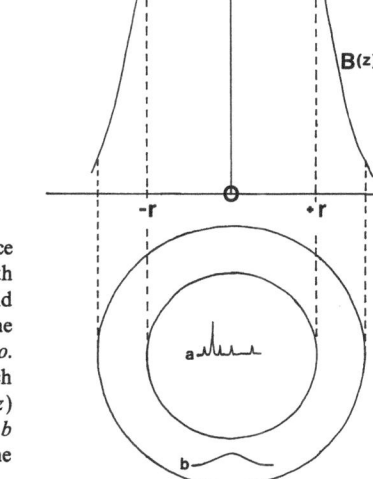

Figure 5.3. The topical magnetic resonance profiled-field localization method. The strength of the profiled field is shown as $B(z)$ (solid curve). ΔB is the field inhomogeneity in the region extending $\pm r$ from the magnetic center o. From this volume, highly resolved spectra such as a are obtained. Outside this region, $B(z)$ decreases rapidly, and very broad lines such as b are obtained, which can be separated from the spectrum by computer processing.

decaying FID from the highly resolved localized spectrum. The technique is limited in application because, although the size of the homogeneous volume can be altered under operator control, localization is restricted to the magnet's center only. Hence the tissue of interest has to be moved to the homogeneous volume, and this can produce positioning problems. An additional disadvantage is that the surface of the homogeneous volume is not well defined and is also irregular, having an approximately star-shaped cross-section.[26] In spite of these handicaps, field profiling has some positive features. The only extra hardware requirements are the additional resistive coils; once these are installed, implementation is relatively easy using simple single-pulse methods. No complicated pulse sequences or gradient switchings are needed.

In addition to the preliminary evaluation of magnetic field profiling in animal studies,[25, 26] the method has found application in clinical studies of the liver,[28, 29] the brain,[28] and tumors,[30] although problems have been encountered in its implementation in studies of the heart and kidneys.[28]

5.2.2.2. Depth-Resolved Surface-Coil Spectroscopy

One of the simplest methods that uses gradients for localization is DRESS.[31] The technique is applied with a surface coil, and spectroscopic information is received only from a tissue slab, parallel to the plane of the coil, with predetermined depth and thickness. Within the tissue slab, sensitivity is defined by the spatial properties of the surface coil. Sensitivity decreases slowly with distance from the coil's axis. The selected volume is sharply defined only perpendicular to the plane of the coil (i.e., at the edges of the slab). A 90° selective pulse is used, and a magnetic field gradient G_x is switched on during the transmit period, as shown schematically in Figure 5.4. The selective pulse is designed to be modulated by a sinc function, which is defined mathematically by

$$\text{sinc}(x) = \frac{\sin(\pi x)}{\pi x} \tag{5.7}$$

The envelope of such a selective pulse on an oscilloscope would be similar to that shown in Figure 5.4c. As shown in Figure 5.2d, the slab of spins which experience a 90° flip will be at a depth governed by the fundamental frequency of the selective pulse and the magnitude of G_x. The thickness of the slab will depend on the selective pulse bandwidth Δv and G_x. Because of the application of G_x, the resonant frequency is also a function of depth within the slab. This means that, during the selective pulse, dephasing will occur due to the different rates of precession at different depths within the slab. In order to rephase the spins, a negative gradient pulse is applied

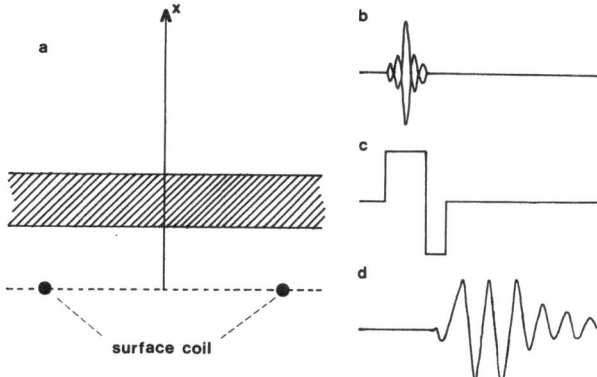

Figure 5.4. The geometry and pulse–gradient sequencing of the depth-resolved surface-coil spectroscopy (DRESS) approach to localization. (a) The geometry of the surface coil and the tissue from which signal is to be obtained (hatched area). The coil is parallel to the yz plane. The lateral extent of the slab of tissue from which signal is localized is defined by the spatial sensitivity of the surface coil. (b) The frequency-selective pulse which, when G_x (a linear magnetic gradient) is switched on, has a frequency bandwidth such that only spins within the hatched area of (a) are flipped. (c) The combination of the positive G_x gradient pulse and the selective RF pulse defines the slab of tissue in which spins are flipped. The negative G_x pulse rephases the magnetization which has spread out in the xy plane due to the variation of resonant frequency across the slab during the positive G_x pulse. (d) The FID, collection of which is commenced as soon as (or just before) rephasing is completed.

immediately after the termination of the selective RF pulse. Acquisition of the localized spectroscopic data is commenced directly after, or shortly before, the end of the refocusing gradient. Because spins outside the tissue slab of interest remain unperturbed by the selective pulse, no signal is obtained from them. In addition to the testing of the technique on phantoms,[31] the method has been successfully applied to the canine heart,[31] human brain,[31] and human liver and heart.[32]

An ingenious variation of this localization modality is slice interleaved DRESS (or SLIT-DRESS).[33] Because nuclei outside the selected slab of tissue remain in their fully relaxed state, it is possible to utilize the pulse repetition time in order to collect data from a range of depths if required. This is done by sequentially changing the RF frequency of a rapid series of selective pulses so that different slabs of tissue experience a 90° flip. The whole series of selective pulses is completed within a normal DRESS pulse-repetition time T_R. Each slab effectively receives only one pulse per series, and hence data can be collected from the whole volume in front of the coil without saturating the sample. This means that the total collection time is the same as that for a single-slab DRESS acquisition. Further resilience against saturation is provided by sequencing the selective pulse-frequency changes so that neighboring slabs do not receive consecutive excitations.

Several variations of the DRESS approach, involving suppression of the signal from tissue water, have been tried to obtain ^{1}H spin–echo spectra from the adult human brain in vivo.[35] Several important metabolites were resolved, including N-acasp, phosphorylcholine, and the combined PCr and Cr resonances.

5.2.2.3. Image Selected in Vivo Spectroscopy

This localization procedure (ISIS) has obtained clinically useful results, especially when applied with more recent refinements. The ISIS[36] and the more recently developed OSIRIS[37] techniques are flexible and can localize the signal to a volume whose position, shape and size can be determined and rapidly changed by the operator with reference to a conventional ^{1}H MR image. The early implementations of ISIS and OSIRIS defined a sensitive volume in one, two, or three dimensions i.e., a slab, a column, or a cube. Recent developments permit flexible orientation of the sensitive volume and the creation of more-complicated sensitive volumes. In order to define a cubic sensitive volume, the original three-dimensional ISIS method requires the co-addition of data collected from eight separate scans, but localization of the signal to a slab[38] or column can be obtained more simply. The procedure depends on the inversion of the magnetization in the selected volume using selective pulses and gradients prior to data acquisition. The pulse sequence for three-dimensional ISIS is shown in Figure 5.5. Up to three consecutive 180° selective pulses are applied during a sample-

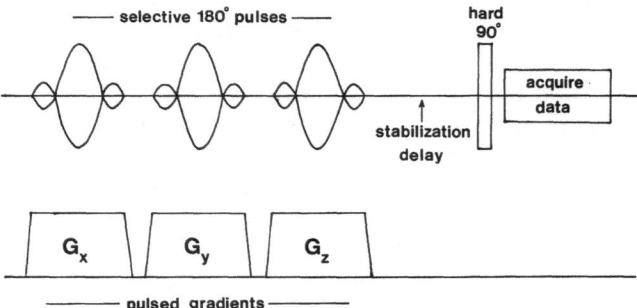

Figure 5.5. The pulse and gradient scheme employed in the ISIS localization technique. The magnet-stabilization delay allows for the decay of eddy currents resulting from the pulsed gradients. The selective pulses are switched on or off and the data are added or subtracted according to the protocol given in Table 5.2.

preparation period prior to each scan. During each pulse, a gradient (G_x, G_y, or G_z) is switched on according to the protocol defined in Table 5.2. Each selective pulse inverts the magnetization in a slab of tissue perpendicular to the applied magnetic gradient. After the necessary number of selective pulses, the magnetization is sampled by applying a 90° hard pulse after a brief period to allow for stabilization of the magnetic field following the gradient switching. If the data are added or subtracted as indicated in Table 5.2, the signal is localized to the cube defined by the intersection of the three mutually orthogonal tissue slabs.

The manner in which ISIS functions is easily understood by considering its operation in one dimension. This defines a sensitive volume which is a slab and requires only two scans. In the first scan only a nonselective 90° pulse is applied (i.e., gradients and selective pulses are not used, as in scan 1 in Table 5.2). After this pulse, the magnetization has been flipped by 90° throughout the entire tissue volume within the field of view of the coil. The flip angle depends on the amplitude and duration of the applied pulse and is uniform or varies in space depending on whether a volume coil (homogeneous RF field) or a surface coil is used. In the second scan the magnetization is prepared by applying a selective inverting 180° pulse before the 90° hard pulse. This inverts the magnetization in the selected slab but leaves spins outside this volume unperturbed. The 90° pulse now flips the spins in the entire volume again. The result of this pulse sequence is that spins in the selected slab have a total flip angle of 270°, whereas spins outside the slab have been flipped by only 90°. If the data acquired from the second scan are subtracted from those obtained in the first scan, the resultant signal originates entirely from the selected slab of tissue. The

Table 5.2. Selective Pulse Sequencing and Protocol for Addition and Subtraction of Data for Three-Dimensional Image-Selected in Vivo Spectroscopy (ISIS)[a]

	Selective pulse			
Scan number	x-axis	y-axis	z-axis	Add/subtract
1	OFF	OFF	OFF	+
2	ON	OFF	OFF	−
3	OFF	ON	OFF	−
4	OFF	OFF	ON	−
5	ON	ON	OFF	+
6	ON	OFF	ON	+
7	OFF	ON	ON	+
8	ON	ON	ON	−

[a] From Ordidge et al.[39]

complete ISIS localization method extends this methodology to three dimensions.

Because ISIS relies on inversion of the magnetization, there is no loss of signal due to transverse relaxation during the preparation period (i.e., before the 90° read pulse) and hence no dependence of signal strength on metabolite T_2s. ^{31}P resonances of ATP and other metabolites with short T_2s are therefore detectable. However, some longitudinal recovery occurs, and hence there is a slight dependence of signal strength on T_1.

An important handicap with the fundamental ISIS technique is its dependence on the addition or subtraction of data collected from eight successive experiments. If the system's gain changes for any reason during this scan cycle, exact cancellation of signals from tissue outside the selected volume will not occur. This can be of particular importance close to the coil, where sensitivity is high. In each scan, moreover, signal is obtained from a large volume of sample. The signal remaining after addition and subtraction originates from a smaller volume of tissue and hence is of reduced amplitude when compared with signal from a single scan. Adequate digitization can therefore be difficult to achieve. Several variations of the fundamental ISIS technique have been suggested in order to avoid these problems. One solution has been to apply a selective RF pre-pulse (constructed from noise and in combination with a suitable pulsed gradient) which randomizes, in amplitude and polarity, the magnetization in regions outside the volume of interest.[39] The pre-pulse is designed so that its spectrum is zero over the range of frequencies corresponding to the localized volume and consists of noise over the remaining bandwidth. Magnetization created in the xy plane by the pulse is allowed to decay by transverse relaxation before the conventional ISIS is performed. Because the net magnetization outside the volume of interest is mostly nulled, signal comes mainly from the localized volume. Hence receiver gains can be increased, thereby improving digitization and reducing subtraction errors. If the full three-dimensional ISIS is to be carried out, then three consecutive, suitably designed pre-pulses must be applied using x, y, and z gradients. The basic ISIS technique is further improved by optimizing the magnetization inversion with the use of Silver–Hoult hyperbolic secant selective RF pulses.[40] The combination of these two variations of ISIS is called OSIRIS.[37]

In applications using surface coils, for which cancellation errors from regions near to the coil can be severe, a combination of ISIS and the HASP method[18] (described briefly in Section 5.2.1.1.), called "improved depth-selective single-surface coil spectroscopy" (IDESS) has been used.[41] In this variation, a two-dimensional ISIS experiment is performed so as to localize signal to a bar-shaped volume coaxial with the surface coil. Silver–Hoult selective pulses[40] are used so that, provided enough RF power can be applied, the magnetization inversion is very efficient over a large volume, thereby largely avoiding problems which could be introduced by the spatial variation of flip angle encountered with surface coils. Signal from tissue close to the coil is greatly reduced by collecting FIDs using pulses of different duration, the shortest of which corresponds to a 90° flip angle in the region of interest. The flip angle is increased in 180° steps by increases in pulse length equal to twice this minimum duration. A minimum of sixtyfour FIDs needs to be collected for the complete cycle of ISIS/HASP combinations, and if cyclically ordered phase sequencing (CYCLOPS) is used, this expands to 256 FIDs. With the appropriate linear combination of these 256 FIDs, the resultant spectrum originates almost entirely from the selected localization volume. The technique has been tested in vivo on the human liver and gives very good suppression of the PCr signal due to abdominal muscle.

Other techniques suggested for the improvement of ISIS involve the use of selective presaturation[42] or composite pulses.[43] Furthermore, the combination of ISIS with DEPTH pulses and water suppression[44] has been used successfully to obtain localized ^1H spectra from the adult human brain and in order to produce a sensitive volume which has an improved fit to the target lesion, a method called conformal ISIS has been suggested.[45] In its simplest form conformal ISIS involves the use of nonorthogonal gradients so as to localize the signal to an operator-defined parallelipiped rather than to a cube. A further sophistication consists of the use of additional slice-defining gradients to produce a more complicated geometric shape for the sensitive volume (e.g., four slices to define a tetrahedron). The sensitive volume can also be oriented in any desired direction. One of the main advantages of conformal ISIS is that, because the sensitive volume is larger and has a better fit to the target tissue, improved SNR can be obtained without risk of spectral contamination from exterior tissue. This permits the collection of better-quality spectra or the use of shorter-duration studies.

5.2.2.4. Phase Encoding

Data collection for phase encoding is also fairly easy to implement, in that it requires only a magnetic field gradient stepped in amplitude over successive pulses, as shown in Figure 5.6. This encodes the depths of elemental sample volumes according to the phases of their magnetization. The technique is based on an earlier imaging scheme which phase-encoded by incrementing the duration of one of three applied magnetic gradients.[46] Phase-encoded spectroscopic localization has been employed using both spin–echo[47] and single-pulse[48] approaches. A pulse is applied to the sample and a phase-encoding gradient G_x directed along the x-axis is switched on immediately. This causes the spins at locations along the x-axis to precess

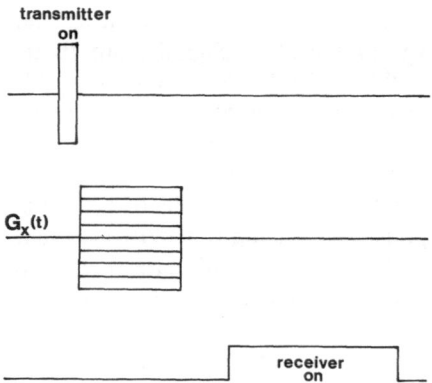

Figure 5.6. A pulse–gradient sequence for the collection of phase encoded spectra in one spatial dimension. $G_x(t)$ is a stepped magnetic field gradient along the x axis. The amplitude of $G_x(t)$ is incremented for each successive scan to provide the phase encoding. After processing (a two-dimensional Fourier transform), spectra are obtained for contiguous, parallel slabs of tissue orthogonal to the x axis.

at increasingly higher rates as the local field gets larger. Hence, when G_x is switched off, the spin magnetization has a relative phase which depends on its position along the x axis, and further differences in precession rate are now due to chemical shift only. After a delay while gradient-induced eddy currents decay, the phase-encoded FID or the spin echo (brought about by an additional 180° pulse) is collected. This process is repeated N times with an accompanying increment of G_x for each successive scan. The whole process can be repeated in order to improve the SNR by signal averaging. The N sets of data are then subjected to a two-dimensional FT which produces spectra as a function of location along the x-axis. As in the case of DRESS, the spectra are obtained from parallel slabs of tissue. However, these can be either parallel to the plane of the surface coil or perpendicular to it (phase-encoding magnetic field gradient along the y- or z-axis).

The spin–echo phase-encoding technique has been demonstrated in ^{31}P studies of rat-leg skeletal muscle with a spatial resolution of 3.3 mm.[47] One leg had occluded circulation and showed an altered metabolism when compared with the control leg. The single-pulse approach has been applied to several normal and pathological human tissues, including liver, brain, muscle, and tumor tissues, with positional resolution of 10 to 20 mm.[48] Typical total data-collection times were on the order of 15 to 20 min.

Phase encoding has several drawbacks as a localization method. In the spin–echo approach, due to the delay between the 90° and 180° pulses, the ATP and phospholipid ^{31}P magnetization undergoes substantial transverse relaxation, and these resonances are greatly reduced. The phase-encoding delay between the pulse and the beginning of data acquisition in the single-pulse method causes spectrum-baseline distortions which can make quantitation difficult. One advantage, though, is that this is a technique in which information is obtained from the whole coil volume at the same time. An extension of this method to two and three dimensions (chemical shift imaging) will be described in the next section.

5.2.2.5. Chemical-Shift Imaging

The chemical-shift imaging (CSI) method of obtaining localized spectra is a multidimensional extension of phase-encoding and is not to be confused with various imaging modalities with similar names. In the imaging procedures either the water signal or the signal originating from fatty acyl groups in the vicinity of, and including, the $(-CH_2)_n$ chain resonance, is greatly reduced by presaturation[49] or other methods including, for example, solvent suppression[50] and spatial phase-encoding in two directions.[51] Using these techniques, images are obtained of the relaxation properties or spin densities of water or fat only, in contrast to what occurs in the conventional modality, which in fact produces two strong overlapping images (one of water and one of fat). At high field strengths, due to the difference in chemical shift, the water- and fat-proton images may not coincide very well, and this can lead to additional interpretational difficulties which are avoided if one of the subimages is suppressed.

The spectroscopic CSI in which we are interested is essentially a two- or three-dimensional version of the phase-encoding method[46] described in the previous section. Spatial information is obtained by three-dimensional phase encoding in which gradients $G_y(t)$ and $G_z(t)$ are incremented scan by scan, in addition to the gradient $G_x(t)$ shown in Figure 5.6. Alternately, the RF pulses can be selective so as to define slices, and phase encoding takes place in two dimensions, with the selective pulse gradient held constant in order to define each slice. (An example of the application of this form of CSI to the human brain is shown in Figure 3.9). No gradient is applied during data acquisition, and hence chemical shift information is preserved. Spectra are obtained for each cube (voxel) in a three-dimensional lattice; in the case of two-dimensional phase-encoding in combination with a selective pulse, the lattice is a layer only one voxel thick. If ^{31}P is the nucleus to be studied, it is necessary to use rather coarse volume resolution (typically greater than $3 \times 3 \times 3$ cm^3) in order to obtain adequate SNR because of the low tissue concentration and sensitivity. One problem with a direct three-dimensional version of the previously described spin–echo phase-encoding method[47] is that signals are greatly reduced from resonances which have short T_2s, such as ATP. An early attempt at three-dimensional CSI used the FIDs rather than spin echoes,[52] and recent improvements have made this method, which avoids the suppression of short T_2 resonances, suitable for application in vivo.[53, 54] Spectroscopic information is obtained simultaneously from the whole slice or volume, and reasonable spectra can be obtained within 15 to 30 min for each of the voxels.[54, 55] This method, like OSIRIS, has the distinct advantage that the tissue volume to be localized can be directly related to a conventional ^{1}H magnetic resonance image. Spectra from voxels can be added together after data collection to produce irregular localized volumes which correspond as closely as possible with the spatial distribution of tissue pathology. Absolute quantitation of metabolite concentrations can be achieved simply by ensuring that a vial containing a standard solution is contained in a voxel within the field of view but exterior to the subject.[54, 55]

5.2.2.6. Stimulated-Echo Methods

Of particular application to ^{1}H localized spectroscopy are techniques which utilize stimulated echoes. These echoes are produced when three RF pulses are applied to a sample at various time intervals.[56] Because three pulses are required, if these are selective and the successive gradients are mutually orthogonal, the stimulated echoes will originate only from a localized cubic region. Two techniques utilizing stimulated echoes have been suggested—stimulated-echo acquisition mode (STEAM)[57] and volume-selective spectroscopy (VOSY).[58] Recent applications have included the use of STEAM with water-resonance suppression to obtain localized ^{1}H spectra of the human brain[59, 60] and the application of VOSY for localized, in vitro spectral editing of lactate.[61]

5.2.2.7. Other Localization Methods

The number of potential techniques is increasing rapidly, and it has been possible to describe in any detail only a few of the methods which have been utilized so far. New localization tools are constantly being developed, and the direction of research is consistently aimed at producing methods which are simple to use and give better sensitive-volume definition and improved rejection of signal from outside this volume. Many suggestions have been made in addition to those described above. Among these are

1. Volume-selective excitation (VSE),[62, 63] which employs three successive "sandwich" pulses consisting of two 45° selective pulses with an intervening 90° hard pulse and accompanying x, y and z gradients for each "sandwich."
2. A variation of VSE called "spatially resolved spectroscopy" (SPARS),[64, 65] which employs a somewhat more complicated pulse–gradient sandwich, has been used in combination with water-suppression techniques to obtain localized ^{1}H spectra of metabolites such as N-acasp, PCr, and Cr from in situ human brain.

3. A novel surface-coil technique called "homogeneity spoil spectroscopy,"[66] in which a thin layer of ferromagnetic particles is interposed between the surface coil and the subject, thereby creating a locally inhomogeneous magnetic field. This destroys the coherence of surface signals while preserving signal from deeper tissue such as liver or brain.

4. "Fast-rotating gradient spectroscopy" (FROGS),[67] which combines the intrinsic localizing properties of a surface coil with a series of selective saturating pulses, thereby destroying the magnetization in a chosen slab of tissue (usually superficial in order to reduce signal from surface muscle, e.g., in studies of brain or liver).

5. A method called "discrete isolation from gradient-governed elimination of resonances" (DIGGER)[68, 69] employs two frequency-shifted sinc pulses and pulsed-field gradients to destroy the coherence of unwanted signals, and then a hard 90° pulse is used to read the z-magnetization in the region of interest, giving complete T_2-independent volume selection in a single acquisition.

6. "Point-resolved rotating-gradient surface-coil spectroscopy" (PROGRESS)[70] uses a slice-selective 90° pulse after the application of a Gaussian-shaped 180° pulse and two orthogonal time-varying magnetic field gradients.

Some of the preceding localization techniques can be combined with water suppression and are, therefore, applicable to in vivo ^1H spectroscopy (permitting the investigation of lactate and amino acids, etc.) as well as to the study of ^{31}P and other nuclei. It is of interest to note that several spectrometer manufacturers have implemented software routines on their spectrometers to run some of the above techniques including ISIS, OSIRIS, VOSY, and DEPTH pulses.

5.3. SPIN–ECHO TECHNIQUES

The underlying principles of spin–echo magnetic resonance spectroscopy have been covered in Chapter 2 and are also well described in the literature.[71, 72] We will now deal with some of the uses of spin echoes in the investigation of in vivo tissue.

5.3.1. Simplification of Spectra

Spin–echo techniques are used quite commonly to simplify spectra (particularly ^1H spectra) by reducing the signals obtained from nuclei which have short T_2s. Many ^1H spectra obtained from samples containing proteins exhibit broad overlapping peaks which make interpretation difficult unless a simplification method is adopted. For this purpose several multipulse spin–echo techniques have been investigated. These include two sequences commonly used to measure spin–lattice relaxation times (T_1s), $180° - \tau - 90°$ and $(90° - \tau - 90°)_n$, as well as the Carr–Purcell methods A and B,[73] which can be used to measure spin–spin relaxation times (T_2s), $90° - \tau - 180° - \tau$ and $90° - (\tau - 180° - \tau)_n$, respectively.[74] In these pulse sequences, τ represents an interpulse delay, n indicates repetition of the bracketed section n times; in the latter sequence the 90° pulse can have its phase shifted by 90° with respect to the 180° pulse.[75] This variation is commonly known as the Carr–Purcell–Meiboom–Gill (CPMG) technique. The total duration of pulse sequence B is $2\tau_n$, and this is typically several hundred msec with a τ of the order of a few msec. The 90° phase shift helps to compensate for inaccurate setting of the 180° pulse, which would otherwise cumulatively shift the magnetization away from the xy plane. The most effective sequence for spectral simplification of the four has been found to be the Carr–Purcell method B. This is because the signals from nuclei with short T_2s decay rapidly and, by choosing a suitable value for τ, resonances from nuclei with longer T_2s are selectively observed. As a rule the signals from large, relatively immobile molecules are suppressed in favor of smaller species. NH resonances are reduced because of their very short T_2s, but resonances from protons that are relatively distant from other protons, such as the C2 and C4 protons of histidine, are preserved due to their high mobility. Sequence A has been used to study human erythrocyte metabolism, with τ ranging from 20 to 60 msec.[76] Sequence B (with the 180° pulse phase shifted by $+90°$ and $-90°$ on alternate scans) has been applied in vivo to obtain ^{1}H spin–echo spectra from rat muscle using $\tau = 1.2$ msec and pulse trains lasting 270 amd 540 msec[77] (see Figure 1.6).

5.3.2. Spin Echoes with Surface Coils

The RF field of a surface coil is inhomogeneous and, as discussed in Chapter 4, the flip angle θ varies as a function of position. Only a small local region of the sample can experience flips close to 90° and 180°. Hence, when used with surface coils, it is better to rewrite the Carr–Purcell sequences A and B as $\theta - \tau - 2\theta - \tau$ and $\theta - (\tau - 2\theta - \tau)_n$. Because of the nonuniform flip angle, unwanted signals, termed "ghosts" and "phantoms," are generated and can distort the spectra.[14] These unwanted contributions to the spectra can be substantially reduced by using a phase-cycling protocol (EXORCYCLE) for the 2θ pulse.[15] This is implemented in a fashion similar to that of the cycling used for localization with the DEPTH-pulse sequence [see section 5.2.1.1. and the pulse sequence in equation (5.3)].

Table 5.3. EXORCYCLE Pulse-Phasing System for the Reduction of Phase and Amplitude Distortions in Spin-Echo Spectra[a,b]

Scan	θ pulse phase	2θ pulse phase	Add/subtract
1	x	x	$+$
2	x	y	$-$
3	x	$-x$	$+$
4	x	$-y$	$-$

[a] From Bodenhausen et al.[13]
[b] The phases x, y, $-$x, and $-$y correspond to phase angles relative to that of the θ pulse of 0°, 90°, 180°, and 270°, respectively.

Table 5.3 gives the protocol required to implement EXORCYCLE phase sequencing for surface coils. More complicated and accurate compensation systems for use with surface-coil spin echoes have been developed.[78, 79]

In certain situations it has been found unnecessary to use phase cycling, particularly when T_2 and τ are several times longer than T_2^*.[80] In this case the unrefocused magnetization, which causes the spectral distortion, decays much faster than the refocused magnetization. For small biological metabolites, T_2 is often much longer than T_2^*, and it is therefore possible, in many cases, to utilize surface coils for spin–echo spectroscopy without phase cycling, providing τ is long enough.

Surface coil ^{1}H spin–echo sequences have been used successfully in vivo for the detection of histidine in the brains of histidinemic mice using

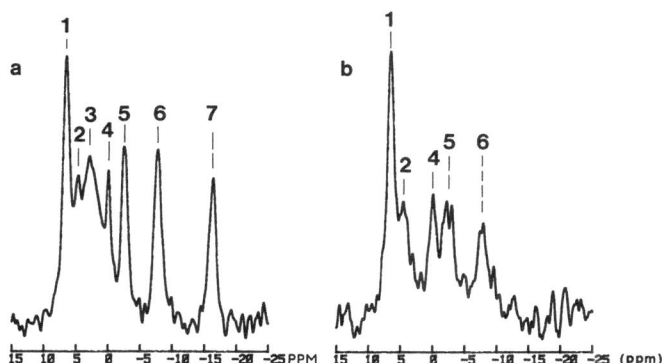

Figure 5.7. A comparison between single-pulse and spin–echo ^{31}P spectra. (a) A ^{31}P single-pulse spectrum of neonatal brain. The FID was the sum of 256 scans at 2.256-sec intervals using a 90° flip angle at the center of a 5-cm-diameter, single-turn surface coil. Peak identifications: 1, phosphomonoesters; 2, inorganic orthophosphate; 3, phosphodiesters; 4, phosphocreatine; 5, 6, and 7, the γ, α, and β phosphates of nucleoside triphosphates. (b) A ^{31}P spin–echo spectrum obtained from the same infant using a τ delay of 10 msec. Peaks 3 and 7 are absent from the spectrum due to their short T_2s. (Spectra obtained in collaboration with P. Hope, P. Hamilton and E.O.R. Reynolds.)

EXORCYCLE[81] and for the detection of lactate in rat brain without the need for phase cycling.[82] ^{31}P spectra from adult human brain have been simplified by using a spin–echo sequence with a τ of 30 msec[83] (see Figure 3.8). The phosphodiester signal was greatly reduced relative to PCr and, due to their very short T_2s, the nucleoside triphosphate (NTP) resonances and the broad features due to phospholipid and bone phosphate, normally prominent in single-pulse brain spectra, were practically indiscernible. If a heteronuclear spin–echo sequence is used (i.e., the coil is tunable for both ^{1}H and ^{31}P and a ^{1}H 2θ pulse is applied immediately preceding the ^{31}P 2θ pulse), then the phosphomonoester and phosphodiester resonances are completely nulled, due to phase modulation of the proton-coupled fine structure, when $\tau = 30$ msec. This leaves the Pi resonance isolated and, because of NOE,[84] the Pi signal is increased by approximately 30%. In this way the employment of ^{31}P spin–echo methods can improve the measurement of Pi chemical shifts, and hence intracellular pHs, because many of the resonances which cause peak overlap are greatly reduced, as shown in the neonatal-brain spectrum displayed in Figure 5.7b.

5.4. SOLVENT SUPPRESSION

One of the greatest problems with the application of ^{1}H magnetic resonance spectroscopy to biological samples is the high concentration of water protons. This concentration can be approximately 80 M in many tissues and presents enormous problems when digitizing FIDs in order to obtain information from metabolites which may have millimolar concentrations. The concentration ratio between water protons and the metabolites of interest will often be of the order of 10,000:1. This means that, if a twelve-bit analogue-to-digital converter (ADC) is used, the resonances required for study may be digitized using only the last one or two bits. The situation can be improved if more bits are available in the ADC. However, it has been found that further techniques are often needed to reduce the water signal while preserving the signals from the metabolites of interest. Various approaches have been suggested[72] including composite pulses, spin echoes, saturation of the solvent magnetization, and binomial pulse sequences. These all attempt either to avoid flipping the solvent magnetization or to arrange that the solvent magnetization is very small during data acquisition.

5.4.1. Composite Pulses

One of the earliest solutions to the problem of solvent suppression was the Redfield pulse[72, 85] (also known as the 214 pulse). This consists of a long RF transmitter pulse, during which the phase is switched by 180° for two segments, each of duration one tenth that of the entire pulse. This pulse can be represented by $2t_x - t_{-x} - 4t_x - t_{-x} - 2t_x$, where nt is the duration of a part of the pulse sandwich and x or $-x$ represents the phase of that part (i.e., either 0° or 180°). It can be arranged that the spectrum of such a pulse has very little power near the frequency of the water resonance, and hence the water spins flip by only a small angle. If the total duration of the 214 pulse is $10t$ (as in the sequence above), then a null in sensitivity is obtained $1/(10t)$ Hz away from the transmitter frequency. The spectrum of the metabolites of interest should be centred on the transmitter frequency and t set so that the chemical shift of the water resonance is at the sensitivity null. The Redfield 214 pulse is capable of achieving solvent suppression of about 300:1 when using a homogeneous RF field (e.g., solenoid, saddle, or other volume coil) but can also give significant suppression when used with a surface coil. Additional solvent reduction can be obtained by fine adjustment of the length of the $4t_x$ section of the composite pulse (in that case called the Redfield $21X$ pulse).[86] Particular care should be taken to monitor tissue-power deposition if Redfield pulses are applied with human subjects.

5.4.2. Spin Echoes

The T_2 of tissue water is often substantially less than that of many metabolites of interest. For instance, in rat muscle the water T_2 has been determined to be approximately 30 msec, whereas T_2s for the resonances of PCr (at 3.03 ppm), taurine (at 3.23 and 3.43 ppm), and anserine (at 7.01 and 8.02 ppm) have been found to be 246, 571, 563, 382, and 473 msec, respectively.[77] The T_2 of water in brain tissue is similarly much shorter than that of metabolites such as lactate and N-acasp.[82] This means that, by judicious selection of spin–echo times, the water resonance can be reduced relative to the signal from other metabolites, and this permits the use of higher receiver gains. This reduces the digitization problem, and in many cases can permit the collection of acceptable ^1H spectra. An example of this can be seen in the ^1H CPMG spin–echo spectra from rat muscle[77] shown in Figure 1.6. The signal from tissue water is greatly reduced. In particular, in the 540 msec spectrum, the water peak has decreased to give a peak area comparable to those of metabolites with millimolar concentrations.

For short spin–echo pulse sequences, the increase in tissue power deposition (relative to that from a singe pulse) may be well within safety limits. Therefore this method for the suppression of the signal from tissue water can, with care, be applied to human subjects.

5.4.3. Saturation Methods

One of the commonest approaches to tissue-water suppression in studies of in vitro or in vivo animal tissues has been to presaturate the water resonance using a long (and hence frequency-selective), low-power RF transmitter pulse at the same frequency as the water resonance.[72, 87, 88] A conventional read pulse is then applied in order to obtain an FID. Solvent-suppression ratios of at least 1000:1 are possible with homogeneous saturation employing a low-power RF pre-pulse of duration 0.1 to 1 sec. Presaturation methods have also been applied in vivo, using surface coils, and in this case a 2-W RF field was swept in frequency over a 200-Hz range centered on water; the sweeping was recycled ten times every 0.4 sec.[89] Great care has to be taken with tissue power deposition and the consequent heating effect. It is probably advisable that presaturation (or any other method requiring long pulses of low but significant power) be avoided when performing studies on human subjects.

5.4.4. Binomial Pulses

The binomial pulse techniques[72, 90–94] are potentially useful for clinical application but have also been employed on intact tissue and dilute solutions. Significant solvent suppression is achieved while using no more power than that of a single pulse producing a comparable flip angle. The method does not rely on a homogeneous RF field, and therefore has potential for use with surface coils. A major drawback is the large variation in signal phase across the observed spectrum, although this can usually be corrected adequately.

A whole family of binomial pulse sequences capable of solvent suppression exists, with a range of suppression and sensitivity-bandwidth properties. The suppression effect depends on the intervals between the constituent pulses. These allow the magnetization of the metabolites of interest to precess about the z axis of the rotating frame, and successive pulses in the sequence flip this magnetization toward the xy plane. The water signal is usually placed on resonance (for some sequences, however, the metabolite resonances of interest are centered on resonance). Successive pulses in the sequence produce negative and positive flips of the water magnetization, which eventually returns to its initial orientation, parallel to the z axis. Hence the water gives only a small signal. Examples of binomial suppression sequences are 11, 121, 1$\bar{1}$, 13$\bar{3}$$\bar{1}$, and 1$\bar{4}6\bar{4}$1. Here 13$\bar{3}$$\bar{1}$ is a shorthand representation of the sequence

$$\theta_x - \tau - 3\theta_{-x} - \tau - 3\theta_x - \tau - \theta_{-x}$$

where θ is the 1 pulse, 3θ is the 3 pulse, τ is the interval between pulses, and x or $-x$ is the pulse phase. The suppression effect of the binomial pulses can be explained visually by use of a series of vector diagrams as shown in Figure 5.8, which shows the flips and precessions experienced by the water and the metabolites of interest during a $1\bar{1}$ pulse. The water signal is arranged to be on resonance and therefore, in the rotating frame, its magnetization does not precess about the z axis during the interval between the θ_x and θ_{-x} pulses. The water magnetization is thus flipped initially away from and then back onto the z axis. Hence little signal is obtained from the water. The metabolites of interest are arranged to be off resonance, and after the θ_x pulse their magnetizations precess about the z axis. τ is set so that, during the interval between the θ_x and θ_{-x} pulses, a metabolite at the center of the spectral band of interest will have precessed by 180°, bringing it directly opposite the water magnetization on the other side of the z axis. The θ_{-x} pulse now flips the metabolite magnetization further away from the z axis and, by giving θ an appropriate value (in this case 45°), the metabolite magnetization can be flipped into the xy plane, allowing the detection of an FID. If the center of the required spectral band is off resonance by v Hz, then the interval τ between the constituent pulses should be set to $1/2v$. At $2v$ Hz off resonance, the metabolite magnetization will precess by 360° during τ and hence will be returned to its starting point and flipped back to the z axis along with the water. This produces a further null in the spectral sensitivity of the sequence. This simple explanation can be easily extended to the more complicated binomial sequences.

The model just described simulates the behavior of the on-resonance

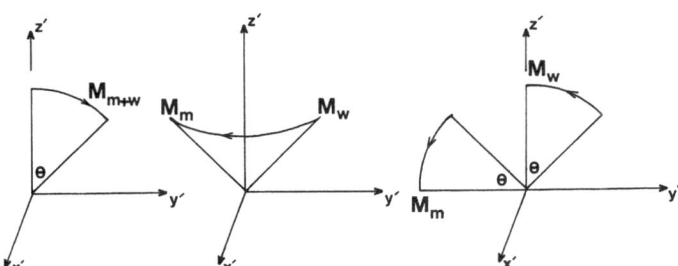

Figure 5.8. A clock diagram illustrating the $1\bar{1}$ binomial solvent-suppression pulse in a rotating frame of reference (x', y', z'). For simplicity only the behavior of the water magnetization $(M_w$, on resonance) and the magnetization of one metabolite $(M_m$, $1/2\tau$ Hz off resonance) are shown. (a) M_{m+w} is flipped about the x' axis through angle θ. (b) During the period τ, M_m precesses about the z' axis through 180° because of its frequency offset $(1/2\tau)$ from resonance. M_w, which is arranged to be on resonance, remains in the $z'y'$ plane. (c) The second pulse in the $1\bar{1}$ sequence flips M_w back to the z' axis and therefore no signal is produced. At the same time, M_m is flipped by $-\theta$ which, if $\theta = 45°$, brings M_m into the $x'y'$ plane, and an FID is detected.

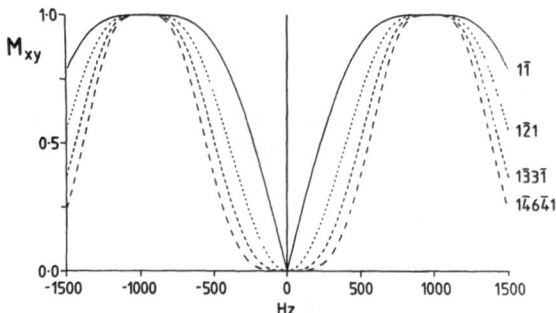

Figure 5.9. The calculated sensitivity as a function of offset frequency from resonance for several binomial solvent-suppression sequences. M_{xy} is the transverse magnetization excited by the sequences. The sensitivity functions were calculated with $\tau = 500$ μsec, $\gamma B_1/2\pi = 5$ kHz, and the flip angle of the first pulse in the sequence equal to 45°, 22.5°, 11.25°, and 5.625° for the $1\bar{1}$, $1\bar{2}1$, $1\bar{3}3\bar{1}$, and $1\bar{4}6\bar{4}1$ sequences, respectively (from Hore,[85] Academic Press Inc.; reprinted with permission).

solvent magnetization and that in which metabolite magnetization is resonating v Hz off resonance. In order to describe the effect of these pulses on the magnetization of resonances at other frequencies, a more complicated approach has to be adopted.[94] The results of such analyses reveal sensitivity functions whose detailed shapes depend on the particular binomial pulse sequence used but which have the following features in common, as shown in Figure 5.9. The sequences which incorporate a 180° phase change on alternate pulses all have a sensitivity null on resonance that becomes broader as the pulse complexity increases. The sensitivity function has a broad maximum, centered on v Hz, which gets narrower as more components are incorporated into the suppression sequence. For pulse combinations which do not involve phase inversion, a sensitivity null is situated v Hz off resonance, while maximum signal is obtained on resonance.

The phase of the signals collected using binomial pulse sequences has a complicated dependence on off set frequency,[94] sometimes making the phase correction of spectra difficult. For pulse sequences involving 180° phase inversions, there is a 180° phase jump on resonance, and substantial phase gradients exist in the main lobes of the sensitivity function.

Recently several improvements to the binomial suppression technique have been suggested in order to reduce the amount of phase correction required and to obtain sensitivity functions with broader maxima.[95–97]

5.5. SPECTRAL EDITING

The spin–echo techniques described in Section 5.3 are very useful for the simplification of spectra; however, they are quite blunt tools in that they have a similar T_2-selective effect on a wide range of metabolite resonances. More precise methods will now be presented that allow the spectroscopist to select a particular resonance and either remove it from the spectrum or preserve it while eliminating the rest of the spectrum.

5.5.1. Reduction of Bone and Phospholipid Signals

Many spectra obtained from mammalian and human brain have a strong, broad underlying hump due to relatively immobile ^{31}P nuclei in phospholipid and bone, as shown in Figures 1.9 and 3.1. This broad hump is particularly strong if the data are collected using surface coils and without further localization methods to reduce signal from superficial tissues such as cranial bone. This common feature of brain spectra can make data analysis difficult because, unless it is removed, peak areas have to be measured above a curved baseline. One approach is to remove the hump after data collection by the use of various computer techniques, which will be described in Chapter 6. Alternately, the broad resonance can be suppressed during data collection, as will now be described.

Several workers have successfully eliminated most of the broad bone–phospholipid component of in vivo spectra by applying selective saturation using a long-duration, narrow-bandwidth RF pulse at a suitable frequency away from the resonances of interest.[98–101] The saturation pulse is applied immediately before the conventional data-acquisition pulse and is of power and duration adequate to saturate the entire phospholipid–bone resonance. The method does not seem to be sensitive to the precise frequency of the RF saturation pulse. Selecting a frequency which positions it outside the spectrum[98, 99, 101] ($+15$ to $+20$ ppm with respect to PCr but still on the wing of the phospholipid–bone hump) has the same effect as using a frequency between the γ and α NTP resonances.[100] Care must be taken when using this technique, because it has been reported that the phosphodiester signal can also be reduced in strength.[101]

An alternate method for the suppression of the bone–phospholipid hump is to use a DANTE pulse sequence.[102] This consists of a finite series of regularly spaced, low-flip-angle hard pulses. In the frequency domain, the power spectrum of such a pulse sequence (with the constituent pulses at intervals of τ sec) consists of an on-resonance peak with sidebands of similar amplitude repeating at a separation of $1/\tau$ Hz. By judicious positioning of the ^{31}P spectrum with respect to the on-resonance frequency and careful selection of τ, it can be arranged that the broad hump is selectively saturated. The main advantage of such an approach is that less power is used to obtain saturation, and this technique may therefore be safer to use in vivo. This method has been successfully used in studies of both piglet brains and those of human neonates.[103, 104]

5.5.2. Spectral Editing in ¹H Spectroscopy

Spectral editing is of particular importance in ¹H spectroscopy due to the narrow bandwidth over which resonances are observed and because of the numerous peaks, many of which are multiplets or broad protein features. At the relatively low field strengths permissible in clinical use, spectral simplification is necessary if the severe problem of peak overlap in ¹H spectra is to be dealt with so as to facilitate the detection and accurate measurement of the areas and positions of individual resonances. The detection of lactate in cerebral tissue is an example of this problem. The easiest lactate resonance to observe in vivo is that due to the methyl protons. This is a doublet and resonates approximately -3.5 ppm from the water. (The methene proton gives a quartet which has a chemical shift closer to that of the water and is, therefore, not so easy to use.) Unfortunately, many fatty acyl-chain protons also resonate in the vicinity of the lactate methyl doublet. Peaks from these protons can be very strong, especially if signal is detected from subcutaneous fat. In addition, many resonances from amino acids lie close by, and at low fields it is often impossible to resolve this part of the spectrum. It has also been reported that following traumatic injury certain resonances closely adjacent to the lactate doublet increase in magnitude, and due to peak overlap, erroneous results could be obtained. However, edited spectra show no excess increase in lactate levels.[105] Obviously, if signals due to resonances other than the lactate methyl protons can be eliminated from the spectrum, then estimates of lactate concentration are both easier to obtain and more reliable.

There are two basic approaches to this type of spectral editing: those which require double irradiation and those which rely on frequency-selective pulses. Both methods depend fundamentally on the phase modulation which occurs when spin–echo techniques are used. (Phase modulation has been described in Section 2.3.2.2b.)

5.5.2.1. Double-Irradiation Methods

A pulse sequence for double-irradiation spectral editing[106, 107] is shown in Figure 5.10. A spin–echo sequence is used with a τ delay of $1/2J$ where J is the proton–proton coupling constant for the resonance under consideration. (This is equivalent to the multiplet component separation.) For the lactate doublet, J is about 7.4 Hz and, due to phase modulation (caused by coupling to the methene proton), if τ is set to $1/2J$ (approximately 68 msec) the resonance is phase inverted. If a decoupling signal (continuous RF of controllable power and duration) at the methene-proton resonant frequency (about 4.1 ppm) is applied to the sample during the spin–echo sequence as shown in Figure 5.10, phase modulation is

Figure 5.10. A pulse sequence for ^{1}H spectral editing, involving double irradiation. The 90° and 180° transmitter pulses produce a spin echo at time 2τ. During the period between the 90° pulse and the commencement of data accumulation, additional power at the specific frequency of the coupled spins is applied on alternate scans. Spin echoes are added to the integration array when decoupling is on and are subtracted when it is off.

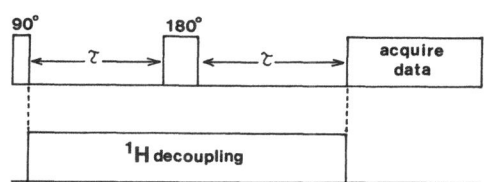

inhibited, and the doublet is observed with normal phase. The sample is observed with the decoupling power on for alternate scans only. The resultant spin echoes are added into the integration array when the decoupling power is on and subtracted when the power is off. The result is that the lactate doublet which had its coupled spin irradiated is preserved, whereas other resonances are nulled. The method can be applied to other multiplets apart from lactate. If the alanine CH proton (at about 3.75 ppm) is irradiated in a sample of rat skeletal or cardiac muscle, glutamate, glutamine, and alanine peaks are preserved because they are all coupled to protons resonating at adjacent chemical shifts.[106] In studies of mouse liver, decoupling irradiation at 4.1 ppm preserves the signal from β-hydroxybutyrate. The method has been applied also to in vivo rat brain and previously unresolved resonances due to alanine, GABA, glutamate, and taurine were observed, in addition to lactate, which was detectable through the intact scalp.[107]

A variation of the above technique employs ^{13}C decoupling and allows the selection of signal from only those protons which are coupled to ^{13}C nuclei.[108] This has been used in ^{13}C glucose and glycogen enrichment studies to investigate lactate and glutamate formation.

5.5.2.2. Methods without Double Irradiation

Spectral editing can also be successfully achieved in vivo without the need to use a decoupler.[109] If a spin–echo method is used, with the θ and 2θ pulses replaced by binomial water-suppression sequences as below,

$$1\bar{3}3\bar{1} - \tau - 2\bar{6}6\bar{2} - \tau - \text{Acquire Data}$$

the phase modulation of a doublet, such as the methyl lactate resonance, is inhibited because the coupled proton resonance (the methene hydrogen resonating at about 4.1 ppm) is suppressed along with the water. In order to edit the spectrum, scans are acquired using alternately the sequence above, which we call (A), and the following sequence, (B):

$$1\bar{3}3\bar{1} - \tau - 2\theta - \tau - \text{Acquire Data}$$

With τ set to approximately 68 or 204 msec, phase inversion of the lactate doublet is obtained using sequence (B), whereas this is inhibited by the second binomial pulse in sequence (A). Data are alternately added or subtracted into the integration array, and resonances apart from the lactate doublet are cancelled from the resulting spectrum. Alternatively the results of the sequences (A) and (B) can be stored separately so that both the sum spectrum (which will contain all the other resonances apart from the lactate doublet) and the difference spectrum (containing only the lactate doublet) can be obtained if needed. The binomial pulses have to be designed so that maximum sensitivity is obtained for the lactate peak, and this means careful setting of the precession delay between the components of the binomial pulse (see Section 5.4.4). If quantitative measurements of lactate are to be obtained, then the peak area has to be multiplied by a phase-modulation factor, approximately 1.2.[110] This is because, in sequence (B), due to the spatial variation of flip angle when a surface coil is used, the 2θ pulse is not 180° throughout the sample, and the phase modulation

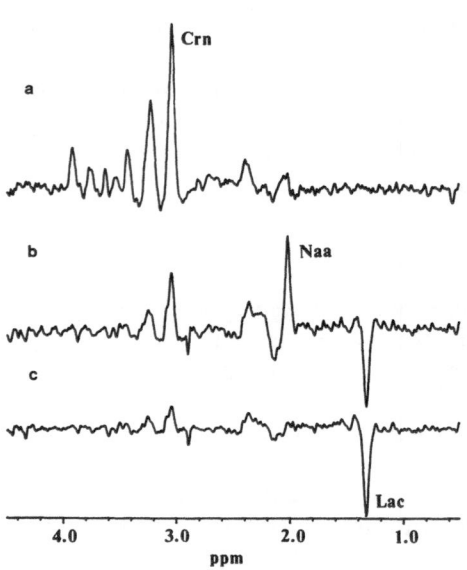

Figure 5.11. ¹H spectral editing for the lactate methyl doublet, using binomial pulses to suppress phase modulation. The spectra were collected from postmortem gerbil brain at 360 MHz using a 10 × 12-mm surface coil placed directly on the exposed cranial bone. (a) The spectrum obtained using a 68-msec spin–echo sequence with the first pulse as a binomial water-suppression pulse ($1\bar{3}3\bar{1} - 180°$). The binomial sequence has been designed to give maximum sensitivity at the resonant frequency of creatine (Crn). (b) As in (a), but the binomial water-suppression sequence has been designed to give maximum sensitivity for N-acetylaspartate (Naa). (c) The spectrum edited for the lactate methyl resonance (Lac) by using alternate $1\bar{3}3\bar{1} - 180°$ and $1\bar{3}3\bar{1} - 2\bar{6}6\bar{2}$ sequences. Note the almost-complete suppression of all other resonances (from Williams, S. R. et al.,[101] Academic Press Inc.; reprinted with permission).

is incomplete. A smaller lactate doublet signal is thus obtained when compared with that detected using sequence (A). Figure 5.11 shows a pair of spectra obtained from postmortem mammalian brain and indicates the efficiency with which this method edits out unwanted resonances. This method of spectral editing has been successfully used to obtain absolute lactate concentrations from brain by comparing the methyl lactate doublet with internal-concentration reference signals (N-acasp, PCr + Cr, and tissue water).[110]

DANTE pulses[102] can also be employed to edit for lactate[111, 112] and using this approach it has been shown that, unless an editing method is used, misleading changes in the strength of resonances adjacent to the lactate doublet can give rise to erroneous results.[105]

5.6. ABSOLUTE QUANTITATION

Although much useful information is obtained from the relative concentrations of metabolites, the ability to obtain absolute values is very important. In the absence of reliable direct methods of absolute quantitation, early attempts to estimate metabolite concentrations in human tissue (mainly in muscle[113, 114] and neonatal brain[117]) were indirect and depended on knowledge of either the total mobile phosphate concentration or the concentration of a particular, normally stable metabolite (usually ATP). For instance, in studies of normal, resting human muscle,[114] metabolite concentrations have been calculated from relative peak areas assuming either an ATP concentration of 5.5 mmole/kg wet wt. or a total mobile phosphate molarity of 49.5 mmole/kg wet wt. Both approaches gave very similar results, and favorable comparison with biopsy data was possible. One problem with the use of ATP as a concentration reference in some tissues, such as brain, is that the γ, α, and β resonances contain significant contributions from other nucleoside triphosphates.[116, 117] Further, if the molarity of ATP or total mobile phosphate changes due to physiological or pathological processes (for instance due to fatiguing exercise in muscle studies), then this absolute quantitation system breaks down and alternative, direct methods have to be sought. Potential applications of direct absolute quantitation methods in the clinic include the sequential monitoring and estimation of damage to brain tissue after a hypoxic or ischemic episode (e.g., birth asphyxia, stroke) or following the progress of a condition such as Duchenne muscular dystrophy.

There are two main approaches to direct absolute quantitation in vivo, and these utilize either external or internal concentration references. An external concentration reference consists of a vial or capillary containing a standard solution located outside the sample but within the sensitive volume of the detection coil. An internal concentration reference is either a stable metabolite occuring naturally in the tissue (endogenous) or a convenient compound which can be introduced safely into the organ being studied (exogenous). A review of the methods currently available for the measurement of metabolite concentrations by MR spectroscopy has recently been published.[118]

The following sections will be almost completely devoted to describing the methods which have been shown so far to have potential for clinical application. Several results from the use of these methods on human tissue in vivo are given in Chapter 3.

5.6.1. External Concentration References

External reference samples have been more commonly employed to determine accurate chemical shifts, but have sometimes also been used for the purpose of estimating metabolite concentrations. Usually a capillary, filled with a standard solution of known volume, is placed close to the sample. If possible, the sample and the reference should both be in a region of locally uniform coil sensitivity. This will not always be possible for in vivo studies, and the standard has often to be placed away from the sample (e.g., immediately on the other side of a surface coil). There are many difficulties with the use of external references, including estimation of the volume of the tissue from which signal originates and the spatial variation of sensitivity and flip angle, which is never exactly uniform (particularly when surface coils are used). Alternatively, the spectrometer response can be calibrated by using a standard phantom solution in place of the tissue to be investigated. Similar problems of tissue volume are encountered. It is also difficult to exactly match the effects of coil loading produced by the sample's conductivity. Successful use of external concentration references has been reported in studies of mammalian cardiac[119, 120] and skeletal muscle.[121] In clinical applications, external references have permitted the determination of absolute concentrations in CSI studies of human brain[54, 55] (see Section 5.2.2.5.). They have also been used as a monitor of coil loading in ISIS studies so that absolute concentrations could be determined using phantom solutions.[122]

5.6.2. Internal Concentration References

In many situations, a method using an internal concentration reference may be the best choice. The reference substance is within the tissue of interest (either naturally or by introduction) and therefore can be located in the same region of the static magnetic field and the coil's sensitive volume. This can avoid many of the quantitation difficulties associated with external concentration references. One problem, however, is that the absolute concentration of the internal reference may not be known as precisely as that of an external reference. Further problems are that the internal reference concentration may change with pathology and that the reference resonance may contain contributions from other substances. There are two main classes of internal concentration reference substances: homonuclear references (in which the substance is observed as part of the spectrum of interest) and heteronuclear references (for which a spectrum of another nuclear species must be collected and therefore the coil must have a multinuclear capability).

5.6.2.1. Internal Homonuclear Concentration References

These have the distinct advantage that they resonate in the same spectrum as the metabolites of interest. The process of quantitation is therefore protected from errors such as those due to system-calibration inaccuracies, gain corrections, or changes in spectrometer sensitivity.

The use of ATP or the total mobile phosphate as a ^{31}P spectroscopy internal reference and the limitations of this technique (dependence on the maintenance of normal physiological conditions) has already been described. It appears that, except in physiologically stable circumstances, a heteronuclear approach may be best for ^{31}P absolute quantitation by internal referencing, although one report on the invasive use of methylphosphonic acid in studies of rat kidney has been published.[123]

One of the major uses of internal homonuclear concentration references has so far been in ^{1}H spectroscopy. This has permitted estimates of amino-acid and lactate concentrations. Successful quantitation of histidine in the brain tissue of histidinemic mice has been achieved using surface-coil spin–echo and binomial water-suppression techniques. Both the PCr + Cr and *N*-acasp signals were utilized as concentration references.[81] To obtain quantitative values, it was necessary to measure the T_2s of the resonances so that the areas of peaks in the spectrum could be related to the tissue concentrations. Good correlation was found between the results obtained by in vivo quantitation and subsequent measurements on brain extracts. One of the reasons for success in quantitating histidine was undoubtedly due to the region of the spectrum in which the C2 and C4 histidine protons resonate. With a spin–echo time of 20 to 60 msec, this region was quite clear of other peaks. The resonance from the methyl protons of lactate is often overlapped by several other resonances, mainly those originating from lipids. In spite of this problem, the application of spectral editing techniques (described in Section 5.5.2.2.) has permitted very convincing in vivo quantitation of brain lactate.[110, 124] This has been achieved using the resonances of Cr, *N*-acasp, and brain-tissue water as concentration references. The N-acasp was found to give lactate concentrations which agreed best with those obtained by measurements of extracts. Further, it has been shown that cerebral lactate is almost 100% detectable, whereas it was once thought that a large fraction was unobservable.[125] Care must be taken with the use of N-acasp as a concentration reference in brain tissue because it has been shown that, its concentration depends not only brain development but also on anatomical location.[126]

5.6.2.2. Internal Heteronuclear Concentration References

The estimation of metabolite concentrations by referencing to the signal from a different nucleus[127] has achieved some degree of success. The main application has been the use of tissue water for the determination of absolute concentrations of phosphorus metabolites[128] although, in principle, the method could be applied to other nuclei. Because concentrations are obtained by comparing signals obtained from different nuclei and hence from separate spectra, great care has to be taken to ensure that the spectrometer's receiver channels have very stable gains. The application of this method depends on knowledge of the tissue-water concentration, calibration of the ^{31}P and ^{1}H channels of the spectrometer using standard solutions, matching the spatial sensitivity of the coil for the two nuclei, and, if necessary, taking into account the loading of the coil by the subject and the calibration solutions.

The nuclear magnetization of a sample containing N spins at thermal equilibrium is given by

$$M_0 = N\gamma^2\hbar^2 I(I+1)B_0/(3k_B T_s) \qquad (5.8)$$

where γ is the gyromagnetic ratio, I is the nuclear angular momentum quantum number (spin), T_s is the sample temperature, $\hbar$ is Planck's constant divided by 2π, and k_B is Boltzmann's constant. Hence the magnetization is directly proportional to N (the number of spins of the nucleus under investigation). If the detection coil is used in transmit–receive mode and has a uniform RF field, then the observed signal strength will be directly proportional to M_0 (and hence to N) provided that the sample volume V_s is small compared with the sensitive volume (V_c) of the coil. The signal intensity obtained from an elemental volume is given by

$$S = [K(B_1)_{xy} \sin \theta] N \qquad (5.9)$$

where θ is the flip angle, $(B_1)_{xy}$ is the component of B_1 in the x, y plane produced by unit current flowing in the coil, and K is a constant that includes (a) the terms apart from N in equation (5.8), (b) spectrometer receiver gains, and (c) a correction factor for saturation if necessary. By substituting $\rho = N/(A_0 V_s)$ in equation (5.9), where ρ is the molar spin concentration and A_0 is Avogadro's number, and taking into account the fact that, in most practical situations (e.g., for a surface coil), neither B_1 nor θ will be uniform throughout the sample, one obtains, on integrating over V_s

$$S = \rho \int_{V_s} [K'(B_1)_{xy} \sin \theta] \, dV \qquad (5.10)$$

where $K' = A_0 V_s K$. If the RF coil is tunable for two different nuclei (e.g., ^{31}P and ^{1}H), then $(B_1)_{xy}$ has an identical spatial distribution for both resonant frequencies.[5, 11] Additionally, if it is arranged that the flip angles at the coil's center are the same for both nuclei (by adjusting either the amplitude or the duration of the transmitter pulse), then θ will have the same spatial dependence at both frequencies. Hence the integral of $(B_1)_{xy} \sin \alpha$ over V_s will also be the same for both nuclei. Considering equation (5.10) applied separately to ^{31}P and ^{1}H, on division one obtains

$$\frac{S_{^{31}P}}{S_{^1H}} = \frac{K'_{^{31}P}}{K'_{^1H}} \times \frac{\rho_{^{31}P}}{\rho_{^1H}} = D \times \frac{\rho_{^{31}P}}{\rho_{^1H}} \qquad (5.11)$$

For this equation to hold true, identical steady-state saturation conditions must apply to both nuclei. It is an easy task to perform calibrations on standard solutions (suitably loaded to imitate tissue, i.e., with an ionic concentration equivalent to approximately 150 mM NaCl) and hence to obtain a value for the calibration constant D. If tissue water is to be used as the internal concentration reference for absolute determinations of a particular phosphorus metabolite, then equation (5.11) becomes

$$\rho_{^{31}P} = \frac{S_{^{31}P}}{S_{H_2O}} \times \frac{111.1 \times f_{H_2O}}{D} \qquad (5.12)$$

in moles/liter, where f_{H_2O} is the tissue-water fraction (0.7 to 0.9 in most body tissues) and 111.1 is the molarity of pure water protons. This approach to direct absolute quantitation has recently been refined to take into account coil loading by conductive samples and the resultant changes in coil quality factor Q at the two frequencies.[128, 129] In these studies the coil was primarily tuned for ^{31}P; coupling to the sample was strong at this frequency, and hence the effect of coil loading was significant. (As loading increases, coil sensitivity is degraded, and a pulse of longer duration or larger amplitude is required to maintain a constant flip angle at the coil's center.) Tuning for ^{1}H was achieved by the incorporation in the coil's circuitry of a selected length of coaxial cable and additional variable tune and match capacitors.[130] This gives only weak coupling with the sample at the ^{1}H frequency; the dependence on loading is minimal, but signal strength is greatly reduced. An RF pulse of large amplitude or of long duration was required to produce the same ^{1}H flip angle as that used for ^{31}P, thereby matching the spatial sensitivity. The effect of coil loading at the ^{31}P frequency was taken into account by calibration of the match capacitance. However, if the coil is strongly coupled to the sample for both phosphorus and protons, loading produces degradation of sensitivity and reduces θ for both nuclei. The ^{31}P signal voltage at the preamplifier input is given by[131]

$$S_{31P} = k_{31P}\, \rho_{31P}\, \frac{C_{31P}}{C_{m,31P}}\, dV \tag{5.13}$$

where k_{31P} is a constant dependent on the nucleus, the geometry of the coil and the sample, and the flip angle; ρ_{31P} is the phosphorus metabolite concentration; $C_{m,31P}$ is the match capacitance; and $C_{31P} = C_{m,31P} + C_{tune}$, where C_{tune} is the tuning capacitance. For a coil that is also strongly coupled for protons, a similar equation applies to the ^{1}H signal (S_{1H}) and the ratio of the ^{31}P and ^{1}H preamplifier input signals is given by

$$\frac{S_{31P}}{S_{1H}} = K\, \frac{C_{31P}}{C_{m,31P}} \times \frac{C_{m,1H}}{C_{1H}} \times \frac{\rho_{31P}}{\rho_{1H}} \tag{5.14}$$

where $K = k_{31P}/k_{1H}$. Further RF coil theory gives the relationship

$$\frac{C_{31P}}{C_{m'31P}} = \frac{4\theta_{31P}\, a Z_1}{\gamma_{31P}\, \mu_0\, V_{31P}\, T_{31P}} \tag{5.15}$$

where a is the coil's radius, Z_1 is the characteristic impedance of the coaxial cable, γ_{31P} is the gyromagnetic ratio, μ_0 is the permeability of free space, V_{31P} is the peak transmitter voltage, and T_{31P} is the length of the RF transmitter pulse required to produce a flip angle θ_{31P} at the coil's center. A similar equation applies to $C_{1H}/C_{m,1H}$. Combination of these equations gives

$$\frac{S_{31P}}{S_{1H}} = K'\, \frac{T_{1H}}{T_{31P}} \times \frac{\theta_{31P}}{\theta_{1H}} \times \frac{\rho_{31P}}{\rho_{1H}} \tag{5.16}$$

where $K' = K(\gamma_{1H}/\gamma_{31P})\,(V_{1H}/V_{31P})$. If the ^{31}P and ^{1}H pulse lengths are held constant and the flip angles at both frequencies vary with loading in a similar fashion, equation (5.16) can be applied directly without correction for loading. Alternately, if loading effects are more severe at one frequency, the pulse lengths required to give matched flip angles in vivo can be measured (by the incorporation of a chemical shift reference sample at the coil's center), and the calibration constant can thereby be corrected.[132] Two factors that can lead to erroneous results are subject movement and consequent coil detuning during the data collection, and the possibility that the spatial distributions of water and phosphorus metabolites are not identical.

The absolute quantitation method (with ^{31}P strongly coupled and ^{1}H weakly coupled) has been applied to the measurement of rat skeletal muscle ATP and the summed nucleoside triphosphates in rat brain (NTP = ATP + guanosine, uridine, and cytidine triphosphates), giving values of 6.79 ± 0.24 and 2.84 ± 0.16 mmole/l, respectively.[128] Both results agree well with published values obtained by biochemical analyses. Results have also been published giving absolute concentrations of ^{31}P metabolites in rat kidney,[133] and values obtained for the ATP (2.33 ± 0.55 mmole/kg wet wt.) correlated well with those from another study in which methylphosphonic acid was used as a concentration reference.[123] In both these studies, the ATP concentration was higher than that given by freeze-clamping, perchloric acid extraction, and enzymatic analysis. This may have been due to metabolites other than ATP contributing to the resonances, or possibly the fact that loading of the coil was not taken into account. However, the last factor would not affect the results from the study in which the methylphosphonic acid was used as an internal concentration reference.

5.6.3. Localized Absolute Quantitation

In the preceding section it was necessary, due to lack of knowledge of the tissue volume and other factors, to relate the signal from a particular nucleus (e.g., ^{31}P) to that from another one with known concentration (e.g., H_2O) and to assume that signal contributions originate only from the tissue of interest. If localization methods are used, alternative approaches can be adopted.[54, 122, 134, 135] If the selected volume is known with some degree of certainty, then it is necessary only to calibrate the spectrometer system with respect to the nucleus of interest using a phantom standard or to apply the same localization procedure to the phantom as would be used in vivo. Coil loading can be taken into account by use of an external reference sample which is present for both in vivo and phantom data collections. Under these circumstances the metabolite concentration ρ is given by the following:[122]

$$\rho = \frac{\rho_s I N_s S_s I_{ref(s)}}{I_s N S I_{ref}} \tag{5.17}$$

where I, $I_{ref(s)}$, I_s, and I_{ref} are the peak integrals of the metabolite resonance to be quantitated, the external reference with the coil loaded by the phantom, the phantom standard, and the external reference with in vivo loading respectively. N, N_s, S, and S_s are the number of scans and the saturation factors used for the in vivo and phantom data collections, respectively; ρ_s is the concentration of the phantom standard.

If three- or four-dimensional chemical-shift imaging is the adopted localization modality, it is necessary only to include a vial containing a standard solution of the appropriate nucleus within the field of view.[54] Because data is collected from the external reference simultaneously with the in vivo spectra, coil loading is the same for both. However, corrections for differential saturation may still have to be applied.

5.7. MEASUREMENT OF RELAXATION CONSTANTS

The relaxation constants T_1 and T_2 have been of particular importance in MR imaging because they give additional information about the state of the tissue over and above that obtained from spin-density images (which are fundamentally maps of tissue-water concentration). It is to be expected that nuclear relaxation will also be of importance in spectroscopy, provided that reliable techniques for localized measurements become available. In addition, knowledge of relaxation times can be very important for quantitation of absolute metabolite concentrations, especially when using spin–echo techniques.

Determinations of T_1 or T_2 using a coil which produces a uniform RF field are carried out using fairly straightforward procedures, and these will be dealt with first. The methods used in vivo are somewhat more complicated, particularly when surface coils are employed, due to the spatial variation of RF amplitude and flip angle. However, an understanding of the techniques used for studies in which the RF amplitude is uniform should make the methods adopted for surface-coil application more comprehensible.

5.7.1. T_1 Measurements in a Uniform RF Field

There are three main approaches to the measurement of T_1 in a uniform RF field: $90° - \tau - 90°$, $180° - \tau - 90°$ (inversion recovery—IR), and saturation recovery (SR).

5.7.1.1. The $90° - \tau - 90°$ Sequence

In this method,[136] an initial $90°$ pulse is applied to the sample, and this flips the magnetization into the xy plane. After a time τ, a further $90°$

pulse is transmitted. During τ, the magnetization in the xy plane dephases due to spin–spin relaxation. The second 90° pulse flips any remaining xy magnetization along the z-axis. Magnetization, which has recovered along the z-axis due to T_1 relaxation, is flipped by this second pulse back into the xy plane and produces a detectable signal as shown in Figure 5.12a. From the fundamental Bloch T_1 relaxation equation (presented in Chapter 2), it is seen that the amplitude $A(\tau)$ is given by

$$A(\tau) = A(\infty)[1 - \exp(-\tau/T_1)] \tag{5.18}$$

where $A(\infty)$ is the amplitude of the signal given by a single 90° pulse. In practice, several values of $A(\tau)$ are measured and, by plotting $\ln[A(\infty) - A(\tau)]$ against τ, a straight line is obtained, from the slope of which T_1 can be determined. This approach has been applied to measure the T_1s of human-brain phosphorylated metabolites in vivo.[28] For this purpose, a coil which surrounded the whole head was used (in order to get a uniform RF field) in combination with the topical magnetic resonance localization method,[25, 26] which has been described in Section 5.2.2.1. The homogeneous volume of the magnet was set to a diameter of 10 cm, and τ values between 0.1 sec and 10 sec were used.

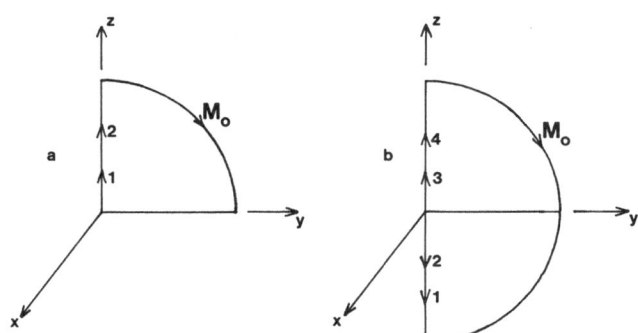

Figure 5.12. The measurement of T_1 in a homogeneous B_1 field. (a) The $90° - \tau - 90°$ sequence (saturation–recovery). M_0 is flipped through 90° into the xy plane, and dephasing commences due to transverse relaxation. The equilibrium magnetization recovers along the z axis by longitudinal relaxation (stages 1 to 2), and a further 90° pulse flips this into the xy plane where it can be detected and its amplitude can be measured. The magnetization measured later at stage 2 of the recovery is greater than that measured earlier at stage 1. (b) The $180° - \tau - 90°$ sequence (inversion–recovery). A 180° pulse inverts M_0, and the magnetization subsequently recovers along the z axis following stages 1, 2, 3, and 4, as shown. A 90° pulse at time τ after the 180° pulse flips the partially recovered magnetization into the xy plane, thereby allowing the measurement of its amplitude.

5.7.1.2. The 180° − τ − 90° Sequence

This method is commonly known as "inversion–recovery."[136] The 180° pulse inverts the sample magnetization so that it is parallel with the negative z-axis. After a period τ, during which partial recovery of the z-axis magnetization takes place, a 90° pulse flips this recovered magnetization into the xy plane, and hence the z magnetization can be sampled, as shown in Figure 5.12b. Because the 180° pulse initially inverts the magnetization, the relationship between $A(\tau)$ and $A(0)$ is different from that given by equation (5.18). Here,

$$A(\tau) = A(\infty)[1 - 2\exp(-\tau/T_1)] \qquad (5.19)$$

For short τ, the magnetization sampled by the 90° pulse is negative but, as τ gets longer, the measured magnetization increases, reaching a null at $\tau = T_1 \ln 2$ (i.e., when $A(\tau) = 0$), and then becomes positive. This means that, in principle, T_1 can be determined simply by finding the null point. In practice, however, more reliable T_1s are obtained by measuring $A(\tau)$ for several values of τ and performing an exponential fit.

5.7.1.3. The Saturation–Recovery Technique

In the saturation–recovery technique,[84] a series of 90° pulses is applied to the sample at intervals t, which are comparable to T_1. After a few pulses, a steady state is reached and the signal amplitude is given by:

$$A(t) = A(\infty)[1 - \exp(-t/T_1)] \qquad (5.20)$$

As for the 90° − τ − 90° method, $A(\infty)$ is the signal amplitude obtained with a single pulse, and T_1 is best determined by using a range of t values.

5.7.2. T_2 Measurements in a Uniform RF Field

All measurements of T_2 are affected to some extent by molecular diffusion, but various techniques have been developed which can minimize the influence of this on results. The basic approach to the measurement of T_2 is to use a spin–echo or a variation thereof. If molecular diffusion takes place during either the dephasing or refocusing periods of the spin–echo sequence, then migrating nuclei will experience a changing magnetic field due to spatial inhomogeneity. Refocusing will be incomplete, and the echo's amplitude will be reduced. In addition to this problem, inaccuracies can also occur if the 180° pulses do not produce exact inversion of the magnetization. Phase modulation of multiplets is also a factor that has to be dealt with by careful selection of spin–echo times, calculation of correction factors, or suppression of the phase modulation. With the former two approaches, knowledge of the appropriate coupling constant J for the multiplet is necessary.

In order to explain the methodology of T_2 measurement, a simple spin–echo technique will be described first, and then some of the more sophisticated methods which help to avoid the problems outlined above will be presented.

5.7.2.1. T_2 Measurements Using a Simple Spin Echo

The basic theory of spin echoes has been dealt with in Chapter 2 and is well covered in the literature,[71, 72] so it will not be repeated here. A simple spin–echo sequence that could be used for T_2 measurement is

$$90° - \tau - 190° - \tau - \text{Acquire Echo}$$

Only the effects of magnetic field inhomogeneities are refocused by the 180° pulse and, as τ is increased, the echo amplitude decreases due to spin–spin relaxation and molecular diffusion. The actual relationship between echo amplitude A and τ is as follows:[74, 136]

$$A(2\tau) = A(0) \exp\left[-\frac{2\tau}{T_2} - \frac{2\gamma^2 G^2 D\tau^3}{3} \right] \qquad (5.21)$$

where G is the gradient of the inhomogeneous static magnetic field and D is the diffusion coefficient. If diffusion were negligible, it would be necessary only to measure $A(2\tau)$ for several values of τ and then to perform an exponential fit to obtain T_2. However, inspection of the second term in equation (5.21) shows that the effect of diffusion becomes pronounced for long spin–echo times due to the τ^3 dependence, and hence this simple approach is not suitable for metabolites with large T_2s.

5.7.2.2. The Carr–Purcell Sequence

In order to reduce the effect of molecular diffusion on the T_2 measurements, a modification of the basic spin–echo method has been introduced which involves replacing the single 180° pulse by a train of 180° pulses at intervals of 2τ.[74] Each subsequent 180° flip causes refocusing of the transverse magnetization, and hence spin echoes are observed at time

τ after each of these pulses, as shown in Figure 5.13. The echo amplitudes in this case are given by:[74, 136]

$$A(t) = A(0) \exp\left[-\frac{t}{T_2} - \frac{\gamma^2 G^2 D \tau^2 t}{3} \right] \tag{5.22}$$

where t is the total duration of the Carr–Purcell sequence which is usually very long when compared with τ. In fact, if τ is short enough, the diffusion term in equation (5.22) can be made negligible, and the echo amplitude then depends on T_2 only.

5.7.2.3. The Carr–Purcell–Meiboom–Gill Sequence

The main difficulty with the Carr–Purcell technique is obtaining the precise setting of the inversion pulses, which have to give an exact 180° flip angle throughout the sample. If the flip angle deviates from 180°, successive pulses leave the magnetization further and further off the xy plane, and the observed signal is reduced. A simple adaptation of the echo-train method, called the Carr–Purcell–Meiboom–Gill (CPMG) technique,[75] has been developed in order to avoid this.

In the previous Carr–Purcell sequence, the pulses all had the same phase and hence produced rotations of the magnetization about one particular axis (e.g., the x-axis). However, if the 180° pulses are phase-shifted by 90°, so that they flip the magnetization about the y-axis instead, then inaccuracies in pulse setting and the consequent deviation of the magnetization from the xy plane are not compounded so severely by successive 180° pulses.

CPMG sequences have been used in conjunction with ^{1}H spectroscopy

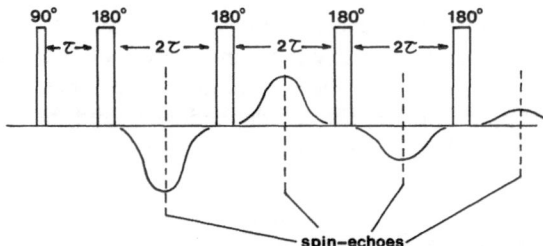

Figure 5.13. A Carr–Purcell spin–echo sequence of 90° and 180° pulses, which flip the magnetization about the x axis. This sequence reduces the effect of molecular diffusion on the determination of T_2s. Echoes are formed at times 2τ, 4τ, 6τ,..., $2n\tau$ where n indicates the nth echo. In the Carr–Purcell–Meiboom–Gill variation of this sequence, the 180° pulses rotate the magnetization about the y axis and reduce the effect of RF-field inhomogeneity and inaccurate flip angles.

in vivo to measure the T_2s of metabolites in rat skeletal muscle.[77] A solenoid coil surrounding the leg was employed in order to obtain a uniform flip angle throughout the tissue, and CPMG pulse trains lasting 270 msec and 540 msec with a delay between consecutive 180° pulses of 1.2 msec were used. The spectra obtained in this study are shown in Figure 1.6.

5.7.3. Measurements of T_1 and T_2 with Surface Coils

The main problem with the use of surface coils for the determination of relaxation constants is the spatial variation of flip angle throughout the sample. By selecting a pulse length or pulse amplitude which gives a flip of 180° at the coil center, it can be arranged that a certain region above the coil will experience a flip angle close to 90°. However, this volume will be poorly defined. Further away from the coil the flip angle will decrease and eventually approach zero. Closer to the plane of the coil and, in particular, adjacent to the windings, flip angles reach values equivalent to many complete rotations of the magnetization vector. Hence the methods described previously can function properly only in a small part of the tissue unless some method of localization is also employed. However, in spite of the spatial variation of flip angle, methods have been developed for use with surface coils, and acceptable results have been obtained.

5.7.3.1. Surface-Coil T_1 Measurements

The methods which have been used successfully for surface-coil T_1 determination are all variations of the basic SR or IR techniques described in Sections 5.7.1.1. and 5.7.1.2. Of the techniques that have been tried, pulse-burst SR,[137] modified fast IR,[138] and a phase-shifted SR method[139] have been found to be the most efficient, especially if the data are fitted to the three-parameter exponential[140]

$$A(\tau) = A + B \exp(-\tau/T_1) \tag{5.23}$$

rather than equations (5.18) or (5.19). In the pulse-burst SR method, the sample is saturated using a cluster of large-flip-angle hard pulses (e.g., twenty 225° pulses at 0.1-msec intervals[140]) prior to sampling the recovered z magnetization with a 90° pulse. A pulsed gradient is applied between the saturating pulse cluster and the sampling pulse in order to prevent echo formation. One advantage of this technique is that successive pulse sequences can be used without having to wait $5T_1$ for the sample to relax, and hence measurements can be made in a short time. For the modified fast IR approach, the relaxation delay between successive sequences is reduced as

τ increases, thereby reducing the experimental time required. In addition, the modified fast IR method circumvents some of the systematic errors caused by inexact setting of the 90° and 180° pulse flip angles which is a problem encountered with the use of surface coils. The phase-shifted SR technique employs saturating radiation, which periodically shifts the phase by 90°. After a delay τ, a 90° pulse is applied in order to sample the recovered z magnetization.

5.7.3.2. Surface-Coil T_2 Measurements

Due to the large spatial variation of RF amplitude and hence of flip angle, it is not possible to use the CPMG method with surface coils. Measured T_2s will therefore be subject to the effects of molecular diffusion. However, useful data can still be obtained, especially if the EXORCYCLE pulse-phasing system[15] is used with a simple spin echo (see Table 5.3). This provides some degree of localization of the signal to the region where the flip angles produced by the first and second pulses in the spin–echo sequence are close to 90° and 180°, respectively. Phase modulation can be a problem, but the use of narrow-band saturation,[106, 107] 180° binomial pulses,[93, 94, 109] or a correction factor[110] can provide solutions.

REFERENCES

1. Chu, A., Delpy, D. T., and Thalaysingam, S.: A transport and life support system for newborn infants during NMR spectroscopy, in Rolfe, P. (ed.): *Neonatal Physiological Measurements*, London, Butterworth, 1986, p. 409.
2. Ceckler, T. L., Bryant, R. G., and Hornak, J. P.: Noise reduction in wide-bore magnets using a patient cage, *Magn. Reson. Med.* 5:173–174, 1987.
3. Iles, R. A., Griffiths, J. R., Stevens, A. N., et al.: Effects of fructose on the energy metabolism and acid-base status of the perfused starved-rat liver, *Biochem. J.* 192:191–202, 1980.
4. Becker, E. D., Feretti, J. A., and Gambhir, P. N.: Selection of optimum parameters for pulse Fourier transform nuclear magnetic resonance, *Anal. Chem.* 51:1413–1420, 1979.
5. Hoult, D. I.: The NMR receiver: A description and analysis of design, *Progress in Nuclear Magnetic Resonance Spectroscopy* 12:41–77, 1978.
6. Hoult, D. I., Chen, C.-N., Eden, H., et al.: Elimination of baseline artifacts in spectra and their integrals, *Journal of Magnetic Resonance* 51:110–117, 1983.
7. Evelhoch, J. L., Crowley, M. G., and Ackerman, J. J. H.: Signal-to-noise optimization and observed volume localization with circular surface coils, *Journal of Magnetic Resonance* 56:110–124, 1984.
8. Haase, A., Hanicke, W., and Frahm, J.: The influence of experimental parameters in surface coil NMR, *Journal of Magnetic Resonance* 56:401–412, 1984.
9. Bottomley, P. A., Edelstein, W. A., Hart, H. R., et al.: Spatial localization in ^{31}P and ^{13}C NMR spectroscopy in vivo using surface coils, *Magn. Reson. Med.* 1:410–413, 1984.

10. Wilkie, D. R., Dawson, M. J., Edwards, R. H. T., et al.: ³¹P NMR studies of resting muscle in normal human subjects, in Pollack, G. H., Sugi, H. (eds.): *Contractile Mechanisms in Muscle*, New York, Plenum Press, 1984, p. 333.

11. Ackerman, J. J. H., Grove, T. H., Wong, G. G., et al.: Mapping of metabolites in whole animals by ³¹P NMR using surface coils, *Nature* 283:167–170, 1980.

12. Tofts, P. S., Cady, E. B., Delpy, D. T., et al.: Surface coil NMR spectroscopy of brain, *Lancet* 1:459, 1984.

13. Laptook, A. R., Corbett, R. J. T., Nguyen, H. T., et al.: Alterations in cerebral blood flow and phosphorylated metabolites in piglets during and after partial ischemia, *Pediatr. Res.* 23:206–211, 1988.

14. Bendall, M. R. and Gordon, R. E.: Depth and refocusing pulses designed for multipulse NMR with surface coils, *Journal of Magnetic Resonance* 53:365–385, 1983.

15. Bodenhausen, G., Freeman, R., and Turner, D. L.: Suppression of artifacts in two-dimensional *J* spectroscopy, *Journal of Magnetic Resonance* 27:511–514, 1977.

16. Bendall, M. R. and Aue, W. P.: Experimental verification of depth pulses applied with surface coils, *Journal of Magnetic Resonance* 54:149–152, 1983.

17. Bendall, M. R., Foxall, D., Nichols, B. G., et al.: Complete localization of in vivo NMR spectra using two concentric surface coils and RF methods only, *Journal of Magnetic Resonance* 70:181–186, 1986.

18. Pekar, J., Leigh, J. S., and Chance, B.: Harmonically analyzed sensitivity profile. A novel approach to depth pulses for surface coils, *Journal of Magnetic Resonance* 64:115–119, 1985.

19. Cox, S. J. and Styles, P.: Towards biochemical imaging, *Journal of Magnetic Resonance* 40:209–212, 1980.

20. Haase, A., Malloy, C., and Radda, G. K.: Spatial localization of high resolution ³¹P spectra with a surface coil, *Journal of Magnetic Resonance* 55:164–169, 1983.

21. Styles, P., Smith, M. B., Briggs, R. W., et al.: A concentric surface coil probe for the production of homogeneous B_1 fields, *Journal of Magnetic Resonance* 62:397–405, 1985.

22. Styles, P., Scott, C. A., and Radda, G. K.: A method for localizing high-resolution NMR spectra from human subjects, *Magn. Reson. Med.* 2:402–409, 1985.

23. Hoult, D. I.: Rotating frame zeugmatography, *Journal of Magnetic Resonance* 33:183–197, 1979.

24. Blackledge, M. J., Rajagopalan, B., Oberhaensli, R. D., et al.: Quantitative studies of human cardiac metabolism by ³¹P rotating-frame NMR, *Proc. Natl. Acad. Sci. USA* 84:4283–4287, 1987.

25. Gordon, R. E., Hanley, P. E., Shaw, D., et al.: Localization of metabolites in animals using ³¹P topical magnetic resonance, *Nature* 287:736–738, 1980.

26. Gordon, R. E., Hanley, P. E., and Shaw, D.: Topical magnetic resonance, *Progress in Nuclear Magnetic Resonance Spectroscopy* 15:1–47, 1982.

27. Campbell, I. D., Dobson, C. M., Williams, R. J. P., et al.: Resolution enhancement in aqueous solutions, *Journal of Magnetic Resonance* 11:172–181, 1973.

28. Oberhaensli, R. D., Galloway, G. J., Hilton–Jones, D., et al.: The study of human organs by phosphorus-31 topical magnetic resonance spectroscopy, *Br. J. Radiol.* 60:367–373, 1987.

29. Oberhaensli, R. D., Galloway, G. J., Taylor, D. J., et al.: Assessment of human liver metabolism by phosphorus-31 magnetic resonance spectroscopy, *Br. J. Radiol.* 59:695–699, 1986.

30. Oberhaensli, R. D., Hilton–Jones, D., and Bore, P. J.: Biochemical investigation of human tumours in vivo with phosphorus-31 magnetic resonance spectroscopy, *Lancet* 2:8–11, 1986.

31. Bottomley, P. A., Foster, T. B., and Darrow, R. D.: Depth-resolved surface-coil spectro-scopy (DRESS) for in vivo ¹H, ³¹P and ¹³C NMR, *Journal of Magnetic Resonance* **59**:338–342, 1984.

32. Bottomley, P. A., Herfkens, R. J., Smith, L. S., et al.: Non-invasive detection and monitoring of regional myocardial ischemia in situ using depth-resolved 31P NMR spectroscopy, *Proc. Natl. Acad. Sci. USA* **82**:8747–8751, 1985.

33. Bottomley, P. A.: Non-invasive study of high-energy phosphate metabolism in human heart by depth-resolved 31P NMR spectroscopy, *Science* **229**:769–772, 1985.

34. Bottomley, P. A., Scott Smith, L., Leue, W. M., et al.: Slice-interleaved depth-resolved surface-coil spectroscopy (SLIT DRESS) for rapid ³¹P NMR in vivo, *Journal of Magnetic Resonance* **64**:347–351, 1985.

35. Bottomley, P. A., Edelstein, W. A., Foster, T. H., et al.: In vivo solvent-suppressed localized hydrogen nuclear magnetic resonance spectroscopy: A window to metabolism? *Proc. Natl. Acad. Sci. USA* **82**:2148–2152, 1985.

36. Ordidge, R. J., Connelly, A., and Lohman, J. A. B.: Image selected in-vivo spectroscopy (ISIS), *Journal of Magnetic Resonance* **66**:283–294, 1986.

37. Connelly, A., Counsell, C., Lohmann, J. A. B., et al.: Outer volume suppressed image related in vivo spectroscopy (OSIRIS), a high-sensitivity localization technique, *Journal of Magnetic Resonance* **78**:519–525, 1988.

38. Tycko, R. and Pines, A.: Spatial localization of NMR signals by narrowband inversion, *Journal of Magnetic Resonance* **60**:156–160, 1984.

39. Ordidge, R. J.: Random noise selective excitation pulses, *Magn. Reson. Med.* **5**:93–98, 1987.

40. Silver, M. S., Joseph, R. I., Chen, C. N., et al.: Selective population inversion in NMR, *Nature* **310**:681–683, 1984.

41. Segebarth, C., Luyten, P. R., and den Hollander, J. A.: Improved depth-selective single surface coil ³¹P NMR spectroscopy using a combination of B_1 and B_0 selection techniques, *Journal of Magnetic Resonance* **75**:345–351, 1987.

42. Sauter, R., Mueller, S., Requardt, H., et al.: Presaturation of superficial slices: An improvement of the ISIS technique for localized in vivo spectroscopy, *Proceedings*, Third Annual Meeting, Society of Magnetic Resonance in Medicine, New York, 1987, p. 944.

43. Matson, G. B., Twieg, D. B., Karczmar, G. S., et al.: Spatially selected ³¹P NMR of human organs with surface coils: Improvement of ISIS with a composite pulse, *Proceedings*, Third Annual Meeting, Society of Magnetic Resonance in Medicine, New York, 1987, p. 605.

44. Hanstock, C. C., Rothman, D. L., Jue, T., et al.: Volume-selected proton spectroscopy in the human brain, *Journal of Magnetic Resonance* **77**:583–588, 1988.

45. Sharp, J. C. and Leach, M. O.: Conformal ISIS, *Proceedings*, Seventh Annual Meeting, Society of Magnetic Resonance in Medicine, San Francisco, 1988, p. 705.

46. Kumar, A., Welti, D., and Ernst, R. R.: NMR Fourier zeugmatography, *Journal of Magnetic Resonance* **18**:69–83, 1975.

47. Haselgrove, J. C., Subramanian, V. H., Leigh, J. S., et al.: In vivo one-dimensional imaging of phosphorus metabolites by phosphorus-31 nuclear magnetic resonance, *Science* **220**:1170–1173, 1983.

48. Bailes, D. R., Bryant, D. J., Bydder, G. M., et al.: Localized phosphorus-31 NMR spectroscopy of normal pathological human organs in vivo using phase-encoding techniques, *Journal of Magnetic Resonance* **74**:158–170, 1987.

49. Bottomley, P. A., Foster, T. H., and Leue, W. M.: Chemical imaging of the brain by NMR, *Lancet* **1**:1120, 1984.

50. Dumoulin, C. L.: A method for chemical shift selective imaging, *Magn. Reson. Med.* **2**:583–585, 1985.

51. Rosen, B. R., Pykett, I. L., Brady, T. J., et al.: NMR proton chemical-shift imaging of experimentally induced fatty liver disease. Enhanced sensitivity over conventional NMR imaging, *Magn. Reson. Med.* 1:238–239, 1984.

52. Brown, T. R., Kincaid, B. M., and Ugurbil, K.: NMR chemical shift imaging in three dimensions, *Proc. Natl. Acad. Sci. USA* 79:3523–3526, 1982.

53. Cox, I. J., Sargentoni, J., Calam, J., et al.: Four-dimensional phosphorus-31 chemical shift imaging of carcinoid metastases in the liver, *NMR in Biomedicine* 1:56–60, 1988.

54. Bottomley, P. A., Charles, H. C., Roemer, P. B., et al.: Human in vivo phosphate metabolite imaging with ^{31}P NMR, *Magn. Reson. Med.* 7:319–336, 1988.

55. Hardy, C. J., Bottomley, P. A., Roemer, P. B., et al.: Rapid ^{31}P spectroscopy on a 4-T whole-body system, *Magn. Reson. Med.* 8:104–109, 1988.

56. Hahn, E. L.: Spin echoes, *Physical Review* 80:580–594, 1950.

57. Frahm, J., Merboldt, K.–D., and Hanicke, W.: Localized proton spectroscopy using stimulated echoes, *Journal of Magnetic Resonance* 72:502–508, 1987.

58. Kimmich, R. and Hoepfel, D.: Volume selective multipulse spin-echo spectroscopy, *Journal of Magnetic Resonance* 72:379–384, 1987.

59. Frahm, J., Bruhn, H., Gyngell, M. L., et al.: Localized high-resolution proton NMR spectroscopy using stimulated echoes: Initial applications to human brain in vivo, *Magn. Reson. Med.* 9:79–93, 1989.

60. Bruhn, H., Frahm, J., Gyngell, M. L., et al.: Cerebral metabolism in man after acute stroke: New observations using localized proton NMR spectroscopy, *Magn. Reson. Med.* 9:126–131, 1989.

61. Knuttel, A. and Kimmich, R.: Multiple-volume-selective proton NMR spectroscopy and spectral editing by spin-echo double resonance (VOSY + SEDOR), *Journal of Magnetic Resonance* 78:205–212, 1988.

62. Aue, W. P., Muller, S., Cross, T. A., et al.: Volume-selective excitation. A novel approach to topical NMR, *Journal of Magnetic Resonance* 56:350–354, 1984.

63. Muller, S., Aue, W. P., and Seelig, J.: Practical aspects of volume-selective excitation (VSE). Compensation sequences, *Journal of Magnetic Resonance* 65:332–338, 1985.

64. Luyten, P. R. and den Hollander, J. A.: Observation of metabolites in the human brain by MR spectroscopy, *Radiology* 161:795–798, 1986.

65. Luyten, P. R., Marien Ad, J. H., Sijtsma, B., et al.: Solvent-suppressed spatially resolved spectroscopy. An approach to high-resolution NMR on a whole-body MR system, *Journal of Magnetic Resonance* 67:148–155, 1986.

66. Hennig, J., Boesch, C., Gruetter, R., et al.: Homogeneity spoil spectroscopy as a tool of spectrum localization for in vivo spectroscopy, *Journal of Magnetic Resonance* 75:179–183, 1987.

67. Sauter, R., Mueller, S., and Weber, H.: Localization in in vivo ^{31}P NMR spectroscopy by combining surface coils and slice selective saturation, *Journal of Magnetic Resonance* 75:167–173, 1987.

68. Doddrell, D. M., Bulsing, J. M., Galloway, G. J., et al.: Discrete isolation from gradient governed elimination of resonances. DIGGER, a new technique for in vivo volume-selected NMR spectroscopy, *Journal of Magnetic Resonance* 70:319–326, 1986.

69. Doddrell, D. M., Galloway, G. J., Brooks, W. M., et al.: The utilization of two frequency-shifted sinc pulses for performing volume-selected in vivo NMR spectroscopy, *Magn. Reson. Med.* 3:970–975, 1986.

70. Bottomley, P. A. and Hardy, C. J.: PROGRESS in efficient three-dimensional spatially localized in vivo ^{31}P NMR spectroscopy using multidimensional spatially selective (ρ) pulses, *Journal of Magnetic Resonance* 74:550–556, 1987.

71. Rabenstein, D. L. and Nakashima, T. T.: Spin–echo Fourier transform nuclear magnetic resonance spectroscopy, *Anal. Chem.* 51:1465A–1474A, 1979.

72. Turner, C. J.: Multipulse NMR in liquids, *Progress in Nuclear Magnetic Resonance Spectroscopy* **16**:311–370, 1984.

73. Campbell, I. D., Dobson, C. M., Williams, R. J. P., et al.: Pulse methods for the simplification of protein spectra, *FEBS Lett.* **57**:96–99, 1975.

74. Carr, H. Y. and Purcell, E. M.: Effects of diffusion on free precession in nuclear magnetic resonance experiments, *Physical Review* **94**:630–638, 1954.

75. Meiboom, S. and Gill, D.: Modified spin–echo method for measuring nuclear relaxation times, *Review Scientific Instruments* **29**:688–691, 1958.

76. Brown, F. F., Campbell, I. D., and Kuchel, P. W.: Human erythrocyte metabolism studies by ^{1}H spin–echo NMR, *FEBS Lett.* **82**:12–16, 1977.

77. Williams, S. R., Gadian, D. G., Proctor, E., et al.: Proton NMR studies of muscle metabolites in vivo, *Journal of Magnetic Resonance* **63**:406–412, 1985.

78. Levitt, M. H. and Freeman, R.: Compensation for pulse imperfections in NMR spin–echo experiments, *Journal of Magnetic Resonance* **43**:65–80, 1981.

79. Hetherington, H. P. and Rothman, D. L.: Phase cycling of composite refocusing pulses to eliminate dispersive refocusing magnetization, *Journal of Magnetic Resonance* **65**:348–354, 1985.

80. Rothman, D. L., Behar, K. L., den Hollander, J. A., et al.: Surface coil spin–echo spectra without cycling the refocusing pulse through all four phases, *Journal of Magnetic Resonance* **59**:157–159, 1984.

81. Gadian, D. G., Proctor, E., Williams, S. R., et al.: Neurometabolic effects of an inborn error of amino acid metabolism demonstrated in vivo by ^{1}H NMR, *Magn. Reson. Med.* **3**:150–156, 1986.

82. Behar, K. L., Rothman, D. L., Shulman, R. G., et al.: Detection of cerebral lactate in vivo during hypoxemia by ^{1}H NMR at relatively low field strengths (1.9 T), *Proc. Natl. Acad. Sci. USA* **81**:2517–2519, 1984.

83. Brindle, K. M., Smith, M. B., Rajagopalan, B., et al.: Spectral editing in ^{31}P NMR spectra of human brain, *Journal of Magnetic Resonance* **61**:559–563, 1985.

84. Freeman, R.: *A Handbook of Nuclear Magnetic Resonance*, New York, John Wiley, 1988.

85. Redfield, A. G., Kunz, S. D., and Ralph, E. K.: Dynamic range in Fourier transform proton magnetic resonance, *Journal of Magnetic Resonance* **19**:114–117, 1975.

86. Redfield, A. G. and Kunz, S. D.: Proton resonance spectrometer for biochemical applications, in Opella, S. J., Lu, P. (eds.): *NMR and Biochemistry*, New York, Dekker, 1979, p. 225.

87. Campbell, I. D., Dobson, C. M., Jeminet, G., et al.: Pulsed NMR methods for the observation and assignment of exchangeable hydrogens: Application to bacitracin, *FEBS Lett.* **49**:115–119, 1974.

88. Hoult, D. I.: Solvent peak saturation with single phase and quadrature Fourier transformation, *Journal of Magnetic Resonance* **21**:337–347, 1976.

89. Behar, K. L., den Hollander, J. A., Stromski, M. E., et al.: High-resolution ^{1}H nuclear magnetic resonance study of cerebral hypoxia in vivo, *Proc. Natl. Acad. Sci. USA* **80**:4945–4948, 1983.

90. Plateau, P. and Gueron, M.: Exchangeable proton NMR without baseline distortion, using new strong-pulse sequences, *Journal of the American Chemical Society* **104**:7310–7311, 1982.

91. Sklenar, V. and Starcuk, Z.: 1-2-1 pulse train: A new effective method of selective excitation for proton NMR in water, *Journal of Magnetic Resonance* **50**:495–501, 1982.

92. Clore, G. M., Kimber, B. J., and Gronenborn, A. M.: The 1-1 hard pulse: A simple and effective method of water resonance suppression in FT ^{1}H NMR, *Journal of Magnetic Resonance* **54**:170–173, 1983.

93. Hore, P. J.: A new method for water suppression in the proton NMR spectra of aqueous solutions, *Journal of Magnetic Resonance* **54**:539–542, 1983.

94. Hore, P. J.: Solvent suppression in Fourier transform nuclear magnetic resonance, *Journal of Magnetic Resonance* **55**:283–300, 1983.

95. Morris, G., Smith, K. I., and Waterton, J. C.: Pulse sequences for solvent suppression with minimal spectral distortion, *Journal of Magnetic Resonance* **68**:526–532, 1986.

96. Hall, M. P. and Hore, P. J.: Computer-optimized solvent suppression, *Journal of Magnetic Resonance* **70**:350–354, 1986.

97. Galloway, J. G., Haseler, L. J., Marshman, M. F., et al.: A simple modification for the elimination of phase distortions, a characteristic of "binomial" solvent suppression pulse sequences, *Journal of Magnetic Resonance* **74**:184–187, 1987.

98. Ackerman, J. J. H., Evelhoch, J. L., Berkowitz, B. A., et al.: Selective suppression of the cranial bone resonance from ^{31}P NMR experiments with rat brain in vivo, *Journal of Magnetic Resonance* **56**:318–322, 1984.

99. Gonzalez-Mendez, R., Litt, L., Koretsky, A. P., et al.: Comparison of ^{31}P NMR spectra of in vivo rat brain using convolution difference and saturation with a surface coil. Source of the broad component in the brain spectrum, *Journal of Magnetic Resonance* **57**:526–533, 1984.

100. Tofts, P. S. and Wray, S.: Changes in brain phosphorus metabolites during the postnatal development of the rat, *J. Physiol.* (London) **359**:417–429, 1985.

101. Cerdan, S., Subramanian, V. H., Hilberman, M., et al.: ^{31}P NMR detection of mobile dog brain phospholipids, *Magn. Reson. Med.* **3**:432–439, 1986.

102. Morris, G. A. and Freeman, R.: Selective excitation in Fourier transform nuclear magnetic resonance, *Journal of Magnetic Resonance* **29**:433–462, 1978.

103. Corbett, R. J. T., Laptook, A. R., Hassan, A., et al.: Quantitation of acidosis in neonatal brain tissue using the ^{31}P NMR resonance peak of phosphoethanolamine, *Magn. Reson. Med.* **6**:99–106, 1988.

104. Corbett, R. J. T., Laptook, A. R., and Nunnally, R. L.: The use of the chemical shift of the phosphomonoester P-31 magnetic resonance peak for the determination of intracellular pH in the brains of neonates, *Neurology* **37**:1771–1779, 1987.

105. Vink, R., McIntosh, T. K., and Faden, A. I.: Nonedited ^{1}H NMR lactate/ N-acetylaspartate ratios and the in vivo determination of lactate concentration in brain, *Magn. Reson. Med.* **7**:95–99, 1988.

106. Rothman, D. L., Arias–Mendoza, F., Shulman, G. I., et al.: A pulse sequence for simplifying hydrogen NMR spectra of biological tissues, *Journal of Magnetic Resonance* **60**:430–436, 1984.

107. Rothman, D. L., Behar, K. L., Hetherington, H. P., et al.: Homonuclear ^{1}H double resonance difference spectroscopy of the rat brain in vivo, *Proc. Natl. Acad. Sci. USA* **81**:6330–6334, 1984.

108. Rothman, D. L., Behar, K. L., Hetherington, H. P., et al.: ^{1}H-observe ^{13}C-decouple spectroscopic measurements of lactate and glutamate in the rat brain in vivo, *Proc. Natl. Acad. Sci USA* **82**:1633–1637, 1985.

109. Williams, S. R., Gadian, D. G., and Proctor, E.: A method for lactate detection in vivo by spectral editing without the need for double irradiation, *Journal of Magnetic Resonance* **66**:562–567, 1986.

110. Williams, S. R., Proctor, E., Allen, K., et al.: Quantitative estimation of lactate in the brain by ^{1}H NMR, *Magn. Reson. Med.* **7**:425–431, 1988.

111. Young, R. S. K., Cowen, B. E., Petrof, O. A. C., et al.: In vivo 31P and in vitro 1H nuclear magnetic resonance study of hypoglycemia during neonatal seizure, *Ann. Neurol.* **22**:622–628, 1987.

112. McIntosh, T. K., Faden, A. I., Bendall, M. R., et al.: Traumatic brain injury in the rat:

Alterations in brain lactate and pH as characterized by ^{1}H and ^{31}P nuclear magnetic resonance, *J. Neurochem.* **49**:1530–1540, 1987.

113. Edwards, R. H. T., Dawson, M. J., Wilkie, D. R., et al.: Clinical use of nuclear magnetic resonance in the investigation of myopathy, *Lancet* **1**:725–731, 1982.

114. Dawson, M. J.: Quantitative analysis of metabolite levels in normal human subjects by ^{31}P topical magnetic resonance, *Biosci. Rep.* **2**:727–733, 1982.

115. Cady, E. B., Costello, A. M. de L., Dawson, M. J., et al.: Non-invasive investigation of cerebral metabolism in newborn infants by phosphorus nuclear magnetic resonance spectroscopy, *Lancet* **1**:1059–1062, 1983.

116. Lolley, R. N., Balfour, W. M., and Samson, F. E.: The high-energy phosphates in developing brain, *J. Neurochem.* **7**:289–297, 1961.

117. Mandel, P. and Edel-Harth, S.: Free nucleotides in the rat brain during post-natal development, *J. Neurochem.* **13**:591–595, 1966.

118. Tofts, P. S. and Wray, S.: A critical assessment of methods of measuring metabolite concentrations by NMR spectroscopy, *NMR in Biomedicine* **1**:1–10, 1988.

119. Billadello, J. J., Gard, J. K., Ackerman, J. J. H., et al.: Determination of intact tissue glycerolphosphorylcholine levels by quantitative ^{31}P nuclear magnetic resonance spectroscopy and correlation with spectrophotometric quantification, *Anal. Biochem.* **144**:269–274, 1985.

120. Garlick, P. B., Radda, G. K., and Seeley, P. J.: Studies of acidosis in the ischaemic heart by phosphorus nuclear magnetic resonance, *Biochem. J.* **184**:547–554, 1979.

121. Dawson, M. J., Gadian, D. G., and Wilkie, D. R.: Contraction and recovery of living muscles studied by ^{31}P nuclear magnetic resonance, *J. Physiol.* (London) **267**:703–735, 1977.

122. Roth, K., Hubesch, B., Meyerhoff, D. J., et al.: Noninvasive quantitation of phosphorus metabolites in human tissue by NMR spectroscopy, *Journal of Magnetic Resonance* **81**:299–311, 1989.

123. Shine, N. R., Xuan, J. A., Koretsky, A. P., et al.: Determination of renal molar concentrations of phosphorus-containing metabolites in vivo using ^{31}P NMR, *Magn. Reson. Med.* **4**:244–251, 1987.

124. Crockard, H. A., Gadian, D. G., Frackowiak, R. S. J., et al.: Acute cerebral ischaemia: Concurrent changes in cerebral blood flow, energy metabolites, pH, and lactate measured with hydrogen clearance and ^{31}P and ^{1}H nuclear magnetic resonance spectroscopy. II. Changes during ischaemia, *J. Cereb. Bl. Flow Metab.* **7**:394–402, 1987.

125. Chang, L. H., Pereira, B. M., Weinstein, P. R., et al.: Comparison of lactate concentration determinations in ischemic and hypoxic rat brains by in vivo and in vitro ^{1}H NMR spectroscopy, *Magn. Reson. Med.* **4**:575–581, 1987.

126. Tallan, H. H.: Studies on the distribution of *N*-acetyl-*L*-aspartic acid in brain, *J. Biol. Chem.* **223**:41–45, 1957.

127. Thulborn, K. R. and Ackerman, J. J. H.: Absolute molar concentrations by NMR in inhomogeneous B_1. A scheme for analysis of in vivo metabolites, *Journal of Magnetic Resonance* **55**:357–371, 1983.

128. Wray, S. and Tofts, P. S.: Direct in vivo measurement of absolute metabolite concentrations using ^{31}P nuclear magnetic resonance spectroscopy, *Biochim. Biophys. Acta* **886**:399–405, 1986.

129. Tofts, P. S. and Wray, S.: Non-invasive measurement of molar concentrations of ^{31}P metabolites in vivo, using surface coil NMR spectroscopy, *Magn. Reson. Med.* **6**:84–86, 1988.

130. Gordon, R. E. and Timms, W. E.: An improved tune and match circuit for B_0 shimming in intact biological samples, *Journal of Magnetic Resonance* **46**:322–324, 1982.

131. Tofts, P. S.: The noninvasive measurement of absolute metabolite concentrations in vivo using surface-coil NMR spectroscopy, *Journal of Magnetic Resonance* **80**:84–95, 1988.

132. Kuan, W. P., Scott, K. N., and Briggs, R. W.: Determination of absolute molar concentrations of ^{31}P metabolites in vivo: Surface coil applications to mouse gluteus muscle, *Proceedings*, Seventh Annual Meeting, Society of Magnetic Resonance in Medicine, San Francisco, 1988, p. 938.

133. Shapiro, J. I. and Chan, L.: In vivo determination of absolute molar concentrations of renal phosphorus metabolites using the proton concentration as an internal standard, *Journal of Magnetic Resonance* **75**:125–128, 1987.

134. Vermeulen, J., Luyten, P. R., and den Hollander, J. A.: Determination of metabolite concentrations from localized ^{31}P NMR spectra of the human brain, *Proceedings*, Third Annual Meeting, Society for Magnetic Resonance in Medicine, New York, 1987, p. 136.

135. Bottomley, P. A. and Hardy, C. J.: Rapid, reliable in vivo assays of human phosphate metabolites by nuclear magnetic resonance, *Clin. Chem.* **35**:392–395, 1989.

136. Farrar, T. C. and Becker, E. D.: *Pulse and Fourier Transform NMR*, New York, Academic Press, 1971.

137. Markley, J. L., Horsley, W. J., and Klein, M. P.: Spin–lattice relaxation measurements in slowly relaxing complex spectra, *Journal of Chemical Physics* **55**:3604–3605, 1971.

138. Gupta, R. K., Ferretti, J. A., Becker, E. D., et al.: A modified fast inversion–recovery technique for spin–lattice relaxation measurements, *Journal of Magnetic Resonance* **38**:447–452, 1980.

139. Matson, G. B., Schleich, T., Serdahl, C., et al.: Measurement of longitudinal relaxation times using surface coils, *Journal of Magnetic Resonance* **56**:200–206, 1984.

140. Evelhoch, J. L. and Ackerman, J. J. H.: NMR T_1 measurements in inhomogeneous B_1 with surface coils, *Journal of Magnetic Resonance* **53**:52–64, 1983.

6

Spectrum Analysis

6.1. INTRODUCTION

The purpose of spectrum analysis as applied to MRS data obtained from in vivo tissue is to ascertain physiologically important parameters such as the relative or absolute concentrations of metabolites and the intracellular pH. These can be determined from the areas and positions of peaks in the spectrum. In order that the measurements reflect the actual tissue values, the processes by which spectra are analyzed must be objective, repeatable, and capable of satisfying quality-control tests. For a valid interpretation of MRS results, the user should be fully aware of the manner in which the raw data has been processed and of any problems that could affect the outcome of the analytical techniques used.

Over the many years that high-resolution Fourier transform MRS has been applied to in vitro samples, a vast armory of digital techniques has been developed for the extraction of useful information from the data.[1] A modern pulse FT spectrometer or its associated workstation should have a comprehensive and flexible selection of software routines, which can be applied to the data in order to optimize the accuracy of the results.

There are two basic approaches to the spectral analysis of MRS data. Ever since the development of FT MRS, a great deal of effort has been put into the measurement of areas and positions from the spectrum itself. More recently, however, methods have been developed by which resonances are fitted to the FID without the need for FT processing. Whichever approach is adopted for data analysis, the user should always be aware that results may depend on both the operator and the technique. To test an analysis method rigorously, it must be applied either to data collected from a phantom containing known metabolite concentrations or to a simulated spectrum with defined peak areas and noise.

In order to fully understand the following sections, it is necessary to describe the digitized spectrometer output. For quadrature detection, this will consist of n real and imaginary pairs of points sampled from the FID at a constant rate (i.e., the time interval τ between successive pairs will be the same). This sampling rate must be at least twice the highest frequency present in the spectrum. The initial raw data will consist of real and imaginary FIDs (the u and v output channels of the receiver, respectively) made up of several overlapping signals. These FIDs, which decay quasi-exponentially, are superimposed on a background of noise. After a certain time, the FID will have decayed so much that it is no longer visible above the noise level. The bandwidth of the FT of the FID will be $1/\tau$, and it must be arranged (by setting appropriate filter bandwidths, data-array sizes, and sampling rates) that this bandwidth is larger than that of the anticipated spectrum. The output from the FT will consist of real and imaginary spectra (also termed u and v), each consisting of n points. The user can work with either or both of these or with the power spectrum, which is the sum of the squared real and squared imaginary spectra. The power spectrum, however, will have broader peaks than the u or v spectra. The resolution of the u and v spectra (the frequency interval between successive points) is equal to the spectral bandwidth divided by the number of points in the spectrum, i.e., equal to $2/n\tau$.

6.2. MEASUREMENT OF THE SPECTRUM

When attempting to obtain objective and accurate measurements of peak areas and positions from spectra, the operator has often to deal with bad SNRs, overlapping peaks, and underlying baselines which are not completely flat. For accurate measurements to be made, the peaks must be adequately resolved and superimposed on a separable baseline. The effect of peak overlap is demonstrated in Figure 6.1. It should be noted that this effect can alter the apparent positions of peaks as well as make area measurement difficult.

Many procedures can be adopted to overcome these problems partially or completely, but there are drawbacks. Methods which improve SNR very often also broaden the lines, thereby increasing peak overlap. Resolution-enhancement techniques often degrade SNR. Thus, there are usually trade-offs to be made among various aspects of spectral analysis. Curve fitting can be applied, in which the spectrum is simulated by super-position of computed peak profiles with various heights, widths, and

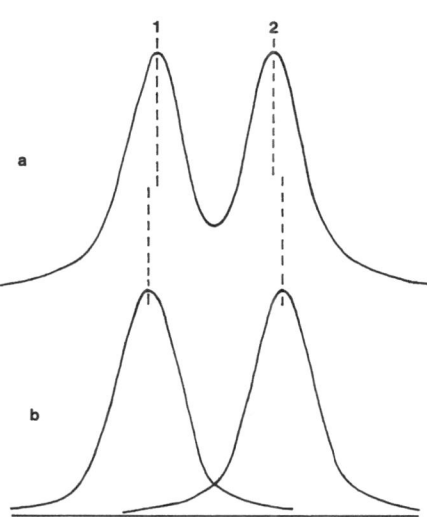

Figure 6.1. The problem of peak overlap in spectrum analysis. (a) Two overlapping peaks in a spectrum. (b) The constituent peak profiles that make up the spectrum in (a). These overlapping profiles demonstrate how difficult it is to measure the areas of peaks 1 and 2 accurately from the spectrum in (a). It is also clear that the apparent chemical shifts are different from the true values due to the superposition of each peak on the wing of its neighbor.

chemical shifts. As is usual with such computer modeling, the more detail that is required (i.e., the greater the number of peaks, fitting of multiplets, etc.) the longer it takes for the processing to be completed. The use of digital techniques is, however, often essential for pulse-FT MRS. A straightforward transform of the FID frequently produces a very ragged and unusable spectrum.

The next sections will deal with some of the methods commonly used to obtain spectra from which reliable results can be acquired. Various approaches that can be adopted to obtain objective measurements from such spectra will also be considered. There are many problems with current analysis techniques, and Section 6.3 will describe some recent developments designed to avoid these.

6.2.1. Raw-Data Baseline Correction

One of the first procedures that has to be applied to the raw (pre-FT) FID data is to correct any constant voltage offset that may be present. This is usually done by calculating the mean value of a section of data consisting mainly of background noise and subtracting this offset from the entire set of data points. If this offset is not removed, the resulting FT output would include a spike in the spectrum at zero frequency. A significant voltage offset in the raw data could also prevent the correct application of other software procedures.

6.2.2. Noise Reduction

The application of various techniques to improve the SNR of spectra is very common. One of the most commonly used techniques is to multiply the FID by an exponentially decaying function, as shown in Figure 6.2. Noise dominates a significant portion of the FID and this factor could also contribute a lot of noise to the spectrum. Often, the data which contains most of the useful information comes at the beginning of the data set. Therefore, multiplication by an exponentially decaying function preserves useful information while reducing noise. Unfortunately there is a trade-off in that application of the technique, in effect, reduces the duration of the FID. The result is that spectral peaks are broadened. For this reason, the method is generally known as "line broadening," and on most spectrometers the operator is able to apply a factor which describes the resultant line broadening in Hz. It is up to the user to decide how much resolution can be sacrificed in order to gain SNR.

Another method which can be used either alone or in conjunction with line broadening is to decide at which point the signal has dropped significantly below the noise level and then to set all the following points to zero. This approach can lose a large amount of noise without significantly reducing the spectral resolution. In applying this approach it is often necessary to smooth the transition from real data to zeroes by "apodizing" the last few points. In this application this process multiplies

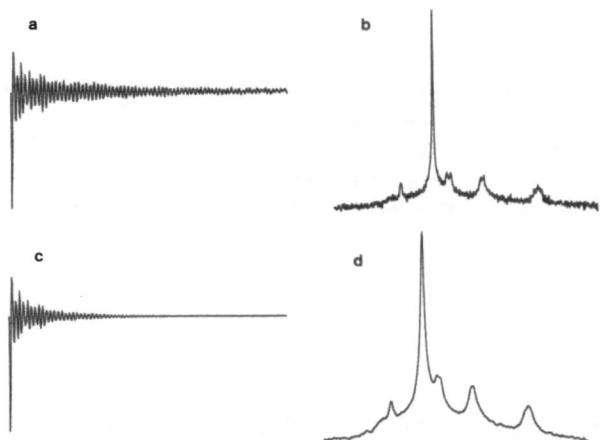

Figure 6.2. The effect of exponential line broadening on a ^{31}P spectrum obtained from skeletal muscle. (a) The raw FID. (b) The spectrum obtained by Fourier transformation of the raw FID. (c) The FID in (a) after multiplication by a decaying exponential with a time constant equivalent to 12-Hz line broadening. Note the reduced signal and noise in the tail of the FID. (d) The spectrum obtained by Fourier transformation of the FID in (c). Note that, in addition to the reduced noise, there is also loss of resolution.

these data by a ramp function which decreases from one to zero and hence removes sudden jumps which could introduce spectral artifacts.

6.2.3. Baseline Smoothing and Resolution Enhancement

In order to obtain a final spectrum, from which peak areas and positions can be measured easily, it is often necessary to use various techniques to improve the resolution and to flatten the underlying baseline. Curvature of the baseline may be due to incorrect setting up of spectrometer delays at collection time, to signals from relatively immobile nuclei (e.g., ^{31}P nuclei in phospholipids or in bone), or to misset phasing. Many methods have been developed in order to obtain spectra which are easy to measure and interpret, and many of these are commonly used in high-resolution MRS.[1] Nearly all the techniques used for achieving the objectives of resolution enhancement and baseline smoothing utilize the inverse relationship between signal duration and frequency bandwidth. FIDs of short duration produce broad spectral features, whereas FIDs which decay slowly give narrow lines. Hence any technique which selectively reduces the amplitude of signals at the beginning of the acquired data will also have a predominant effect on the broader features in the spectrum.

Before collecting data, the spectroscopist should always check that the small delays between the transmitter pulse, the receiver switch-on, and the commencement of digitization are correctly set in order to minimize baseline roll caused by the spectrometer filters (see Chapter 5). This is quite easy to do, given the pulse length and filter bandwidth.[2] Having reduced filter-induced baseline roll as much as possible at collection time, the operator has available several techniques for further baseline smoothing, including apodization,[1,3] convolution differencing,[1,4] profile correction,[1,5,6] and polynomial fitting.[7]

6.2.3.1. Apodization

The apodization approach to baseline smoothing depends on the fact that broad spectral features decay rapidly in the FID, whereas signals producing narrow spectral peaks are of much longer duration. In the time domain (i.e., the FID), broad spectral features have a large amplitude only in the first few digitized points. Therefore, if the FID is apodized by being multiplied by a linear ramp function which increases from zero at its first digitized point to unity a few points later, broad signals in the spectrum are selectively suppressed relative to the narrow resonances. Although proven useful in some applications,[3] apodization is quite a crude approach to baseline smoothing. Optimum results are not always obtained and line-shape distortion can be produced.[1]

6.2.3.2. Convolution Differencing

The convolution-difference technique[1,4] basically consists of obtaining a raw spectrum which includes unwanted underlying broad features. From this spectrum is subtracted the same spectrum, but with enough exponential line broadening (described in Section 6.2.2) to obliterate the spectral peaks. The method can be tailored to some degree by varying the amount of the broadened spectrum which is subtracted. The process is mathematically identical to multiplying the FID by the function

$$A(t) = 1 - a \exp(-bt/T) \qquad (6.1)$$

where a is the fraction of the broadened spectrum to be subtracted, b is the line-broadening time constant, and T is the acquisition time. Convolution differencing has the advantage over apodization that two parameters are available to the operator for fine adjustment of the baseline smoothing. Baseline flattening using the convolution-difference technique is demonstrated in Figures 6.3a and 6.3c.

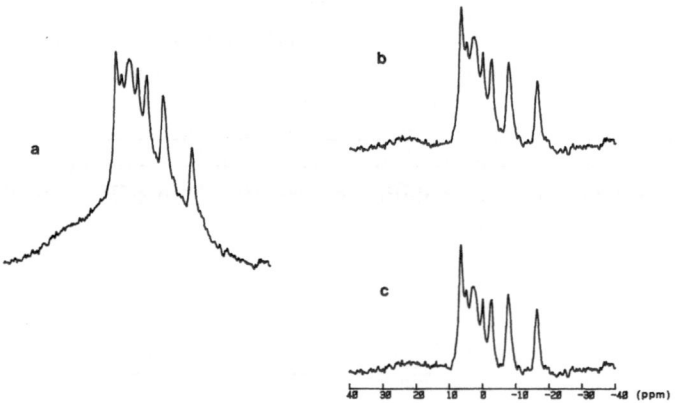

Figure 6.3. Removal of the broad hump due to relatively immobile phosphorus nuclei in bone and phospholipids from a ^{31}P brain spectrum using the LIRE[5] and convolution-difference[4] methods. (a) The spectrum including the bone–phospholipid signal. Exponential line broadening of 12 Hz has been applied to the FID in order to reduce noise. (b) The bone–phospholipid hump has been removed using the LIRE procedure with a time constant T_2^P equivalent to a spectral feature width of 1000 Hz (about 30 ppm) and a factor n of 4. (c) Bone–phospholipid hump removal using the convolution-difference method, in which the subtracted spectrum was exponentially broadened by 250 Hz (about 7 ppm).

6.2.3.3. Profile Correction

Profile correction has been introduced on two occasions for different reasons. It was first developed as a resolution-enhancement technique called the "limited rising exponential" method (LIRE),[5] but was subsequently employed in the "topical magnetic resonance" (TMR) localization procedure (described in Chapter 5) for the removal of broad spectral signals originating in the inhomogeneous part of the profiled magnetic field.[6] In the TMR formulation, the FID is multiplied by the following function:

$$A(t) = \frac{1}{1 + n \exp(-t/T_2^P)} \tag{6.2}$$

The parameter n controls the amount of smoothing, while T_2^P is the decay-time constant of the broad features in the FID. This method also gives the operator more subtle control over the baseline smoothing conditions than that provided by apodization. This approach can give quite good results, as demonstrated by the spectra shown in Figures 6.3a and 6.3b.

6.2.3.4. Polynomial Fitting

Another method for flattening the spectrum baseline is to fit either a single polynomial function or multiple polynomial functions to the regions between the peaks, using a least-squares procedure. This curve is then subtracted from the spectrum. In practice, it has been demonstrated that a quadratic fit can be quite adequate because the coefficients of powers higher than two are not significantly different from zero.[7] This is true only in the absence of features with more complicated profiles (e.g., ^{31}P phospholipid or bone humps); otherwise, more complicated polynomials may have to be considered.[1]

6.2.3.5. Resolution Enhancement

Many of the preceding techniques for baseline smoothing have also been used for resolution enhancement simply by applying them with different parameter values, as shown in Figure 6.4. However, other numerical methods can also be considered. It should be noted that there is usually a trade-off between resolution and signal strength. Most resolution-enhancement procedures reduce SNRs.

6.2.3.5a. Zero Filling. The number of points in the spectrum can be increased simply by adding an equal number of zero points at the end of

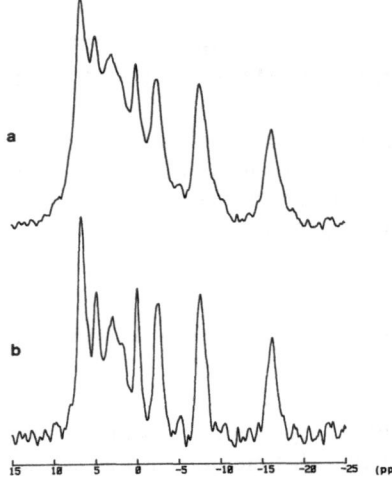

Figure 6.4. Resolution enhancement of a ^{31}P spectrum using LIRE.[5] (a) A neonatal brain spectrum with the bone–phospholipid hump removed and 12-Hz line broadening applied to optimize the signal-to-noise ratio. (b) The same spectrum as in (a), but resolution enhanced using LIRE with a time constant T_2^P equivalent to a spectral linewidth of 60 Hz (about 2 ppm) and a factor n of 3.

the FID before performing the FT.[1] This can improve the interpolation in the FT and bring out detail in the spectrum which would otherwise be lost (i.e., the resolution may be improved). In addition to revealing finer structure in the spectrum, zero filling permits more precise determination of the positions and heights of resonances because more points are available to define the peaks. Figure 6.5 shows the effect of zero filling on a doublet, which would otherwise be unresolved.

6.2.3.5b. Gaussian Transformation. One method of enhancing the resolution in the spectrum is to transform the original lineshapes to quasi-Gaussian profiles, which do not have such broad "wings."[1] This can be done by first multiplying the FID by a positive exponential, which cancels the decay of the signal. The data are then multiplied by a negative squared exponential before Fourier transformation and this produces lines with quasi-Gaussian profiles of preselected width. The usual trade-off of SNR for resolution prevails.

6.2.3.5c. Sine Bell. In this method, the FID is multiplied by a sine function of period twice the data acquisition time and of zero phase (i.e., the "sine-bell" is zero at the beginning of the FID, has a maximum value of unity halfway through the FID, and is zero again at the end of the FID).[1] This approach is similar to others in that it selectively reduces spectral components at the beginning of the FID that decay rapidly with time. By shifting the sine bell along the time axis it is possible to achieve optimization of its effect.

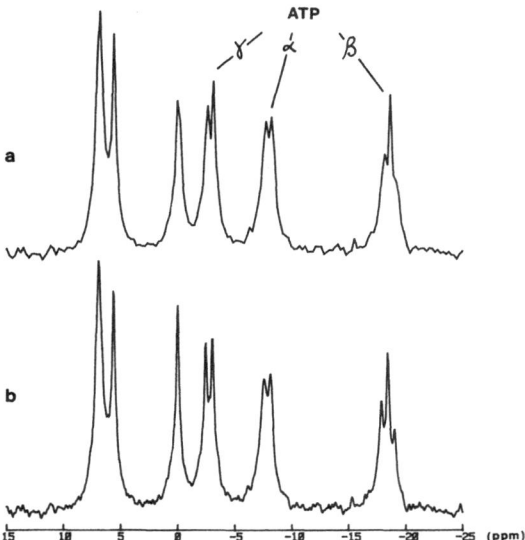

Figure 6.5. The effect of zero filling on a ^{31}P spectrum obtained from phosphorus metabolites in vitro. From left to right: phosphoethanolamine, Pi, PCr, and ATP. (a) The spectrum obtained by direct Fourier transformation of a 1024-point FID. (b) The result obtained by zero filling from 1024 to 4096 points before Fourier transformation. Other processing is the same as for the spectrum in (a). Note the increased resolution obtained particularly for the γATP doublet and the βATP triplet. The relative heights of these peaks are also closer to those expected from their multiplet structures.

6.2.4. Phasing the Spectrum

The outputs from the Fourier transform of the real and imaginary FID inputs consist of two spectra commonly called u and v. Due mainly to the delay between the end of the transmitter pulse and the beginning of data acquisition and to the receiver filters and other spectrometer hardware, but possibly also to the pulse sequence used for data collection (e.g., binomial solvent suppression), these spectra have phases which depend on frequency. In order to obtain spectra in which each point has the same phase and which are superimposed on flat baselines, this frequency dependence has to be corrected. Most of the phase effects can be accounted for by the application of either a phase adjustment that is constant with respect to frequency or a correction which has a linear frequency dependence. Spectra are phase-corrected by applying the following transformations:

$$\left.\begin{array}{l} u'_i = u_i \cos \phi - v_i \sin \phi \\ v'_i = u_i \sin \phi - v_i \cos \phi \end{array}\right\} \tag{6.3}$$

where u_i' and v_i' are the new ith spectrum points calculated from the original unphased spectra u_i and v_i, ϕ is a linear phase-correction angle calculated for each ith point from a combination of a constant phase offset and a phase gradient, both of which are under operator control. Using this approach it is possible to phase most spectra manually, as shown in Figure 6.6. Phase corrections can be entered on the spectrometer keyboard or by knobs on the console. The latter method is, in the author's opinion, the most rapid way of correcting the phase of spectra manually, especially if real-time visual monitoring of the effect of the phase adjustment is available.

In general, manual phasing of spectra is a chore which the spectroscopist could do without, particularly if batches of spectra have to undergo identical processing. Manual phasing is also one of the several areas of spectrum analysis in which results can become operator-dependent. In order to provide automatic and consistent phase correction, several methods have recently been developed. The spectrometer filters can be calibrated using a strong, sharp resonance line (such as that from water).[8] This allows for the correction of all amplitude and phase distortions caused by both the receiver and phase-detection anomalies. Purely computational techniques have been tried also. These include the application of a modified "simplex" procedure, an iterative optimization routine[9] in which all the variables are changed simultaneously. Of three optimization criteria which have been tested (maximization of the minimum spectral amplitude,

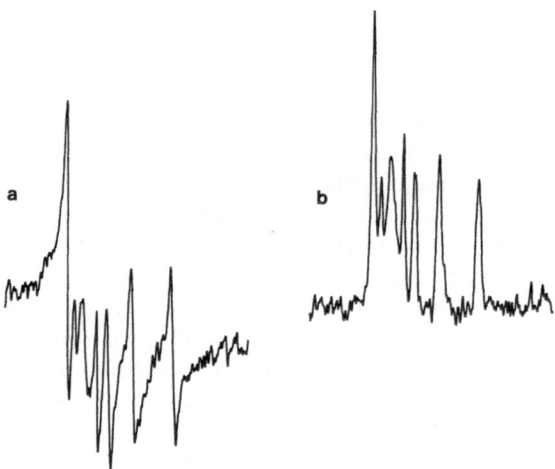

Figure 6.6. Manual phasing of a ^{31}P spectrum. (a) The uncorrected spectrum exhibiting the dependence of phase on frequency due mainly to instrumental factors. (b) The spectrum in (a), with its phase adjusted by the application of a combination of constant and linear frequency-dependent corrections.

maximization of the summed spectral amplitudes below the baseline, and maximization of the spectral area), the first two methods give by far the best results.[10]

A more recent development is a phase correction based on the "dispersion versus absorption" relationship (DISPA).[11] A plot of the dispersion mode signal (the u or real spectrum) against the absorption mode signal (the v or imaginary spectrum) for a single Lorentzian spectral peak gives a circle, one point of which passes through the origin, of diameter equal to the absorption-mode peak height. A peak misphased by angle ϕ gives a circle rotated about the origin by the same angle ϕ. The phase angles for several peaks can be measured in this way from the raw spectral data to give a phase spectrum from which constant and first-order corrections can then be determined.

6.2.5. Baseline Flattening

Even if receiver and digitization delays have been correctly set, broad features from relatively immobile nuclei have been removed computationally, and the spectrum has been optimally phased, there may still be a residual slope to the baseline. In order to facilitate the measurement of peak areas, this has to be removed. The traditional, manual method to do this has utilized the integral of the spectrum. The amplitude of each point of the integral corresponds to the sum of the amplitudes of all the preceding points in the spectrum. A computer can be used to calculate hypothetical baselines for the spectrum. From these, residual integrals can be generated and displayed. The calculations can be based on operator-controlled baseline offsets. When the optimum baseline is found, the regions of the spectrum in between peaks should be flat in the integral. This is because the spectrum between the peaks consists of random noise, and the averaged contribution to the integral over these regions is small. The optimum baseline found in this way can then be subtracted from the spectrum. The technique can also be applied purely computationally, provided a method for testing optimum fit is available. Figure 6.7 shows a ^{31}P spectrum of human muscle and its integral before and after this form of baseline correction. In addition to this approach, the methods which have already been mentioned in Section 6.2.3.4 for baseline smoothing by polynomial fitting[1,7] can be applied with good effect.

6.2.6. Area Measurement

The task now before the spectroscopist is to measure the areas of the peaks present in the spectrum in order to calculate either relative or absolute metabolite concentrations. Some researchers have presented

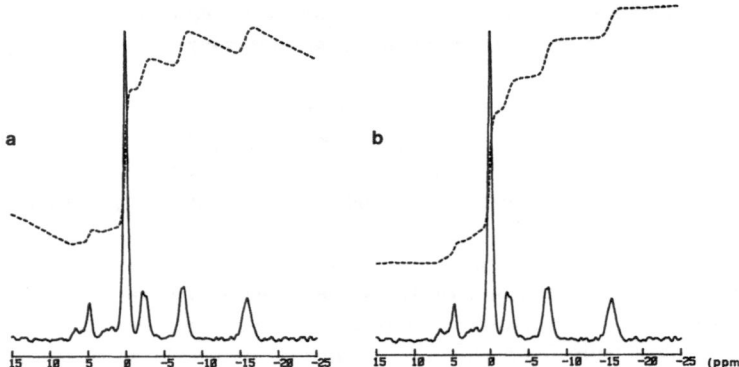

Figure 6.7. Baseline correction of a ^{31}P skeletal-muscle spectrum using the integral (dashed). The integral is the running sum of the data amplitudes across the spectrum. (a) The original spectrum in which the baseline, although almost linear, has a slope, and the background noise is therefore asymmetrically distributed about zero. (b) The spectrum with the baseline corrected by subtraction of a linear function. The integral steps between the resonances are now horizontal.

results in which relative concentrations have been deduced from peak heights. This approach will give good results only if the widths of all the peaks are the same. As this is rarely true, areas are far more reliable as well as being preferable on a theoretical basis. Ideally, the resolution obtained in the spectrum will produce peaks which are distinct. However, for most in vivo studies, many peaks will overlap with their neighbors. As shown in Figure 6.1, this overlap not only makes the measurement of peak areas difficult but can also alter the apparent chemical shifts. Many approaches to spectral area measurement have been used including cutting out the peaks from the plotter paper and then weighing them! With the availibility of modern digital methods, more sophisticated approaches can be chosen.

6.2.6.1. Spectrum Gating

This is the digital equivalent of cutting and weighing. The method consists simply of defining the local spectral bandwidth which contains a particular peak. The computer then calculates the area below the spectrum and between the bandwidth limits, as shown in Figure 6.8a. This approach does not take peak overlap into account and hence may work well only for isolated resonances. If peaks do overlap, then the area measured for each resonance contains a contribution from its neighbor. If care is taken in setting the limits between which the peak areas are measured, then reasonable results can be obtained. For specific tissues, the user should use the same local bandwidth settings for each peak so that operator dependence is

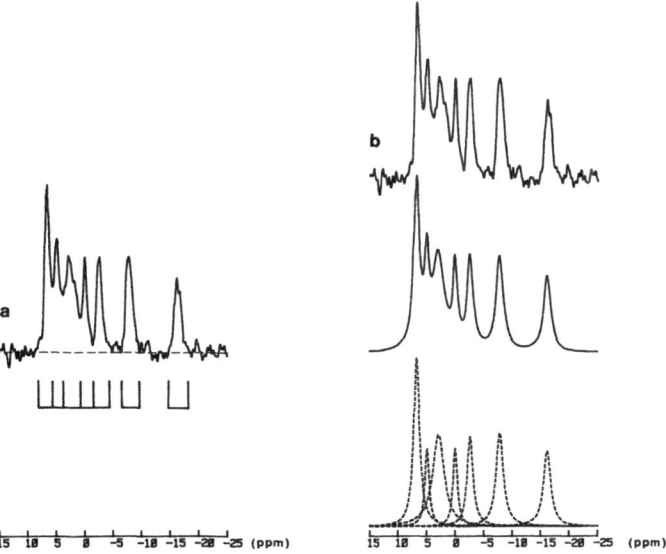

Figure 6.8. Two digital methods of spectrum analysis. (a) Spectrum gating. The peak areas are measured between the spectrum and the baseline (dashed) using the gate limits as shown. In this method, severe peak overlap can cause errors in measurement of both area and position. The chemical shifts of resonances can be measured using statistical routines. (b) Curve fitting as applied to an in vivo spectrum (top). A simulated spectrum (middle) is synthesized from peak profiles generated from a controlled combination of Lorentzian and Gaussian functions. The individual constituents of the simulated spectrum are shown (dashed lines). A best fit to the in vivo data is obtained using iterative computer procedures in which the profile positions, widths, and heights are varied. This approach reduces problems related to peak overlap.

avoided. Optimum gate settings can be determined by the inspection of a few spectra of exceedingly good quality. Care should be taken when spectra containing shifted peaks are measured (e.g., those collected during acidosis).

6.2.6.2. Curve Fitting

A more sophisticated approach, which requires more computational time, is the fitting of hypothetical curves to the peaks by a least-squares technique.[12-15] This is called "curve fitting" and allows each peak in the spectrum to be fitted using its own particular operator-designed profile. The raw spectrum is then simulated by the superposition of the required number of fitted profiles, as shown in Figure 6.8b. Because the characteristics of each fitted component are known, the user can rapidly obtain areas and positions once the best fit has been computed. Most software implementations allow the operator to decide which mathematical curve is to be fitted to the peaks. It is most common to use either Lorentzian or Gaussian profiles or a mixture of both. It is also possible to fit multiplets, if the splitting is known, although this increases the length of the computation. Processing time can be greatly reduced if some information is fed into the software routine in advance. This can consist of the number of peaks to be fitted; their approximate positions, widths, and heights; and whether they are singlets, doublets, or triplets. The fitting routine can then use these values as starting points for its iterative procedures.

The main advantage of curve fitting is that the effects of peak overlap are avoided. Peak areas and positions measured in this way should give a more accurate picture of physiological conditions present in the tissue. In most of the applications of this technique so far, assumptions are made about the shape of the peak profile which may be incorrect when compared with the actual one. This can mean that, although a least-square fit has been found for the hypothetical profile, the selected curve does not accurately represent the actual line shape.

6.2.7. Position Measurement

Estimations of intracellular pH and metal-ion concentrations depend on chemical shift measurements and therefore on the positions of peaks in the spectrum. Obviously the results obtained will be better if the measurement technique takes peak overlap into account and if the method is relatively insensitive to noise.

If curve fitting has been used to analyze a particular spectrum, then the positions of the peaks are given directly by the various fitted profile functions. The effects of peak overlap are avoided because the raw spectrum is simulated by the superposition of the individual fitted profiles. The method should also be insensitive to noise spikes because the fitting is performed over a large spectral band.

Simpler digital techniques can have problems, which the user should be aware of. One common way to measure the position of a resonance is to find the chemical shift of the maximum amplitude point. This can then be adopted as the chemical shift of the peak, or a statistical mean can be calculated from this point and its neighbors. If the SNR is large and there is no close overlap between peaks, this approach can be adequate. In other situations, however, noise spikes or other significant nearby peaks can lead to erroneous results because the computer may start its calculations using these spectral features rather than the true resonance maximum.

6.3. OTHER ANALYSIS METHODS

There have lately been several developments in the analysis of MRS data. Until recently, results have been obtained by the measurement of spectra produced by Fourier transformation, line broadening, and the application of the various techniques described in Section 6.2 for baseline smoothing and resolution enhancement. The new techniques fit decaying exponentials to the FID [linear least-squares procedure (LPSVD)[16] and variable projection method (VARPRO)[17]], use an iterative method of producing the spectrum by maximizing the entropy,[18–20] or adopt a different approach to filtering the spectrum (PIQABLE).[21] Further novel methods for spectral analysis appear regularly in the literature, and the reader is advised to examine the relevant journals to keep up to date.

6.3.1. FID Fitting

In addition to the idea of fitting profiles to the spectral peaks, it has been suggested that a better approach is to fit the FID with a function consisting of exponentially decaying sinusoids. This avoids the necessity of Fourier transformation and other processing. One of the main advantages of these techniques is that the time-domain fitting is not significantly handicapped by the absence of part of the FID. Thus, if the first few FID points contain large signals predominantly from rapidly decaying immobile nuclei (e.g., ^{31}P nuclei in bone), this section of data can be omitted. It is claimed that FID fitting gives better resolution than Fourier transformation. Two methods have been suggested, both of which use matrix computations to find solutions.

6.3.1.1. A Linear Least Squares Procedure

The Linear Least-Squares Procedure (LPSVD)[16] is based on the Kumaresan–Tufts method for estimating complex damped exponentials.[22] The procedure simulates the FID with a function consisting of exponentially decaying sine waves of arbitrary phase, using the principle of linear prediction in combination with linear least-squares fitting. No starting values are required. The method depends on a process called single-value decomposition, which differentiates between signal and noise in the FID. The outputs from LPSVD consist of the frequency (chemical shift), damping factor, amplitude, and phase for each resonance detected.

6.3.1.2. The Variable Projection Method

The Variable Projection Method (VARPRO)[17] uses an iterative procedure, devised by Golub and Pereyra.[23] The technique can be applied without any starting values. However, because it is an iterative method, prior knowledge of the spectral components can be supplied to the routine, and this gives substantial improvements in accuracy. Details of multiplet structure, such as number of components, relative amplitudes, and splitting, are the sort of information that can be usefully input to the routine. Figure 6.9 shows results from the application of VARPRO, with and without prior knowledge, to a ^{31}P human brain spectrum and compares this with LPSVD and conventional Fourier transformation.

6.3.2. The Maximum-Entropy Method

In the maximum-entropy method (MEM) approach,[18–20] of all the possible spectra which could fit the data, the one selected is that which has the minimum structure and maximizes the entropy, defined as:

$$S = -\int f(v) \log f(v) \, dv \qquad (6.4)$$

where $f(v)$ is the spectrum (normalized so that $\int f(v) \, dv = 1$). In order to apply the MEM technique, a featureless trial spectrum is selected as an

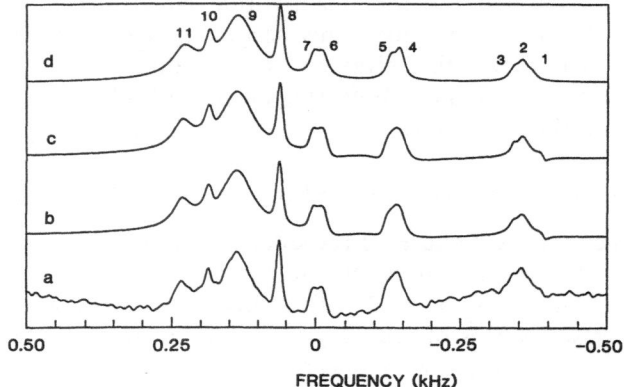

Figure 6.9. Various analysis procedures applied to ^{31}P data obtained from human brain in vivo. (a) A direct Fourier transform of the FID. (b) The Fourier transform of the FID fit obtained using linear prediction and single-value decomposition (LPSVD).[16] (c) The Fourier transform of the FID fit using the variable-projection method (VARPRO) without prior knowledge of resonance parameters. (d) As in (c), but using information about phases, multiplet structure, and linewidths. Peak identifications: 1, 2, and 3, the βNTP triplet; 4 and 5, the αNTP doublet; 6 and 7, the γNTP doublet; 8, PCr; 9, phosphodiesters; 10, Pi; and 11, phosphomonoesters (from Van der Veen et al.,[17] Academic Press Inc.; reprinted with permission).

initial fit. The best fit to the data is found by using an iterative procedure in which the trial spectrum is successively modified in a methodical manner at each iteration until the entropy is maximized.

6.4. PEAK IDENTIFICATION

The correct identification of the metabolites contributing to peaks in the spectrum is vital to the physiological interpretation of the data. The process by which identifications are made is not easy. There are several considerations to be taken into account. These include the relative concentration of the unknown metabolite, the chemical shift of the peak at normal physiological pH and the detailed chemical-shift titration against pH, the multiplet structure of coupled resonances and their associated J coupling constants, and the dependences of the chemical shift and peak amplitude on temperature.[13,24–26] A vast literature exists concerning the absolute concentrations of various metabolites in living tissues as assayed using conventional biochemical techniques. Therefore the researcher has some idea of the substances which ought to show up in a particular spectrum, and these can be adopted as trial metabolites before proceeding with further tests. Extracts of tissues can be made and examined using high-resolution MRS so that precise chemical shifts, multiplet structures, and coupling constants can be obtained for the unknown metabolites. These can be compared directly with the MRS properties observed for solutions of trial substances, provided care is taken to match the pH, metal-ion concentrations, and temperature. However, there may be more than one trial metabolite that satisfies these tests within experimental error. In this case it is often possible to eliminate incorrect trial substances by performing a titration against pH. Trial metabolites considered as candidates for the unknown substance should have chemical shifts that match the unknown in the tissue extract over the pH range examined. It should be noted that this test can still be open to error. All that has been achieved is that the researcher has managed to find a metabolite whose chemical shift titration and multiplet structure matches the unknown substance. However these characteristics and prior knowledge of the expected concentration provide substantial evidence of identity.

6.5. DATA REDUCTION

Regardless of the analysis technique used, the output will consist of a relative area and chemical shift for each of the peaks detected in the spectrum. Further reduction of the data is necessary in order to provide clinically useful information. If necessary, the areas have to be corrected for partial saturation effects (i.e., if the pulse-repetition interval was less than $5T_1$), and the results have to be presented in a physiologically meaningful manner.

6.5.1. Correcting for Partial Saturation

If the spectrum has been collected using a relatively short interval between pulses, then the measured areas will be smaller than those that would be obtained with complete relaxation. The amount by which the areas will be reduced varies from peak to peak and depends on the T_1s pertaining to the metabolites contributing to that particular peak. The usual approach is to multiply each area by its saturation correction factor before proceeding with any further data reduction. The saturation factor for each peak can be determined in a separate series of measurements by comparing the signal obtained under completely relaxed pulsing conditions with that measured using the pulse interval giving partial saturation. Alternatively, if T_1 information is available, a correction factor can be calculated.[27]

6.5.2. Data Presentation

In order to obtain the maximum benefit from the analysis of MR spectra, the results must be presented in an easily assimilable form. This is often done by converting the areas into physiologically significant quantities such as concentration ratios. Some commonly used ratios are PCr/ATP, PCr/Pi, and PCr/(PCr + Pi). It can also be useful to present the relative metabolite concentrations as fractions of the total mobile (i.e., MRS-visible) phosphate (P_{tot}), because under many circumstances this may be assumed to be constant. As an example, for a ^{31}P spectrum obtained from human skeletal muscle, P_{tot} is often approximated by

$$P_{tot} = ME + Pi + PD + PCr + 3 \times \beta ATP \qquad (6.5)$$

where ME, Pi, PD, PCr, and βATP refer to the saturation-corrected areas under the peaks for phosphomonoester, inorganic phosphate, phosphodiester, phosphocreatine, and the β phosphate resonance of ATP. The individual metabolite concentrations can then each be presented as a fraction of the total mobile phosphate simply by dividing each area by P_{tot}. If absolute quantitation has been carried out for one of the metabolites

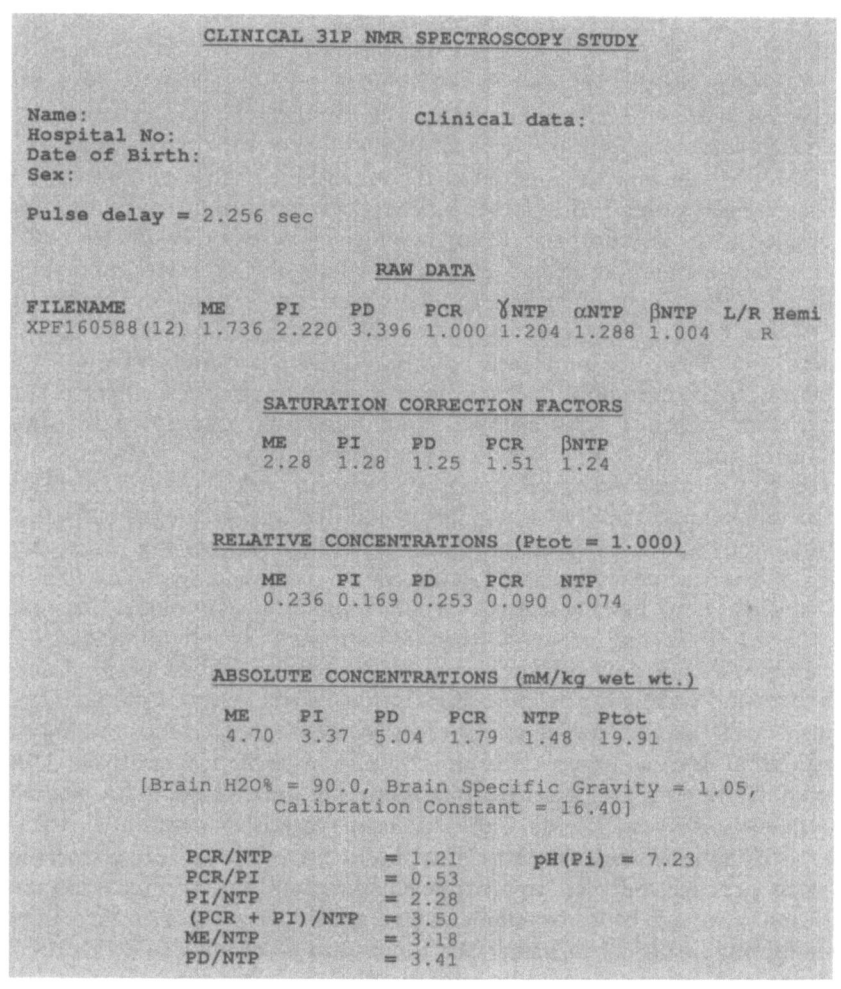

Figure 6.10. A clinical report form for a ^{31}P spectroscopic study of neonatal brain.

(e.g., ATP), then it is a simple matter to calculate absolute concentrations for the remaining resonances, including P_{tot}, from the relative concentrations.

A convenient clinical report form is shown in Figure 6.10. This figure presents basic patient data and raw and processed MRS results including absolute quantitations along with essential collection parameters, correction factors, and calculation constants.

6.6. TESTING THE ANALYSIS METHODS

The usefulness of an analysis technique is directly related to its ability to give relative and absolute metabolite concentrations, intracellular pHs, and metal-ion concentrations comparable with the real tissue values. One way of testing the analysis method is to compare the results with previously published values obtained by other biochemical techniques. Caution should be applied with this approach. Many biochemical assays rely on techniques such as freeze-clamping in order to arrest metabolic processes, and these methods vary in their efficacy. Pi and PCr are particularly susceptible to errors due to the finite time taken for the freezing of biopsy or in situ tissue. In addition, the published data may not be comprehensive enough to take into account anatomical or developmental variations in concentration. More rigorous tests apply the analysis technique to either data obtained from a solution which contains known concentrations of metabolites or a simulated spectrum in which the peak areas and positions are predefined and the SNR is similar to that obtained in the clinical study. If a phantom solution is used, great care must be taken both in its production and when collecting the data. Many proprietary chemicals are hydrated to some extent, leading to lower phantom concentrations than anticipated. In order to avoid this, an alternate supply of anhydrous metabolite should be obtained, or slow desiccation of the original substance can be carried out. Conventional biochemical assay of the phantom solution can provide useful corroboration. Resonances from metabolites dissolved at low concentrations in ordinary water have relatively long relaxation times. Unless the phantom is doped with a suitable paramagnetic ion (Mn^{2+} or Cu^{2+}) or is made up as a gel (in order to reduce relaxation times), data must be collected using very long intervals between pulses; otherwise, relative peak areas will be distorted. The test phantom also needs to be stable with respect to bacterial and fungal degradation, and this can usually be taken care of by storage at low temperature and the inclusion of a preservative. As a monitor of long-term stability, it should be possible to sample the phantom solution from time to time so that conventional biochemical assays can be performed in order to check the metabolite concentrations.

REFERENCES

1. Lindon, J. C. and Ferrige, A. G.: Digitisation and data processing in Fourier transform NMR, *Progress in Nuclear Magnetic Resonance Spectroscopy* 14:27–66, 1980.
2. Hoult, D. I., Chen, C–N., Eden, H., et al.: Elimination of baseline artifacts in spectra and their integrals, *J. Mag. Res.* 51:110–117, 1983.

3. Gadian, D. G., Proctor, E., Williams, S. R., et al.: Neurometabolic effects of an inborn error of amino acid metabolism demonstrated in vivo by ^{1}H NMR, *Magn. Reson. Med.* **3**:150–156, 1986.

4. Campbell, I. D., Dobson, C. M., Williams, R. J. P., et al.: Resolution enhancement of protein PMR spectra using the difference between a broadened and a normal spectrum, *J. Mag. Res.* **11**:172–181, 1973.

5. Akitt, J. W.: A new kind of function for resolution enhancement, *J. Mag. Res.* **32**:311–324, 1978.

6. Gordon, R. E., Hanley, P. E., and Shaw, D.: Topical magnetic resonance, *Progress in Nuclear Magnetic Resonance Spectroscopy* **15**:1–47, 1982.

7. Dietrich, W. and Gerhards, R.: A simple method for the fitting of baselines and resonance peaks in NMR spectroscopy, *J. Mag. Res.* **44**:229–237, 1981.

8. Neff, B. L., Ackerman, J. L., and Waugh, J. S.: Fully automatic software correction of Fourier transform NMR spectra, *J. Mag. Res.* **25**:335–340, 1977.

9. Nelder, J. A. and Mead, R.: A simplex method for function minimization, *The Computer Journal* **7**:308–313, 1965.

10. Siegel, M. M.: The use of the modified simplex method for automatic phase correction in Fourier transform nuclear magnetic resonance spectroscopy, *Analytica Chimica Acta* **133**:103–108, 1981.

11. Craig, E. C. and Marshall, A. G.: Automated phase correction of FT NMR spectra by means of phase measurement based on dispersion versus absorption relation (DISPA), *J. Mag. Res.* **76**:458–475, 1988.

12. Hilberman, M., Subramanian, V. H., Haselgrove, J. H., et al.: In vivo time-resolved brain phosphorus nuclear magnetic resonance, *J. Cereb. Blood Flow Metab.* **4**:334–342, 1984.

13. Gyulai, L., Bolinger, L., Leigh, J. S., et al.: Phosphorylethanolamine—the major constituent of the phosphomonoester peak observed by ^{31}P-NMR on developing brain, *FEBS Lett.* **178**:137–142, 1984.

14. Maris, J. M., Evans, A. E., McLaughlin, A. C., et al.: ^{31}P nuclear magnetic resonance spectroscopic investigation of human neuroblastoma in situ, *N. Engl. J. Med.* **312**:1500–1505, 1985.

15. Bergmann, G., Dietrich, W., Gunther, U., et al.: Fast parameter fit of two randomly overlapping Lorentzians, *J. Mag. Res.* **76**:193–199, 1988.

16. Barkhuijsen, H., de Beer, R., Bovee, W. M. M. J., et al.: Retrieval of frequencies, amplitudes, damping factors, and phases from time-domain signals using a linear least-squares procedure, *J. Mag. Res.* **61**:465–481, 1985.

17. Van der Veen, J. W. C., de Beer, R., Luyten, P. R., et al.: Accurate quantification of in vivo ^{31}P NMR signals using the variable projection method and prior knowledge, *Magn. Reson. Med.* **6**:92–98, 1988.

18. Sibisi, S., Skilling, J., Brereton, R. G., et al.: Maximum entropy signal processing in practical NMR spectroscopy, *Nature* **311**:446–447, 1984.

19. Laue, E. D., Skilling, J., Staunton, J., et al.: Maximum entropy method in nuclear magnetic resonance spectroscopy, *J. Mag. Res.* **62**:437–452, 1985.

20. Waller, M. L. and Tofts, P. S.: Edited ^{31}P brain spectra using maximum entropy data processing, *Magn. Reson. Med.* **4**:385–392, 1987.

21. Nelson, S. J. and Brown, T. R.: A method for automatic quantification of one-dimensional spectra with low signal-to-noise ratio, *J. Mag. Res.* **75**:229–243, 1987.

22. Kumaresan, R. and Tufts, D. W.: Estimating the parameters of exponentially damped sinusoids and pole-zero modeling in noise, *Transactions on IEEE Acoustic Speech and Signal Processing ASSP-30* (No. 6) **30**:833–840, 1982.

23. Golub, G. H. and Pereyra, V.: The differentiation of pseudo-inverses and nonlinear least

squares problems whose variables separate, *Society for Industrial and Applied Mathematics Journal on Numerical Analysis* **10**:413–432, 1973.

24. Glonek, T., Kopp, S. J., Kot, E., et al.: P-31 nuclear magnetic resonance analysis of brain: The perchloric acid extract spectrum, *J. Neurochem.* **39**:1210–1219, 1982.

25. Arus, C., Chang, Y., and Barany, M.: Proton nuclear magnetic resonance spectra of excised rat brain. Assignment of resonances, *Physiol. Chem. Phys. Med. NMR* **17**:23–33, 1985.

26. Pettegrew, J. W., Kopp, S. J., Dadok, J., et al.: Chemical characterization of a prominent phosphomonoester resonance from mammalian brain. ^{31}P and ^{1}H NMR analysis at 4.7 and 14.1 tesla, *J. Mag. Res.* **67**:443–450, 1986.

27. Roth, K., Hubesch, B., Meyerhoff, D. J., et al.: Noninvasive quantitation of phosphorus metabolites in human tissue by NMR spectroscopy, *J. Mag. Res.* **81**:299–311, 1989.

Index